Methods in Molecular Biology

Series Editor
John M. Walker
School of Life and Medical Sciences
University of Hertfordshire
Hatfield, Hertfordshire, UK

For further volumes:
http://www.springer.com/series/7651

For over 35 years, biological scientists have come to rely on the research protocols and methodologies in the critically acclaimed *Methods in Molecular Biology* series. The series was the first to introduce the step-by-step protocols approach that has become the standard in all biomedical protocol publishing. Each protocol is provided in readily-reproducible step-by-step fashion, opening with an introductory overview, a list of the materials and reagents needed to complete the experiment, and followed by a detailed procedure that is supported with a helpful notes section offering tips and tricks of the trade as well as troubleshooting advice. These hallmark features were introduced by series editor Dr. John Walker and constitute the key ingredient in each and every volume of the *Methods in Molecular Biology* series. Tested and trusted, comprehensive and reliable, all protocols from the series are indexed in PubMed.

Plant Biotic Stress

Methods and Protocols

Edited by

Muthappa Senthil-Kumar

BRIC-National Institute of Plant Genome Research, New Delhi, Delhi, India

Hema Ramanna

BRIC-National Institute of Plant Genome Research, New Delhi, Delhi, India

Humana Press

Editors
Muthappa Senthil-Kumar
BRIC-National Institute of Plant Genome Research
New Delhi, Delhi, India

Hema Ramanna
BRIC-National Institute of Plant Genome Research
New Delhi, Delhi, India

ISSN 1064-3745 ISSN 1940-6029 (electronic)
Methods in Molecular Biology
ISBN 978-1-0716-4745-5 ISBN 978-1-0716-4746-2 (eBook)
https://doi.org/10.1007/978-1-0716-4746-2

This Humana imprint is published by the registered company Springer Science+Business Media, LLC, part of Springer Nature.
The registered company address is: 1 New York Plaza, New York, NY 10004, U.S.A.

Preface

The book *Plant Biotic Stress* comprises 22 chapters covering insect pests, various plant-pathogen interactions, viruses, phytoplasmas (PLOs), fungi, oomycetes, insects, and nematodes across a diverse range of plant species. These include model plants, agriculturally and horticulturally relevant crops, as well as species from ecological and forestry contexts. Additionally, the book explores how abiotic environmental factors influence plant-biotic interactions. This book serves as a valuable resource, providing both a knowledge base and practical protocols for researchers, including graduate students, postdoctoral researchers, senior faculty, and lab technicians working in the fields of agriculture and plant stress biology. It offers a balanced coverage of well-established and routinely used protocols alongside novel, recently developed tools that facilitate in-depth and reproducible research for crop protection.

The 22 chapters include 1 review article and 21 methodology chapters. Together, they present a comprehensive collection of protocols and techniques commonly used in research on plant-biotic interactions. These methodologies address a broad spectrum of experimental requirements and are organized thematically based on the type of biotic stress involved. The *general plant–pathogen interaction* category includes chapters such as ectopic gene expression in *Arabidopsis*, enabling researchers to analyze gene roles in disease resistance or susceptibility. In the *virus studies* section, RNA interference (RNAi) in *Okra* is explored as a gene-silencing tool that targets viral genes to inhibit replication and spread. Another chapter details the spot inoculation of begomoviruses using whitefly vectors, which mimics natural transmission and allows for reliable symptom development studies. Detection methods for RNA viruses are thoroughly described, encompassing both traditional and advanced molecular techniques. These include symptom observation for initial assessment, enzyme-linked immunosorbent assay (ELISA) for detecting viral proteins, reverse transcription polymerase chain reaction (RT-PCR) and loop-mediated isothermal amplification (RT-LAMP) for amplifying viral RNA with high sensitivity, northern blot hybridization for analyzing RNA levels, and next-generation sequencing (NGS) for in-depth analysis of viral populations and mutation patterns.

In the *PLO/Virus* category, the detection of Sesame Phyllody Disease is elaborated using quantitative PCR (qPCR), which enables detection even in low concentrations within host tissues. Another chapter describes the transfection of *Arabidopsis* protoplasts as a means to perform sucrose feeding assays, where isolated plant cells are introduced with plasmid DNA to study sugar-responsive gene expression and metabolic responses. Additionally, bacterial twitching motility is examined, providing insights into surface colonization and biofilm formation, which are critical for understanding plant-pathogenic bacterial interactions. A substantial portion of the book focuses on *fungal pathogens*, which pose major threats to agricultural productivity. Chapters detail screening methodologies for chickpea genotypes against dry root rot, typically involving inoculation with *Macrophomina phaseolina*, followed by observation of root damage and disease severity. Similarly, for Fusarium wilt, caused by *Fusarium oxysporum* f.sp. *ciceris*, techniques such as sick-plot screening or hydroponic inoculation are employed to distinguish resistant and susceptible genotypes.

CRISPR-mediated genome editing in rice is also discussed, providing step-by-step guidance on designing guide RNAs, constructing vectors, and regenerating gene-edited plants to develop disease-resistant lines. Another chapter introduces oligo-dT anchored cDNA-SCoT-based differential gene expression in chickpea to identify genes differentially regulated in response to fungal stress, aiding in marker development and stress biology. The extracellular acidification assay is also described, measuring pH shifts in the surrounding medium to evaluate the effectiveness of antifungal agents against plant pathogens. Additionally, leaf inoculation methods for *Sclerotinia sclerotiorum* in mustard are outlined to quantify infection severity and assess host resistance. Under the *oomycete* category, a protocol using differential-display reverse transcription-PCR (DDRT-PCR) is introduced to investigate gene expression differences in plants.

The *insect pest* section includes protocols for estimating foliar volatiles in black gram using gas chromatography-mass spectrometry (GC-MS), which helps identify compounds associated with insect resistance or attraction. Additional chapters focus on CRISPR/Cas9-mediated genome editing of the oriental fruit fly, a major pest species, detailing how targeted gene disruption can aid in population control. SSR markers are utilized in another chapter for identifying and breeding black gram genotypes resistant to bruchid pests, supporting the development of durable resistance. Moreover, RNAi-based methods for managing leafhopper infestation are outlined, involving the design of double-stranded RNA to silence essential pest genes, thereby reducing survival and reproduction. In the *nematode* category, chickpea germplasm is evaluated for resistance to parasitic nematodes. The methodology includes assessing root gall formation, gall index scoring, and potential molecular validation of resistance genes, facilitating the selection of resistant lines. Finally, the *combined stress* category addresses the challenge of multiple simultaneous stresses by providing a protocol for rapid screening of chickpea genotypes under dry root rot and osmotic stress.

This collection of methodologies not only supports ongoing research but also provides a strong foundation for developing improved crop protection strategies.

New Delhi, India

Muthappa Senthil-Kumar
Hema Ramanna

Contents

Contributors

S. L. Abhang • *Biotechnology Centre, Dr. PDKV, Akola, Maharashtra, India*

Krishna Ramakrishnan Acharya • *BRIC-National Institute of Plant Genome Research, New Delhi, Delhi, India*

Selvaraj Anandakumar • *Department of Agricultural Microbiology, Tamil Nadu Agricultural University, Coimbatore, Tamil Nadu, India*

Anjali Anjali • *BRIC-National Institute of Plant Genome Research, New Delhi, Delhi, India*

Shanthinie Ashokkumar • *Department of Plant Biotechnology, Centre for Plant Molecular Biology & Biotechnology, Tamil Nadu Agricultural University, Coimbatore, Tamil Nadu, India*

E. Ashwini • *ICAR-Indian Institute of Oilseeds Research, Hyderabad, Telangana, India*

Shuhada Begum • *Department of Molecular Biology and Biotechnology, Tezpur University, Tezpur, Assam, India*

C. Bharadwaj • *Pulse Research Laboratory, Division of Genetics, ICAR-Indian Agricultural Research Institute, New Delhi, Delhi, India*

Chikmagalur Nagaraja Bhargava • *Division of Basic Sciences, ICAR-Indian Institute of Horticultural Research, Bengaluru, India; Tata Institute for Genetics and Society , Bengaluru, India*

Shuvam Bhuyan • *Department of Molecular Biology and Biotechnology, Tezpur University, Tezpur, Assam, India*

Naveen C. Bisht • *BRIC-National Institute of Plant Genome Research, New Delhi, India*

T. Boopathi • *ICAR-Indian Institute of Oilseeds Research, Hyderabad, Telangana, India*

V. Celia Chalam • *ICAR-National Bureau of Plant Genetic Resources, New Delhi, India*

Chaitali Narendra Chavan • *BRIC-National Institute of Plant Genome Research, New Delhi, India*

Aswin Reddy Chilakala • *BRIC-National Institute of Plant Genome Research, New Delhi, Delhi, India*

Joy Das • *ICAR-Central Institute for Cotton Research, Nagpur, India*

Rutupurna Das • *School of Applied Sciences, Centurion University of Technology and Management, Bhubaneswar, Odisha, India*

S. N. Datarkar • *Biotechnology Centre, Dr. PDKV, Akola, Maharashtra, India*

V. Dinesh Kumar • *ICAR-Indian Institute of Oilseeds Research, Hyderabad, Telangana, India*

Shikha Dixit • *ICAR-National Institute of Biotic Stress Management, Raipur, Chhattisgarh, India*

Sudhakar Duraialagaraja • *Department of Plant Biotechnology, Centre for Plant Molecular Biology & Biotechnology, Tamil Nadu Agricultural University, Coimbatore, Tamil Nadu, India*

P. Duraimurugan • *ICAR-Indian Institute of Oilseeds Research, Hyderabad, Telangana, India*

K. B. Durga Bhavani • *ICAR-Indian Institute of Oilseeds Research, Hyderabad, Telangana, India*

ATHIMOOLAM DURGADEVI • *BRIC-National Institute of Plant Genome Research, New Delhi, Delhi, India*
LUKAPRIYA DUTTA • *Department of Molecular Biology and Biotechnology, Tezpur University, Tezpur, Assam, India*
KOKILADEVI ESWARAN • *Department of Plant Biotechnology, Centre for Plant Molecular Biology & Biotechnology, Tamil Nadu Agricultural University, Coimbatore, Tamil Nadu, India*
BHARAT H. GAWADE • *ICAR-National Bureau of Plant Genetic Resources, New Delhi, India*
GAYACHARAN • *ICAR-National Bureau of Plant Genetic Resources, New Delhi, India*
SHUBHRA JYOTI GIRI • *Department of Molecular Biology and Biotechnology, Tezpur University, Tezpur, Assam, India*
V. GOWTHAM • *ICAR-Central Institute for Cotton Research, Coimbatore, India*
K. S. ILIGER • *ICAR-IARI's Regional Research Centre, Dharwad, Karnataka, India; Technology Development, Advanata Seeds, Bengaluru, India*
A. P. INGLE • *Biotechnology Centre, Dr. PDKV, Akola, Maharashtra, India*
FEBA JACOB • *Centre for Plant Biotechnology and Molecular Biology, Kerala Agricultural University, Thrissur, India*
P. V. JADHAV • *Biotechnology Centre, Dr. PDKV, Akola, Maharashtra, India*
MONIKA JAIN • *Department of Molecular Biology and Biotechnology, Tezpur University, Tezpur, Assam, India*
THANGAVEL KALAISELVI • *Department of Agricultural Microbiology, Tamil Nadu Agricultural University, Coimbatore, Tamil Nadu, India*
ASHOK KARUPPANNASAMY • *Division of Basic Sciences, ICAR-Indian Institute of Horticultural Research, Bengaluru, India; Division of Entomology , ICAR-Indian Agricultural Research Institute, New Delhi, India*
VIPUL GANESH KELKAR • *Centre for Plant Biotechnology and Molecular Biology, Kerala Agricultural University, Thrissur, India; Seagull Biosolutions Pvt. Ltd, Pune, India*
ZAKAULLAH KHAN • *ICAR-National Bureau of Plant Genetic Resources, New Delhi, India*
KUMAR K. KRISH • *Department of Plant Biotechnology, Centre for Plant Molecular Biology & Biotechnology, Tamil Nadu Agricultural University, Coimbatore, Tamil Nadu, India*
H. KUDAPA • *Biotechnology Centre, Dr. PDKV, Akola, Maharashtra, India; International Crops Research Institute for The Semi-Arid Tropics, Hyderabad, Telangana, India*
SHIVAJI AJINATH LAVALE • *Centre for Plant Biotechnology and Molecular Biology, Kerala Agricultural University, Thrissur, India*
ARUL LOGANATHAN • *Department of Plant Biotechnology, Centre for Plant Molecular Biology & Biotechnology, Tamil Nadu Agricultural University, Coimbatore, Tamil Nadu, India*
SANJOY MAJUMDER • *School of Applied Sciences, Centurion University of Technology and Management, Bhubaneswar, Odisha, India*
BHANU MALHOTRA • *BRIC-National Institute of Plant Genome Research, New Delhi, India*
MANABENDRA MANDAL • *Department of Molecular Biology and Biotechnology, Tezpur University, Tezpur, Assam, India*
ASHISH MARATHE • *ICAR-National Institute of Biotic Stress Management, Raipur, Chhattisgarh, India*
DEEPU MATHEW • *Centre for Plant Biotechnology and Molecular Biology, Kerala Agricultural University, Thrissur, India*

K. Midhuna Madhu • *Centre for Plant Biotechnology and Molecular Biology, College of Agriculture, Vellanikkara, Thrissur, India*
M. P. Moharil • *Biotechnology Centre, Dr. PDKV, Akola, Maharashtra, India*
Smita Nair • *Centre for Plant Biotechnology and Molecular Biology, College of Agriculture, Vellanikkara, Thrissur, India*
Manivannan Narayana • *Centre for Plant Breeding and Genetics, Tamil Nadu Agricultural University, Coimbatore, India*
Soundararajan Raga Palanisamy • *Department of Agricultural Entomology, Tamil Nadu Agricultural University, Coimbatore, India*
Prachi Pandey • *BRIC-National Institute of Plant Genome Research, New Delhi, Delhi, India*
Gagan Kumar Panigrahi • *School of Applied Sciences, Centurion University of Technology and Management, Bhubaneswar, Odisha, India*
B. S. Patil • *ICAR-IARI's Regional Research Centre, Dharwad, Karnataka, India*
Basil Babu Paul • *Centre for Plant Biotechnology and Molecular Biology, Kerala Agricultural University, Thrissur, India*
Vignesh Ponnurangan • *Department of Plant Pathology, Centre for Plant Protection Studies, Tamil Nadu Agricultural University, Coimbatore, Tamil Nadu, India*
K. P. Raghavendra • *ICAR-Central Institute for Cotton Research, Nagpur, India*
Pandian Rajendran • *ICAR-National Institute of Biotic Stress Management, Raipur, Chhattisgarh, India*
Asokan Ramasamy • *Division of Basic Sciences, ICAR-Indian Institute of Horticultural Research, Bengaluru, India*
Shubhashish Ranjan • *BRIC-National Institute of Plant Genome Research, New Delhi, India*
Maniraj Rathinam • *ICAR-National Institute for Plant Biotechnology, LBS Building, Pusa Campus, New Delhi, India*
Suvendra Kumar Ray • *Department of Molecular Biology and Biotechnology, Tezpur University, Tezpur, Assam, India*
Revanasiddappa • *ICAR-National Institute of Biotic Stress Management, Raipur, Chhattisgarh, India*
Annapurna Sahoo • *School of Applied Sciences, Centurion University of Technology and Management, Bhubaneswar, Odisha, India*
Bhimeshwari Sahu • *ICAR-National Institute of Biotic Stress Management, Raipur, Chhattisgarh, India*
Kunja Bihari Satapathy • *School of Applied Sciences, Centurion University of Technology and Management, Bhubaneswar, Odisha, India*
Dhandapani Senthamilselvi • *Department of Agricultural Microbiology, Tamil Nadu Agricultural University, Coimbatore, Tamil Nadu, India*
Muthappa Senthil-Kumar • *BRIC-National Institute of Plant Genome Research, New Delhi, Delhi, India*
K. Shankarganesh • *ICAR-Central Institute for Cotton Research, Coimbatore, India*
Varanavasiappan Shanmugam • *Department of Plant Biotechnology, Centre for Plant Molecular Biology & Biotechnology, Tamil Nadu Agricultural University, Coimbatore, Tamil Nadu, India*

P. SHARMA • *Biotechnology Centre, Dr. PDKV, Akola, Maharashtra, India; Department of Biotechnology, CSIR-IHBT, Palampur, Himachal Pradesh, India*
PALAIYUR NANJAPPAN SIVALINGAM • *ICAR-National Institute of Biotic Stress Management, Raipur, Chhattisgarh, India*
ROHINI SREEVATHSA • *ICAR-National Institute for Plant Biotechnology, LBS Building, Pusa Campus, New Delhi, India*
J. SRIDHAR • *ICAR-National Institute of Biotic Stress Management, Raipur, Chhattisgarh, India*
RAGUL SUBRAMANIYAN • *Centre for Plant Breeding and Genetics, Tamil Nadu Agricultural University, Coimbatore, India*
RUCHI TIWARI • *BRIC-National Institute of Plant Genome Research, New Delhi, India*
SHAILESH TRIPATHI • *AICRP on Rabi Pulses, ICAR-Indian Institute of Pulses Research, Kanpur, Uttar Pradesh, india*
PARANIDHARAN VAIKUNTAVASAN • *Department of Plant Pathology, Centre for Plant Protection Studies, Tamil Nadu Agricultural University, Coimbatore, Tamil Nadu, India*
MOHAN VARGHESE • *BRIC-National Institute of Plant Genome Research, New Delhi, India*
VIJAYKUMAR K. N. • *ICAR-IARI's Regional Research Centre, Dharwad, Karnataka, India; Plant pathology, VNR Seeds Pvt Ltd, Raipur, India*

Chapter 1

Plant Biotic Stress: Tools and Techniques for Crop Protection

Krishna Ramakrishnan Acharya, Aswin Reddy Chilakala, and Muthappa Senthil-Kumar

Abstract

As climate change continues to impact crop yields, developing strategies to enhance plant tolerance to biotic stress has become increasingly important. This requires a thorough evaluation of the tools and methodologies used to manipulate and study biotic stress tolerance. It is crucial to comprehensively understand both conventional and modern techniques, as well as their effectiveness in addressing the specific needs of the crop under study. Detecting diseases at the early stages of plant development can prevent significant losses in large-scale cultivations. Two broad approaches commonly used to mitigate biotic stresses are eliminating causative agents such as fungi, bacteria, nematodes, viruses, or pests, and imparting resistance to the plant. Although there are similarities in the tools and techniques used to address different biotic stresses, each scenario requires dedicated case studies. It is also essential to stay up to date with the latest developments in plant biotechnology to incorporate a cross-disciplinary approach in conducting and validating experiments. This chapter provides an overview of methods covered in this book ranging from molecular breeding to nondestructive techniques that help achieve the goal of safeguarding plant health.

Key words Integrated pest management, Disease detection, Germplasm screening, Plant-pathogen interaction, Gene editing, Advanced crop protection

1 Introduction

Pests and diseases have long threatened agricultural productivity and product preservation. Food and Agriculture Organization estimated a global crop loss of 20–40% annually due to pests. Annual monetary costs of about US$220 billion and US$70 billion occur due to plant diseases and invasive pests, respectively. Additionally, the risks of food shortages and loss in floral biodiversity are expected to increase with climate change [1, 2]. Also, the altered disease pressure brought on by known and emerging pathogens due to climate change could nullify the effective yield

Muthappa Senthil-Kumar and Hema Ramanna (eds.), *Plant Biotic Stress: Methods and Protocols*, Methods in Molecular Biology, vol. 2966, https://doi.org/10.1007/978-1-0716-4746-2_1,

improvements made over the next 50 years [3], and such interactions are well compiled in the recently developed database on stress combinations and their interactions in plants (SCIPdb; http://www.nipgr.ac.in/scipdb.php) [4].

2 Disease Detection and Estimation

Scientific advancements made in the detection of plant disease from modern plant pathology have been crucial leading to scientific and effective pest control strategies. An understanding of the type and severity of plant diseases is crucial for timely management decisions and for addressing fundamental questions in plant stress biology and the genetics of disease resistance. Further, this knowledge allows for the precise and cost-effective application of pesticides. Conventional disease identification methods often occur at late stages, when plants are already significantly affected, limiting management options. While technological advancements have improved traditional visual assessment of plant symptoms, expert identification remains time-consuming and expensive. Finding an expert and obtaining their assistance can be challenging, delaying disease management, particularly in large areas [5, 6]. In many cases, infections begin through a vector or infected plant material, spreading to other plants and causing significant harm before becoming noticeable to humans. Consequently, there is a pressing need for methods beyond conventional approaches. The list of the detection methods according to the disease stages in their applications is schematically shown in Fig. 1.

While traditional visual scouting methods have been used for centuries, advancements in molecular biology have provided powerful tools for pathogen detection. DNA-based and serological methods offer significant advantages and complement conventional approaches. These methods have numerous advantages, including sensitivity, specificity, and rapid detection. Serological assays, like ELISA, use antibodies to detect pathogens usually viral pathogens that are not culturable. Both polyclonal and monoclonal antibodies are used in ELISA, each with respective advantages and disadvantages. Serological methods have successfully been used to detect pathogens including *rice tungro bacilliform virus* [8], *citrus yellow vein clearing virus* [9], *citrus yellow mosaic virus* [10], *Xanthomonas axonopodis* [11].

DNA-based methods, including fluorescence in situ hybridization and polymerase chain reaction (PCR) variants, and RNA-based methods like reverse transcription-PCR and nucleic acid sequence-based amplification, are essential for accurate and rapid pathogen detection. PCR-based diagnostics offer several advantages, including the ability to detect multiple targets in complex samples and unculturable pathogens. However, they can be susceptible to

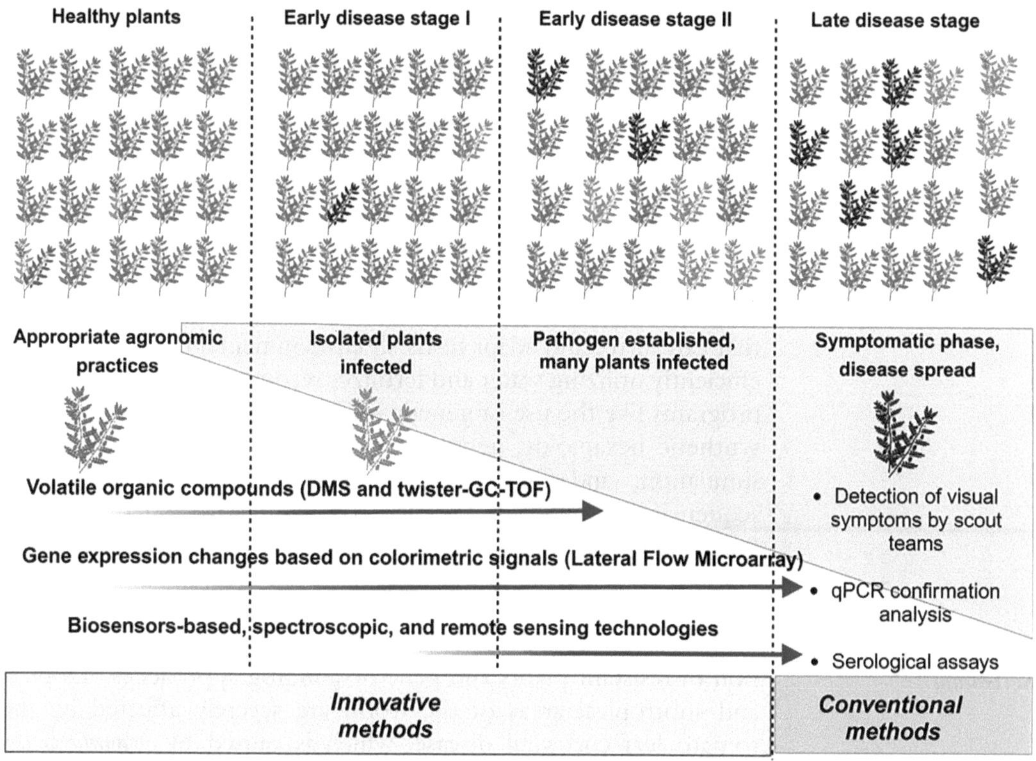

Fig. 1 Various surveillance and detection methods for crop pests and diseases with their appropriate time of applicability. Various methods described in the literature [7] are depicted here to show the type of method that can be used at a particular growth and disease occurrence stage. *DMS* differential mobility spectrometry, *GC-TOF* gas chromatography-time of flight

contamination and inhibitors. Real-time PCR allows for pathogen quantification but requires optimization for specific conditions. Loop-mediated isothermal amplification (LAMP) utilizes a DNA polymerase and multiple primers to amplify a target DNA sequence under isothermal conditions. For RNA viruses, RT-LAMP combines reverse transcription with LAMP to amplify the RNA target. Positive LAMP results can be visually detected by turbidity or colorimetric methods. The high amplification rate and isothermal conditions of LAMP make it a valuable tool for early plant disease detection in plants and can be applied in crop management [12]. One of the important cases of sesame phyllody was one where four leafhopper species were identified as potential vectors. Among these, *Hishimonus phycitis* was confirmed to transmit the phytoplasma to healthy sesame plants. Molecular analysis of 16S rDNA sequences from infected plants verified the presence of the phytoplasma, which clustered with aster yellows phytoplasmas [13]. This study underscores the broad significance of DNA-based methods in disease detection. An exhaustive list of the applications of DNA-based methods is reviewed in one of the recent reviews [14]. Nucleic acid-based techniques are highly sensitive and effective for confirming visual scouting, but may not be reliable for screening asymptomatic plants due to their requirement for careful evaluation for each pathogen-plant combination and designing needs in oligonucleotides and probes production.

3 Strategies to Identify Pathogen-Resistant Genetic Stocks

The diverse genetic resources enable breeders to develop cultivars that can thrive and adapt to harsh environmental conditions while efficiently utilizing water and fertilizer resources. The pre-breeding programs like the use of genetic self-incompatibility, inter-crosses, synthetic hexaploidy, genomic selection coupled with stochastic simulation, and PCR-based genotyping assays like amplicon sequencing (AmpSeq and rAmpSeq) can effectively identify, transfer the desirable traits, and incorporate the genotypic information in the germplasm [15].

3.1 Genotypic Screening

Molecular markers for resistance alleles greatly enhance the selection of resistant plants and gene pyramiding approaches. Tropical and subtropical areas of the world are severely affected by the tomato leaf curl viral disease, which is caused by *begomoviruses* through whiteflies. Gene introgressions have been demonstrated by sequence-characterized amplified region and cleaved amplified polymorphic sequence marker-assisted detection of resistance quantitative trait loci (QTLs) against t*omato yellow leaf curl virus* [16]. In addition, gene-targeted, functional, resistance gene-based, and RNA-based markers were developed to avoid false positive and negative results during breeding and post-resistance-breaking events [17]. Conventional single nucleotide polymorphism (SNP) screening involves hybridization, enzymatic or post-amplification-based molecular beacons, SNP microarray, cleaved amplified polymorphic sequence, restriction fragment length polymorphism, single-strand conformation polymorphism, and denaturing gradient gel electrophoresis. In contrast, SNP genotyping and diversity analysis using genic-SNPs through high-resolution melting analysis was performed in black gram (*Vigna mungo* (L.) Hepper) [18]. This method is appropriate for high-throughput screening, since it monitors the change in fluorescence that happens when double-stranded DNA melts in the presence of a saturated DNA intercalating dye in real-time PCR [19].

3.2 Phenotypic Screening

Future crop breeding research needs phenomics, or phenotyping as a strategy that can be integrated with genotypic and environmental data. The approaches are broadly classified into fields: natural sick plot, artificial sick plot, controlled environment: sick pot, seed soaking, split root design, root dipping, etc. Nematode infestation can cause crop damage that is difficult to diagnose, as symptoms that appear aboveground are not specific. To accurately conduct phenotyping experiments, a robust statistical design is necessary in a controlled environment where plants inoculated with a known initial nematode population are available. Resistance to root-knot nematode is determined through visual inspection and rating of infected roots using root-galling index and calculating root damage index using gall size evaluation and per cent galled area [20]. The pot screening technique can identify 90% of *Fusarium* wilt in susceptible chickpea lines. The process involves raising chickpea seedlings in sterilized soil, dipping their roots in inoculum, transplanting them into pots filled with sterilized soil, and the measurement of disease incidence [21]. Whereas the blotting paper technique for dry root rot disease screening under drought stress in chickpea requires liquid fungal inoculum, seedling preparation, and disease assessment via root scanner-based scoring [22]. *Sclerotinia sclerotiorum* damages field-grown mustard crops by affecting the stems. Inoculating mycelial mat from a pure culture of surface-sterilized sclerotia to the stem and measurement of lesion length is done to quantify such infections [23].

4 Genetic Modification and Gene-Editing Approaches for Disease Resistance

4.1 Gene-Silencing Methods

The RNA interference (RNAi) mechanism provides plants with a defense against viruses by reducing the amount of viral genome multiplication and/or expression. The two different mechanisms are posttranscriptional gene silencing (PTGS), where degradation of transcripts or translational inhibition by small RNAs occurs, and transcriptional gene silencing (TGS), where DNA methylation and chromatin remodeling occur. Methods used for PTGS include hairpin RNAi, virus-induced gene silencing, microRNA-induced gene silencing (MIGS), artificial microRNAs, and host-induced gene silencing, while TGS include viral vectors and inverted repeat constructs with sequence homology to the target gene promoter [24]. These techniques are used to knockdown, that is, decrease the expression of a gene rather than gene knockout, where eradication of the gene expression occurs. The advent of short interfering RNA (siRNA) and expression of short hairpin RNA in vivo from plasmid vectors has reduced the off-target effects caused by the conventional delivery of longer double-stranded RNA (dsRNA).

Single-stranded RNA serves as the genetic material for most plant viruses. Replicative intermediates of dsRNA formed during RNA-dependent RNA polymerase-based replication of the viral genome cause RNAi in the host. Thus, transgenic overexpression of viral RNA frequently results in the production of dsRNA, which in turn causes RNAi [25]. This works better against RNA viruses than DNA viruses and can be used to develop resistance for viruses lacking potent viral suppressors of RNA silencing [26]. Researchers have developed okra lines that are resistant to *yellow vein mosaic virus* using universal intron-containing hairpin RNA construct, transforming plants in vitro using the pRNAi-ligation-independent cloning vector, and checking transgenic lines for resistance using sense strand PCR amplification [27].

Varying the duration of exposure of the germlings to *Agrobacterium* infection-cocultivation, and manipulating virulence induction conditions are done to optimize transformation efficiency and generate marker-free transgenic pigeon pea resistant to *Helicoverpa armigera* [28]. The application of RNAi in generating *Begomovirus*-resistant okra plants was developed recently with an effective *Agrobacterium*-mediated tissue culture-dependent transformation procedure [29]. The significant advancement in constructing gene cassettes with many traits against disease has led to the effective use of resistance genes. New technological developments, such as gene assembly in *Agrobacterium* by nucleic acid transfer using recombinase technology, can successfully target one or two major diseases [30].

4.2 Gene-Editing Methods

Initially, the CRISPR/Cas (CRISPR: clustered regularly interspaced short palindromic repeat, Cas: CRISPR-associated) system was recognized as an adaptive immunity route in prokaryotes that offered defense against bacteriophages and was comparable to the RNA interference mechanism observed in eukaryotes. A short synthetic guide RNA and a nonspecific double-stranded endonuclease known as Cas9 make up the system. A distinct 20-nucleotide spacer/targeting sequence on the guide RNA is connected to a trans-activating sequence to stabilize and attract the Cas9 nuclease, which in turn causes the recognition and cleavage of genomic targets. The last steps in the knockout process include homologous recombination and homology-directed repair performed by the host cell in the presence of a donor DNA fragment.

A two-step technique that involves modification of body segmentation, mating time and partner, detoxification enzymes in target insects, and releasing them back into the natural habitat can be utilized to effectively deploy genome editing in insects. In contrast, plant phenotypic changes, such as alterations in volatile blends, visual traits, gustatory cues, and oviposition positions, can avoid recognition by pests [31]. Producing male sterile lines through conventional breeding or marker-assisted selection involves a series of complex and time-consuming procedures, ranging from the selection of appropriate genetic materials to meticulous laboratory work. The process requires significant investments in terms of resources, time, and expertise, making it a challenging but critical aspect of plant breeding and genetic engineering. The YK17S cultivar generated by the CRISPR/Cas9 system was proved to be a new thermosensitive genic male sterility line with superior yield potential which can be widely utilized in two-line hybrid breeding of indica rice [32].

The oriental fruit fly, *Bactrocera dorsalis*, is a highly destructive and polyphagous insect pest species that poses a significant threat to agriculture worldwide. Native to Southeast Asia, this fly has now spread across the globe and caused extensive economic losses in over 150 fruit crops, including citrus, mango, and guava [33]. They have a remarkable ability to adapt to different climates and environments, making it a formidable foe for farmers and agricultural authorities. Insects rely on olfaction for foraging, mating, and predator avoidance. The odorant-binding proteins and odorant interactions in insects can be studied using in vitro microscale thermophoresis binding assays, in vivo electroantennogram response, and behavioral attraction using an olfactometer [34]. The female fruit fly lays its eggs inside the fruit, and the larvae feed on the fruit, causing it to rot and drop prematurely. Sex determination pathways are diverse and of interest for both fundamental knowledge and developing genetic methods for population control of insect pests. Hence, in vivo CRISPR/Cas-based precision-guided sterile insect technique was developed as a potential strategy to manipulate transformer2 gene that disrupts female sex development and male fertility [35].

Understanding the immune system of plants has made it possible to exploit the pathogen recognition stage. Researchers are now able to develop transgenic plants containing modified node-like receptors (NLR), interspecies NLRs, NLR helpers, and NLR protease traps that identify the multiple plant proteins cleaved by pathogen proteases [30].

5 Recent Advancements in Pathogen Management Strategies

RNAi technology can be an effective and sustainable way to control pests. It has been proven to work against various types of insects, including dipterans, coleopterans, hemipterans, lepidopterans, hymenopterans, and isopterans. The development of genetically modified crops that express dsRNA could revolutionize insect pest management and have significant commercial potential. A novel transacting siRNA-based MIGS approach was used to combat leaf hopper (*Amrasca biguttula*) and root-knot nematode (*Meloidogyne incognita*) simultaneously in cotton [36]. The implementation of start codon targeted polymorphism marker-based targeted fingerprinting in many commercially viable and underutilized plant species for differential gene expression is a simple, novel, cost-effective, highly polymorphic, and reproducible technique [37].

5.1 Plant-Insect Interactions

While *Agrobacterium*-mediated delivery of viruses was previously considered to be less efficient than using whiteflies, 100% transmission efficiency of *mungbean yellow mosaic India virus* in mungbean was demonstrated using *Agrobacterium* culture [38]. The tritrophic interaction of insect-transmitted plant viruses requires a better understanding and management of the insect vectors. To reach high transmission rates, it is necessary to rear vectors in large quantities and manipulate them accordingly. This limitation can be surpassed by the use of the microcage-based technique, which provides more inoculation and acquisition rates for *begomoviruses*, a high retention rate of individual whiteflies compared to larger cages and keeps the leaf surfaces intact [39].

Aphids and certain other homopterans feed on phloem sieve elements, delicately maintaining the viability of these cells. They achieve this by preventing the coagulation of phloem proteins and inhibiting the subsequent callose formation. The electrical penetration graph technique records signal waveforms reflecting different insect activities, such as mechanical stylet work, saliva secretion, and sap ingestion [40]. Meta-analysis and data synthesis methods are applied to such electrophysiological studies from literature to evaluate behavioral experiments describing host-plant and nonhost resistance [41]. A novel role of a cathepsin-type protease in aphid saliva, triggering robust host plant defenses supports the theory of host specialization in generalist aphids [42]. Conversely, to avoid the activation of plant defenses and ensure sustained feeding, they use mouthparts like acrostyle having salivary effector binding capacity that supports its role in the delivery of aphid effectors into plant cells [43].

5.2 Plant-Bacteria Interaction

Ralstonia solanacearum, a rod-shaped phytopathogenic bacterium, causes severe wilt disease in plants. Exploring its invasion and colonization mechanisms is necessary to predict crop protection targets. The regulatory genes in the chemotaxis pathway actively control various pili-related phenotypes, including twitching motility, natural transformation, and biofilm formation. Moreover, these genes play a direct and active role in virulence, particularly in the initial stages of plant infection [44]. In vitro time-lapse photography was employed to visualize its motility dynamics [45]. The bacterial colony had intermittent fingerlike projections extending in all directions, with successive layers of bacteria stacking on top of one another, forming a visually distinctive multilayered filmlike structure. Interestingly, the use of bacteriophage was shown to cause the emergence of mucoid phage-resistant colonies that lack virulence and displayed reduced twitching [46].

5.3 Interaction of Plants with Beneficial Microbes

In response to stress, plants exhibit an augmented defense strategy characterized by an accelerated synthesis of phenolics and enhanced activity of superoxide radical-quenching enzymes. Such expressions are pronounced in the presence of a symbiotic microbial consortium. The findings from [47] underscore the priming of defence response in black gram after infection with arbuscular mycorrhizal fungi, *Glomus intraradices* against insect pest, *Spodoptera litura*. The microbe-enriched farm yard manure approach using antagonistic *Bacillus* strains was devised for the suppression of bacterial wilt causing *Ralstonia solanacearum* in eggplant (*Solanum melongena* L.) [48]. Furthermore, preparation of microbe-enriched manure requires the identification of microbial inoculants using molecular techniques and biochemical characterization. Other factors include investigating the direct mechanisms of plant growth-promoting rhizobacteria (PGPR), like phosphate solubilization and phytohormone production; indirect mechanisms, like cyanide production and siderophore-mediated biocontrol; induced systemic resistance; plant genes modulated by PGPR; volatiles; and antimicrobial peptides [49].

6 Approaches to Decode Mechanistic Plant-Pathogen Interactions

Plants are affected by numerous pathogens including viruses, bacteria, fungi, oomycetes, nematodes, and insects. In nonhost interaction, pathogens fail to invade the plant due to preexisting physical and chemical barriers, such as cuticles, cell walls, and constitutively generated antimicrobial chemicals. Meanwhile, in host interaction, pathogens surpass these layers of defense and elicit the immune response. Thus, the interactions are classified as compatible, where the pathogen is virulent to a nonresistant plant, and incompatible, where the pathogen is avirulent to a resistant plant. Early signaling processes, such as protein phosphorylation or activation of plasma membrane proteins, are triggered by plant defense elicitors. These processes either directly or indirectly mobilize or produce different signaling molecules, such as free calcium, nitric oxide, and reactive oxygen species. By linking branch pathways that amplify and define the physiological response through transcriptional and metabolic changes, these signaling molecules govern several activities [50].

6.1 Conventional Approaches

Most plants accumulate proline under drought, salinity, extreme temperatures, UV radiation, or pathogen infection and once the stress is released, proline undergoes oxidative degradation, providing nitrogen and energy for the recovery process [51]. It was postulated that the hypersensitive response (HR) reaction, a crucial part of plant defense against the spread of invasive pathogens is influenced by proline-P5C cycle-induced ROS burst in mitochondria [52]. L-proline and DL-P5C in the form of

Table 1
Applications of DDRT-PCR

Host	Pathogen	Investigation	References
Apple	*Penicillium expansum*	Post-harvest pathology	[54]
Apricot	European stone fruit yellows phytoplasma	Plant metabolic pathways	[55]
Wheat	*Puccinia striiformis*	Yellow rust resistance	[56]
Rice	*Magnaporthe grisea*	Elicitor-responsive genes	[57]
Lemon	*Capnodium citri*	Host defense mechanism	[58]
Sugarcane	*Colletotrichum falcatum*	Red rot resistance genes	[59]
Rubber tree	*Oidium heveae*	Powdery mildew resistance	[60]
Potato	*Globodera rostochiensis*	Systemic acquired immunity	[61]
Banana	*Odoiporus longicollis*	Secondary metabolism	[62]

2,4-dinitrophenylhydrazine-hydrochloride-double salt are standards for estimating proline and P5C concentrations in plant extracts. The O-amino-benzaldehyde assay can determine the concentration of the metabolic intermediate P5C at 440 nm [53].

Over the years, the advancements in displaying differential gene expression among different cell types have facilitated rapid and efficient gene identification and the utilization of automated analytical tools (*see* Table 1). Two main phases make up the differential-display reverse transcription-PCR (DDRT-PCR) process: (i) cDNA pools are created by reverse transcription of RNAs isolated from various cell populations using a set of degenerate, anchored oligo-dT primers, and (ii) random partial sequences from the cDNA pools are amplified by PCR using the original anchored dT primer and an upstream arbitrary primer [63]. PCR results from various cell types are compared using gel electrophoresis and documentation to identify and isolate differentially expressed transcripts. When it is known that transcription plays a significant part in molecular pathogenesis, the improvements and changes made to DDRT-PCR make it a viable alternative [64]. Moreover, gene expression validations can be performed via northern blot, subtractive cloning, and sequencing to determine their biological role [65, 66].

Proteomic studies of plant-pathogen interactions have revealed several pathogen and plant proteins expressed in different pathosystems. Some plant varieties express resistance proteins that trigger a hypersensitive response, leading to programmed cell death and containment of the pathogen [67]. Cloning the target genes into appropriate binary vectors, followed by *Agrobacterium*-mediated transformation into the plant under investigation, is the typical procedure for ectopic expression. Table 2 represents the potential

Table 2
Emerging ectopic/transient expression studies in *Arabidopsis thaliana*

Sl. no.	Donor	Protein	Vector	Inference	References
1.	Wild Chinese grapevine	VpCN (R protein with NB-ARC domain)	pCAMBIA 2300	Resistance against *Erysiphe necator* powdery mildew	[68]
2.	*Cucumber Mosaic Virus* (CMV)	Movement protein (MP)	pFAST-G01	Virus effector suppresses host immune signaling	[69]
3.	Parasitic nematode	MiMIF-2 (macrophage migration inhibitory factor)	Super1300-FLAG	Promote parasitism by interfering with the annexin-mediated plant immune responses	[70]
4.	Peanut	AhGLK1b (GOLDEN2-like transcription factor)	pK7WG2.0	Dual resistance to fungal and bacterial pathogens	[71]
5.	Downy mildew fungus	HaRxLL470 (conserved RxLR effector)	pHB-YFP and pER8-3 × flag	Enhances pathogenicity and suppresses transcriptional activation of defense-related genes	[72]
6.	*Sclerotinia sclerotiorum*	SsPINE1 (polygalacturonases inhibiting proteins-INactivating effector 1)	pCNG and pCNF	Effector-based enhancement of necrotrophic virulence	[73]
7.	*Arabidopsis thaliana*	AT5G51630 DRP (disease-related protein)	pPTN289	Resistance against pathogen infection and drought stress	[74]

approaches using plant and pathogen effector genes to discover new plant immunity regulators. The protein expression can be confirmed through western blotting and GUS staining [75]. Functional validation of the expressed protein can be done through subcellular localization of GFP tagged protein, transactivation assay, co-localization of interacting proteins, yeast two-hybrid systems, co-immunoprecipitation assay, and dual-luciferase assay [69, 72, 76, 77].

6.2 Technological Perspectives

Understanding complicated biological networks and interactions requires the use of quantitative representation and computational modeling of biological data from protein-protein interactions, metabolic networks, differential gene expression, structural and comparative genomics, etc. [78]. RNA-based sequencing analysis enabled the identification of sequential, temporally coordinated activation and suppression of a suite of immune response regulators in wheat during pathogen invasion [79]. Current proteomic studies implement gel-based (2D-PAGE, DIGE) and non-gel-based methods (iTRAQ, MudPIT, ICAT, SILAC) for isolation and quantification, followed by modified mass spectrometry analyzers (time-of-flight quadrupole, ion trap, orbitrap, Fourier transform ion cyclotron resonance) and database analysis systems including iProXpress from protein information resource [80].

Pathogen effectors are great cellular probes that provide an understanding of biological processes as well as critical inadequacies in plant immunological signaling networks [81]. Transcription activator-like effectors can directly modify the expression of host plant genes to help bacterial colonization, namely the induction of members of the SWEET and sulphate transporters, binding to transcription factors and regulation of microRNA stabilization [82]. Phytopathogens are subjected to reverse genetics techniques such as targeted gene disruption/replacement (knockout), gene silencing (knockdown), insertional mutagenesis, and targeting-induced local lesions in genomes to understand the function of genes [83].

Long noncoding RNAs have been identified as a critical component in the remodeling of chromatin and genome architecture, RNA stability, and transcription control, along with enhancer-associated activities toward insect, phytoplasma, fungal, and viral resistance [84]. Genome digestion by methylation-sensitive enzymes during methylation-sensitive amplified polymorphism, identification of methylated cytosines by bisulfite sequencing, shotgun bisulfite-sequencing, chromatin immunoprecipitation sequencing, and methylated-DNA immunoprecipitation are molecular techniques that can be used to identify the epigenetically regulated genes during plant-virus interactions [85].

Integrating genetic, molecular, biochemical, physiological, and morphological data of an organism is important for a systems biology approach. Combining multi-omics information to identify qualitative and quantitative connections between various biological components offers a comprehensive method to study the static and dynamic emergent characteristics on a global level. We may decipher how a gene or a group of genes and their products engage in a plant-pathogen relationship using a hybrid strategy combining network topology and machine learning [86]. By merging molecular techniques with microscopy, methods like bimolecular fluorescence complementation, fluorescence localization after photobleaching, and deterministic and stochastic functional super-resolution have developed alongside fluorescence and confocal microscopes [87]. Noninvasive disease detection tools like hyperspectral sensors and thermal imaging technologies for assessing compatible and incompatible plant-pathogen interactions are employed in precision agriculture and plant phenotyping [88].

By understanding the diverse and exceptional phyto-biomes of the rhizosphere, phyllosphere, and endosphere, microorganisms that enhance crop growth, yield, and resistance against biotic and abiotic stresses can be discovered. Analyzing the alpha and beta diversity of non-culturable microbes includes appropriate plant and soil sampling, metagenomic next-generation sequencing, taxonomic assignment of amplicon sequence variants using DADA2 pipeline with SILVA and/or UNITE databases, and statistical analysis performed using different packages in R statistical software [89–92]. Taking advantage of recombination events that have accumulated over thousands of generations, Genome-Wide Association Study (GWAS) uses natural linkage disequilibrium to identify polymorphisms that are associated with phenotypic variation in both plants and pathogens [93]. Co-GWAS consider the pathobiota context essential to address multiple strains of the same pathogen or multiple pathogens encountered in natural environments and crop fields. Subsequent decoding of the associated molecular network by cloning plant QTLs can identify a comprehensive list of candidate genes for future GWAS by network-assisted genomic association and functional validation without affecting other members of the microbiota [94].

6.3 Emerging Approaches

A range of volatile organic molecules are released by plants in response to biotic and abiotic stresses. These molecules serve as info-chemicals for ecological communication in the phytobiome. Plant volatiles produced by microbes directly limit pathogen development while indirectly promoting resistance to upcoming pathogen assault [95]. Ca^{2+} sensors and their targets can be at the intersection of several signaling pathways, making them potential candidates to serve as key integrators in the fine-tuning of plant physiological responses to pathogens under changing environmental conditions [96]. Live-cell imaging with fluorescent probes, genome editing, and multi-omics approaches are widely used to study calcium sensors and signaling [97]. The role of pH regulation in the plant-pathogen interface is becoming increasingly important. Many necrotrophic fungi create an acidic environment using host machinery and metabolic by-products like organic acids [98]. At the early stage of infection, *Fusarium oxysporum* exhibited enhanced virulence upon rapid apoplastic acidification [99]. Appropriate assays that monitor the extracellular pH modulations caused by pathogens can effectively screen a library of antifungal compounds. Furthermore, there will be an increased demand for research on combined stress in the coming years [100]. A chapter in this book explains a novel method to screen chickpea genotypes at the laboratory level for combined dry root rot and osmotic stress.

Extracellular vesicles (EVs) operate as plant-pathogen mediators by transferring bioactive proteins, nucleic acids, lipids, and other tiny molecules. They are classified as exosomes, microvesicles, and apoptotic bodies. Isolation of EV is precise through immunoaffinity capture, but due to the unavailability of specific biomarkers and respective antibodies, methods such as infiltration-centrifugation technique with high-speed centrifugation for plants and a combination of ultra- and differential centrifugation for fungi and bacteria are being used [101]. The secondary metabolites help plants cope with many stressful pathogens and contribute to their survival. Decoding the metabolic reprogramming in the plant-pathogen system requires processing data collected from nuclear magnetic resonance, gas chromatography, liquid chromatography-mass spectrometry, and mass spectrometry imaging techniques through feature extraction and deconvolution, metabolite annotation, molecular networking, multivariate analysis, and biological annotation [102].

7 Conclusion and Future Perspectives

The defense mechanisms of plants are significantly influenced by biotic, abiotic, and edaphic conditions in the field. Conducting advanced research is imperative to formulate effective strategies for mitigating emerging diseases. The research confined to controlled laboratory conditions and lacking validation in real-world agricultural settings may fail to adequately address the complexities encountered in the field, thereby limiting its practical applicability. Historically, studies have predominantly focused on plant responses to individual stress factors. However, there is a growing awareness of the studies on multifactorial stress combinations, which more accurately explain the intricate nature of environmental challenges in agriculture. A comprehensive understanding of the physiological, biochemical, and molecular response mechanisms under multiple stress conditions is fundamental for developing climate-resilient crops that can withstand fluctuating environments. Consequently, transdisciplinary research on stress combinations is pivotal in ensuring sustainable agriculture and global food security.

Acknowledgments

Projects at M. S. K. lab are funded by the National Institute of Plant Genome Research core funding. We acknowledge DBT-eLibrary and NIPGR library for providing access to e-resources.

References

1. Burdon JJ, Zhan J (2020) Climate change and disease in plant communities. PLoS Biol 18:e3000949. https://doi.org/10.1371/journal.pbio.3000949
2. Velásquez AC, Castroverde CDM, He SY (2018) Plant–pathogen warfare under changing climate conditions. Curr Biol 28:R619–R634. https://doi.org/10.1016/j.cub.2018.03.054
3. Chaloner TM, Gurr SJ, Bebber DP (2021) Plant pathogen infection risk tracks global crop yields under climate change. Nat Clim Chang 11:710–715. https://doi.org/10.1038/s41558-021-01104-8
4. Priya P, Patil M, Pandey P, Singh A, Babu VS, Senthil-Kumar M (2023) Stress combinations and their interactions in plants database: a one-stop resource on combined stress responses in plants. Plant J 116:1097–1117. https://doi.org/10.1111/tpj.16497
5. Liu L, Dong Y, Huang W, Du X, Ren B, Huang L, Zheng Q, Ma H (2020) A disease index for efficiently detecting wheat Fusarium head blight using sentinel-2 multispectral imagery. IEEE Access 8:52181–52191. https://doi.org/10.1109/ACCESS.2020.2980310
6. Mishra P, Polder G, Vilfan N (2020) Close range spectral imaging for disease detection in plants using autonomous platforms: a review on recent studies. Curr Robot Rep 1:43–48. https://doi.org/10.1007/s43154-020-00004-7
7. Martinelli F, Scalenghe R, Davino S, Panno S, Scuderi G, Ruisi P, Villa P, Stroppiana D, Boschetti M, Goulart LR, Davis CE, Dandekar AM (2015) Advanced methods of plant disease detection. A review. Agron Sustain Dev 35:1–25. https://doi.org/10.1007/s13593-014-0246-1
8. Henry Sum MS, Yee SF, Eng L, Poili E, Lamdin J (2017) Development of an indirect ELISA and dot-blot assay for serological detection of rice tungro disease. Biomed Res Int 2017:3608042. https://doi.org/10.1155/2017/3608042
9. Liu Z, Sunzhu Y, Zhou X, Hong J, Wu J (2017) Monoclonal antibody-based serological detection of *Citrus yellow vein clearing virus* in citrus groves. J Integr Agric 16:884–891. https://doi.org/10.1016/S2095-3119(16)61475-2
10. Kumar PV, Sharma SK, Rishi N, Baranwal VK (2017) Efficient immunodiagnosis of Citrus yellow mosaic virus using polyclonal antibodies with an expressed recombinant virion-associated protein. 3 Biotech 8:39. https://doi.org/10.1007/s13205-017-1063-4
11. Alvarez AM (2004) Integrated approaches for detection of plant pathogenic bacteria and diagnosis of bacterial diseases. Annu Rev Phytopathol 42:339–366. https://doi.org/10.1146/annurev.phyto.42.040803.140329
12. Ali MM, Bachik NA, Muhadi NA, Tuan Yusof TN, Gomes C (2019) Non-destructive techniques of detecting plant diseases: a review. Physiol Mol Plant Pathol 108:101426. https://doi.org/10.1016/j.pmpp.2019.101426
13. Jutimala P, Manoj Kumar K, Sahidur R, Shankar GH, Deb NP (2019) Identification of sesame phyllody transmitting insect vectors in Assam, India. Phytopathogenic Mollicutes 9:107–108. https://doi.org/10.5958/2249-4677.2019.00054.9
14. Attaluri S, Dharavath R (2023) Novel plant disease detection techniques – a brief review. Mol Biol Rep 50:9677–9690. https://doi.org/10.1007/s11033-023-08838-y
15. Byrne PF, Volk GM, Gardner C, Gore MA, Simon PW, Smith S (2018) Sustaining the future of plant breeding: the critical role of the USDA-ARS national plant germplasm system. Crop Sci 58:451–468. https://doi.org/10.2135/cropsci2017.05.0303
16. Mangal M, Hussain Z, Lata S, Gosavi G, Tomar BS (2021) Marker assisted detection of TYLCV and late blight resistance in tomato (Solanum lycopersicum). Indian J Agric Sci 91:1466–1469. https://doi.org/10.56093/ijas.v91i10.117431
17. Nevame AYM, Xia L, Nchongboh CG, Hasan MM, Alam MA, Yongbo L, Wenting Z, Yafei H, Emon RM, Ismail MR, Efisue A, Gang S, Wenhu L, Longting S (2018) Development of a new molecular marker for the resistance to tomato yellow leaf curl virus. Biomed Res Int 2018:e8120281. https://doi.org/10.1155/2018/8120281
18. Raizada A, Souframanien J (2021) SNP genotyping and diversity analysis based on genic-SNPs through high resolution melting (HRM) analysis in blackgram [Vigna mungo (L.) Hepper]. Genet Resour Crop Evol 68:1331–1343. https://doi.org/10.1007/s10722-020-01064-6
19. Wittwer CT, Reed GH, Gundry CN, Vandersteen JG, Pryor RJ (2003) High-resolution genotyping by amplicon melting

analysis using LCGreen. Clin Chem 49:853–860. https://doi.org/10.1373/49.6.853

20. Zwart RS, Thudi M, Channale S, Manchikatla PK, Varshney RK, Thompson JP (2019) Resistance to plant-parasitic nematodes in chickpea: current status and future perspectives. Front Plant Sci 10. https://doi.org/10.3389/fpls.2019.00966
21. Yadav RK, Tripathi MK, Tiwari S, Tripathi N, Asati R, Patel V, Sikarwar RS, Payasi DK (2023) Breeding and genomic approaches towards development of Fusarium wilt resistance in chickpea. Life 13:988. https://doi.org/10.3390/life13040988
22. Irulappan V, Senthil-Kumar M (2021) Dry root rot disease assays in chickpea: a detailed methodology. J Vis Exp:e61702. https://doi.org/10.3791/61702
23. Singh M, Avtar R, Lakra N, Hooda E, Singh VK, Bishnoi M, Kumari N, Punia R, Kumar N, Choudhary RR (2021) Genetic and proteomic basis of sclerotinia stem rot resistance in Indian mustard [Brassica juncea (L.) Czern & Coss.]. Genes 12:1784. https://doi.org/10.3390/genes12111784
24. Pandey P, Senthil-Kumar M, Mysore KS (2015) Advances in plant gene silencing methods. In: Mysore KS, Senthil-Kumar M (eds) Plant gene silencing: methods and protocols. Springer, New York, pp 3–23
25. Galvez LC, Banerjee J, Pinar H, Mitra A (2014) Engineered plant virus resistance. Plant Sci 228:11–25. https://doi.org/10.1016/j.plantsci.2014.07.006
26. Zhao Y, Yang X, Zhou G, Zhang T (2020) Engineering plant virus resistance: from RNA silencing to genome editing strategies. Plant Biotechnol J 18:328–336. https://doi.org/10.1111/pbi.13278
27. Ganesh KV, Mathew D, Augustine R, Soni KB, Alex S, Shylaja MR, Cherian KA (2022) Development of transgenic okra (Abelmoschus esculentus L. Moench) lines having RNA mediated resistance to yellow vein mosaic virus (Geminiviridae). J Virol Methods 301:114457. https://doi.org/10.1016/j.jviromet.2022.114457
28. Sankara Rao K, Sreevathsa R, Sharma PD, Keshamma E, Udaya Kumar M (2008) In planta transformation of pigeon pea: a method to overcome recalcitrancy of the crop to regeneration in vitro. Physiol Mol Biol Plants 14:321–328. https://doi.org/10.1007/s12298-008-0030-2
29. Sharma N, Sarao NK, Mohanpuria P, Sharma A (2023) Effective agrobacterium-mediated genetic transformation of okra (Abelmoschus esculentus L.) and generation of RNAi plants resistant to Begomovirus infecting okra. J Hortic Sci Biotechnol 99:1–14. https://doi.org/10.1080/14620316.2023.2251978
30. van Esse HP, Reuber TL, van der Does D (2020) Genetic modification to improve disease resistance in crops. New Phytol 225:70–86. https://doi.org/10.1111/nph.15967
31. Tyagi S, Kesiraju K, Saakre M, Rathinam M, Raman V, Pattanayak D, Sreevathsa R (2020) Genome editing for resistance to insect pests: an emerging tool for crop improvement. ACS Omega 5:20674–20683. https://doi.org/10.1021/acsomega.0c01435
32. Barman HN, Sheng Z, Fiaz S, Zhong M, Wu Y, Cai Y, Wang W, Jiao G, Tang S, Wei X, Hu P (2019) Generation of a new thermo-sensitive genic male sterile rice line by targeted mutagenesis of TMS5 gene through CRISPR/Cas9 system. BMC Plant Biol 19:109. https://doi.org/10.1186/s12870-019-1715-0
33. Yuan J, Zhang J, Zhang Y, QiQiGe W, Liu W, Yan S, Wang G (2022) Protocols for CRISPR/Cas9 mutagenesis of the oriental fruit fly bactrocera dorsalis. J Vis Exp. https://doi.org/10.3791/64195
34. Chen X, Lei Y, Li H, Xu L, Yang H, Wang J, Jiang H (2021) CRISPR/Cas9 mutagenesis abolishes odorant-binding protein BdorOBP56f-2 and impairs the perception of methyl eugenol in Bactrocera dorsalis (Hendel). Insect Biochem Mol Biol 139:103656. https://doi.org/10.1016/j.ibmb.2021.103656
35. Wishard R, Karuppannasamy A, Asokan R, Chikmagalur Nagaraja B, Chalapathi P, Dhawane Y, Kumar SS, Maligeppagol M, Rai A (2023) CRISPR/Cas9 editing of transformer2 gene of the oriental fruit Fly, Bactrocera dorsalis (Hendel) (Diptera: Tephritidae) leads to intersex phenotype. J Asia Pac Entomol 26:102105. https://doi.org/10.1016/j.aspen.2023.102105
36. Karthik K, Hada A, Bajpai A, Patil BL, Paraselli B, Rao U, Sreevathsa R (2022) A novel tasi RNA-based micro RNA-induced gene silencing strategy to tackle multiple pests and pathogens in cotton (Gossypium hirsutum L.). Planta 257:20. https://doi.org/10.1007/s00425-022-04055-2
37. Rai MK (2023) Start codon targeted (SCoT) polymorphism marker in plant genome analysis: current status and prospects. Planta 257:34. https://doi.org/10.1007/s00425-023-04067-6

38. Sivalingam PN, Dokka N, Mahajan MM, Sahu B, Marathe A, Kaushal P, Ghosh PK (2021) Achieving maximum efficiency of Mungbean yellow mosaic India virus infection in mungbean by agroinoculation. 3 Biotech 12:29. https://doi.org/10.1007/s13205-021-03088-w
39. Sivalingam PN, Sahu B, Mahajan MM, Sridhar J, Dokka N, Marathe A, Kaushal P, Ghosh PK (2022) Enhanced transmission efficiency of begomoviruses by a single whitefly (Bemisia tabaci) using a microcage technique. J Gen Plant Pathol 88:378–382. https://doi.org/10.1007/s10327-022-01086-1
40. Tjallingii WF (2006) Salivary secretions by aphids interacting with proteins of phloem wound responses. J Exp Bot 57:739–745. https://doi.org/10.1093/jxb/erj088
41. Leybourne DJ, Aradottir GI (2022) Common resistance mechanisms are deployed by plants against sap-feeding herbivorous insects: insights from a meta-analysis and systematic review. Sci Rep 12:17836. https://doi.org/10.1038/s41598-022-20741-3
42. Guo H, Zhang Y, Tong J, Ge P, Wang Q, Zhao Z, Zhu-Salzman K, Hogenhout SA, Ge F, Sun Y (2020) An aphid-secreted salivary protease activates plant defense in phloem. Curr Biol 30:4826–4836.e7. https://doi.org/10.1016/j.cub.2020.09.020
43. Deshoux M, Monsion B, Pichon E, Jiménez J, Moreno A, Cayrol B, Thébaud G, Mugford ST, Hogenhout SA, Blanc S, Fereres A, Uzest M (2022) Role of acrostyle cuticular proteins in the retention of an aphid salivary effector. Int J Mol Sci 23:15337. https://doi.org/10.3390/ijms232315337
44. Corral J, Sebastià P, Coll NS, Barbé J, Aranda J, Valls M (2020) Twitching and swimming motility play a role in Ralstonia solanacearum pathogenicity. mSphere 5. https://doi.org/10.1128/msphere.00740-19
45. Bhuyan S, Dutta L, Begum S, Giri SJ, Jain M, Mandal M, Ray SK (2023) A study on twitching motility dynamics in Ralstonia solanacearum microcolonies by live imaging. J Basic Microbiol 64:42. https://doi.org/10.1002/jobm.202300272
46. Wang J, Wang X, Yang K, Lu C, Fields B, Xu Y, Shen Q, Wei Z, Friman V-P (2023) Phage selection drives resistance–virulence trade-offs in Ralstonia solanacearum plant-pathogenic bacterium irrespective of the growth temperature. Evol Lett qrad056:253. https://doi.org/10.1093/evlett/qrad056
47. Selvaraj A, Thangavel K, Uthandi S (2020) Arbuscular mycorrhizal fungi (Glomus intraradices) and diazotrophic bacterium (Rhizobium BMBS) primed defense in blackgram against herbivorous insect (Spodoptera litura) infestation. Microbiol Res 231:126355. https://doi.org/10.1016/j.micres.2019.126355
48. Sakthivel K, Manigundan K, Gautam RK, Singh PK, Balamurugan A, Kumar A, Sharma SK (2023) Microbe-enriched farm yard manure (MFYM) approach for the suppression of Ralstonia solanacearum Yabuuchi (Smith) inciting bacterial wilt disease in eggplant (Solanum melongena L.). Plant Soil 491:303–315. https://doi.org/10.1007/s11104-023-06119-y
49. Kashyap AS, Manzar N, Meshram S, Sharma PK (2023) Screening microbial inoculants and their interventions for cross-kingdom management of wilt disease of solanaceous crops – a step toward sustainable agriculture. Front Microbiol 14. https://doi.org/10.3389/fmicb.2023.1174532
50. Garcia-Brugger A, Lamotte O, Vandelle E, Bourque S, Lecourieux D, Poinssot B, Wendehenne D, Pugin A (2006) Early signaling events induced by elicitors of plant defenses. Mol Plant-Microbe Interact 19: 711–724. https://doi.org/10.1094/MPMI-19-0711
51. Rizzi Y, Monteoliva M, Fabro G, Grosso C, Laróvere L, Alvarez M (2015) P5CDH affects the pathways contributing to Pro synthesis after ProDH activation by biotic and abiotic stress conditions. Front Plant Sci 6:572
52. Qamar A, Mysore K, Senthil-Kumar M (2015) Role of proline and pyrroline-5-carboxylate metabolism in plant defense against invading pathogens. Front Plant Sci 6:503
53. Miller G, Honig A, Stein H, Suzuki N, Mittler R, Zilberstein A (2009) Unraveling Δ1-Pyrroline-5-carboxylate-proline cycle in plants by uncoupled expression of proline oxidation enzymes. J Biol Chem 284:26482–26492. https://doi.org/10.1074/jbc.M109.009340
54. Sánchez-Torres P, González-Candelas L (2003) Isolation and characterization of genes differentially expressed during the interaction between apple fruit and Penicillium expansum. Mol Plant Pathol 4:447–457. https://doi.org/10.1046/j.1364-3703.2003.00190.x
55. Carginale V, Maria G, Capasso C, Ionata E, La Cara F, Pastore M, Bertaccini A, Capasso A (2004) Identification of genes expressed in

response to phytoplasma infection in leaves of Prunus armeniaca by messenger RNA differential display. Gene 332:29–34. https://doi.org/10.1016/j.gene.2004.02.030

56. Bozkurt O, Unver T, Akkaya MS (2007) Genes associated with resistance to wheat yellow rust disease identified by differential display analysis. Physiol Mol Plant Pathol 71: 251–259. https://doi.org/10.1016/j.pmpp.2008.03.002
57. Kim CY, Lee S-H, Park HC, Bae CG, Cheong YH, Choi YJ, Han C, Lee SY, Lim CO, Cho MJ (2007) Identification of rice blast fungal elicitor-responsive genes by differential display analysis. Mol Plant-Microbe Interact 13: 470–474. https://doi.org/10.1094/MPMI.2000.13.4.470
58. De Felice B, Wilson RR (2010) Molecular characterization of a novel pathogen-responsive receptor kinase-like in citrus limon. Tree Genet Genomes 6:47–56. https://doi.org/10.1007/s11295-009-0227-8
59. Prathima PT, Raveendran M, Kumar KK, Rahul PR, Kumar VG, Viswanathan R, Sundar AR, Malathi P, Sudhakar D, Balasubramaniam P (2013) Differential regulation of defense-related gene expression in response to red rot pathogen Colletotrichum falcatum infection in sugarcane. Appl Biochem Biotechnol 171:488–503. https://doi.org/10.1007/s12010-013-0346-4
60. Li X, Bi Z, Di R, Liang P, He Q, Liu W, Miao W, Zheng F (2016) Identification of powdery mildew responsive genes in Hevea brasiliensis through mRNA differential display. Int J Mol Sci 17:181. https://doi.org/10.3390/ijms17020181
61. Aseel DG, Elkobrosy DH, Abdelsalam NR, El-Saedy MA, Shama S, Hafez EE (2020) The effect of cyst nematode (Globodera rostochiensis) isolate DDH1 on gene expression in systemic leaves of potato plant: cyst nematode and gene expression. J Microbiol Biotechnol Food Sci 10:93–97. https://doi.org/10.15414/jmbfs.2020.10.1.93-97
62. Sibina L, Kavitha KJ, Shabithraj K, Evans DA (2023) Differential display reverse transcription and variance on protein profile between susceptible and resistant musa cultivars in relation to infestation by Odoiporus longicollis. Proc Natl Acad Sci India Sect B Biol Sci 93: 851. https://doi.org/10.1007/s40011-023-01457-4
63. Benito EP, Prins T, van Kan JAL (1996) Application of differential display RT-PCR to the analysis of gene expression in a plant-fungus interaction. Plant Mol Biol 32:947–957. https://doi.org/10.1007/BF00020491
64. Sturtevant J (2000) Applications of differential-display reverse transcription-PCR to molecular pathogenesis and medical mycology. Clin Microbiol Rev 13:408–427. https://doi.org/10.1128/cmr.13.3.408
65. Yang T, Zhang M, Zhang N (2022) Modified northern blot protocol for easy detection of mRNAs in total RNA using radiolabeled probes. BMC Genomics 23:66. https://doi.org/10.1186/s12864-021-08275-w
66. Chavez-Navarrete T, Sanchez-Timm L, Pacheco-Coello R, Baisakh N, Santos-Ordóñez E (2023) Identification of differential-expressed genes in banana-biostimulant interaction using suppression subtractive hybridization. Agronomy 13:415. https://doi.org/10.3390/agronomy13020415
67. Mehta A, Brasileiro ACM, Souza DSL, Romano E, Campos MA, Grossi-de-Sá MF, Silva MS, Franco OL, Fragoso RR, Bevitori R, Rocha TL (2008) Plant–pathogen interactions: what is proteomics telling us? FEBS J 275:3731–3746. https://doi.org/10.1111/j.1742-4658.2008.06528.x
68. Wen Z, Yao L, Wan R, Li Z, Liu C, Wang X (2015) Ectopic expression in Arabidopsis thaliana of an NB-ARC encoding putative disease resistance gene from wild Chinese Vitis pseudoreticulata enhances resistance to phytopathogenic fungi and bacteria. Front Plant Sci 6:1087. https://doi.org/10.3389/fpls.2015.01087
69. Kong J, Wei M, Li G, Lei R, Qiu Y, Wang C, Li Z-H, Zhu S (2018) The cucumber mosaic virus movement protein suppresses PAMP-triggered immune responses in Arabidopsis and tobacco. Biochem Biophys Res Commun 498:395–401. https://doi.org/10.1016/j.bbrc.2018.01.072
70. Zhao J, Li L, Liu Q, Liu P, Li S, Yang D, Chen Y, Pagnotta S, Favery B, Abad P, Jian H (2019) A MIF-like effector suppresses plant immunity and facilitates nematode parasitism by interacting with plant annexins. J Exp Bot 70:5943–5958. https://doi.org/10.1093/jxb/erz348
71. Ali N, Chen H, Zhang C, Khan SA, Gandeka M, Xie D, Zhuang W (2020) Ectopic expression of AhGLK1b (GOLDEN2-like transcription factor) in Arabidopsis confers dual resistance to fungal and bacterial pathogens. Genes 11:343. https://doi.org/10.3390/genes11030343
72. Chen S, Ma T, Song S, Li X, Fu P, Wu W, Liu J, Gao Y, Ye W, Dry IB, Lu J (2021) Arabidopsis downy mildew effector

HaRxLL470 suppresses plant immunity by attenuating the DNA-binding activity of bZIP transcription factor HY5. New Phytol 230:1562–1577. https://doi.org/10.1111/nph.17280
73. Wei W, Xu L, Peng H, Zhu W, Tanaka K, Cheng J, Sanguinet KA, Vandemark G, Chen W (2022) A fungal extracellular effector inactivates plant polygalacturonase-inhibiting protein. Nat Commun 13:2213. https://doi.org/10.1038/s41467-022-29788-2
74. Sahoo A, Satapathy KB, Panigrahi GK (2023) Ectopic expression of disease resistance protein promotes resistance against pathogen infection and drought stress in Arabidopsis. Physiol Mol Plant Pathol 124:101949. https://doi.org/10.1016/j.pmpp.2023.101949
75. Bouwmeester K, de Sain M, Weide R, Gouget A, Klamer S, Canut H, Govers F (2011) The lectin receptor kinase LecRK-I.9 is a novel phytophthora resistance component and a potential host target for a RXLR effector. PLoS Pathog 7:e1001327. https://doi.org/10.1371/journal.ppat.1001327
76. He Y, Li W, Lv J, Jia Y, Wang M, Xia G (2012) Ectopic expression of a wheat MYB transcription factor gene, TaMYB73, improves salinity stress tolerance in Arabidopsis thaliana. J Exp Bot 63:1511–1522. https://doi.org/10.1093/jxb/err389
77. Qu Z-L, Zhong N-Q, Wang H-Y, Chen A-P, Jian G-L, Xia G-X (2006) Ectopic expression of the cotton non-symbiotic hemoglobin gene GhHbd1 triggers defense responses and increases disease tolerance in Arabidopsis. Plant Cell Physiol 47:1058–1068. https://doi.org/10.1093/pcp/pcj076
78. Pinzón A, Barreto E, Bernal A, Achenie L, González Barrios AF, Isea R, Restrepo S (2009) Computational models in plant-pathogen interactions: the case of Phytophthora infestans. Theor Biol Med Model 6:24. https://doi.org/10.1186/1742-4682-6-24
79. Dobon A, Bunting DCE, Cabrera-Quio LE, Uauy C, Saunders DGO (2016) The host-pathogen interaction between wheat and yellow rust induces temporally coordinated waves of gene expression. BMC Genomics 17:380. https://doi.org/10.1186/s12864-016-2684-4
80. Jain A, Singh HB, Das S (2021) Deciphering plant-microbe crosstalk through proteomics studies. Microbiol Res 242:126590. https://doi.org/10.1016/j.micres.2020.126590
81. Toruño TY, Stergiopoulos I, Coaker G (2016) Plant-pathogen effectors: cellular probes interfering with plant defenses in spatial and temporal manners. Annu Rev Phytopathol 54:419–441. https://doi.org/10.1146/annurev-phyto-080615-100204
82. Perez-Quintero AL, Szurek B (2019) A decade decoded: spies and hackers in the history of TAL effectors research. Annu Rev Phytopathol 57:459–481. https://doi.org/10.1146/annurev-phyto-082718-100026
83. Bhadauria V, Banniza S, Wei Y, Peng Y-L (2009) Reverse genetics for functional genomics of phytopathogenic fungi and oomycetes. Int J Genom 2009:e380719. https://doi.org/10.1155/2009/380719
84. Zaynab M, Fatima M, Abbas S, Umair M, Sharif Y, Raza MA (2018) Long non-coding RNAs as molecular players in plant defense against pathogens. Microb Pathog 121:277–282. https://doi.org/10.1016/j.micpath.2018.05.050
85. Sahu PP, Sharma N, Puranik S, Muthamilarasan M, Prasad M (2014) Involvement of host regulatory pathways during geminivirus infection: a novel platform for generating durable resistance. Funct Integr Genomics 14:47–58. https://doi.org/10.1007/s10142-013-0346-z
86. Mishra B, Kumar N, Mukhtar MS (2019) Systems biology and machine learning in plant–pathogen interactions. Mol Plant-Microbe Interact 32:45–55. https://doi.org/10.1094/MPMI-08-18-0221-FI
87. Zeilinger S, Gupta VK, Dahms TES, Silva RN, Singh HB, Upadhyay RS, Gomes EV, Tsui CK-M, Nayak SC (2016) Friends or foes? Emerging insights from fungal interactions with plants. FEMS Microbiol Rev 40:182–207. https://doi.org/10.1093/femsre/fuv045
88. Mahlein A-K, Kuska MT, Behmann J, Polder G, Walter A (2018) Hyperspectral sensors and imaging technologies in phytopathology: state of the art. Annu Rev Phytopathol 56:535–558. https://doi.org/10.1146/annurev-phyto-080417-050100
89. Barroso-Bergadà D, Massot M, Vignolles N, Faivre d'Arcier J, Chancerel E, Guichoux E, Walker A-S, Vacher C, Bohan DA, Laval V, Suffert F (2023) Metagenomic next-generation sequencing (mNGS) data reveal the phyllosphere microbiome of wheat plants infected by the fungal pathogen Zymoseptoria tritici. Phytobiomes J 7:281–287. https://doi.org/10.1094/PBIOMES-02-22-0008-FI
90. Cambon MC, Cartry D, Chancerel E, Ziegler C, Levionnois S, Coste S, Stahl C, Delzon S, Buée M, Burban B, Cazal J,

Fort T, Goret J-Y, Heuret P, Léger P, Louisanna E, Ritter Y, Bonal D, Roy M, Schimann H, Vacher C (2023) Drought tolerance traits in neotropical trees correlate with the composition of phyllosphere fungal communities. Phytobiomes J 7:244–258. https://doi.org/10.1094/PBIOMES-04-22-0023-R

91. Hassani MA, Gonzalez O, Hunter SS, Holmes GJ, Hewavitharana SS, Ivors K, Lazcano C (2023) Microbiome network connectivity and composition linked to disease resistance in strawberry plants. Phytobiomes J 7:298–311. https://doi.org/10.1094/PBIOMES-10-22-0069-R
92. Huo D, Malacrinò A, Lindsey LE, Benitez M-S (2023) Subtle responses of soil bacterial communities to corn-soybean-wheat rotation. Phytobiomes J 7:392–400. https://doi.org/10.1094/PBIOMES-05-22-0032-R
93. Bartoli C, Roux F (2017) Genome-wide association studies in plant Pathosystems: toward an ecological genomics approach. Front Plant Sci 8:763
94. Demirjian C, Vailleau F, Berthomé R, Roux F (2023) Genome-wide association studies in plant pathosystems: success or failure? Trends Plant Sci 28:471–485. https://doi.org/10.1016/j.tplants.2022.11.006
95. Sharifi R, Lee S-M, Ryu C-M (2018) Microbe-induced plant volatiles. New Phytol 220:684–691. https://doi.org/10.1111/nph.14955
96. Aldon D, Mbengue M, Mazars C, Galaud J-P (2018) Calcium signalling in plant biotic interactions. Int J Mol Sci 19:665. https://doi.org/10.3390/ijms19030665
97. Negi NP, Prakash G, Narwal P, Panwar R, Kumar D, Chaudhry B, Rustagi A (2023) The calcium connection: exploring the intricacies of calcium signaling in plant-microbe interactions. Front Plant Sci 14:1248648. https://doi.org/10.3389/fpls.2023.1248648
98. Vylkova S (2017) Environmental pH modulation by pathogenic fungi as a strategy to conquer the host. PLoS Pathog 13:e1006149. https://doi.org/10.1371/journal.ppat.1006149
99. Kesten C, Gámez-Arjona FM, Menna A, Scholl S, Dora S, Huerta AI, Huang H, Tintor N, Kinoshita T, Rep M, Krebs M, Schumacher K, Sánchez-Rodríguez C (2019) Pathogen-induced pH changes regulate the growth-defense balance in plants. EMBO J 38:e101822. https://doi.org/10.15252/embj.2019101822
100. Ramegowda V, Senthil A, Senthil-Kumar M (2024) Stress combinations and their interactions in crop plants. Plant Physiol Rep 29:1–5. https://doi.org/10.1007/s40502-024-00785-5
101. Wang Z, Zeng J, Deng J, Hou X, Zhang J, Yan W, Cai Q (2023) Pathogen-derived extracellular vesicles: emerging mediators of plant-microbe interactions. Mol Plant-Microbe Interact 36:218–227. https://doi.org/10.1094/MPMI-08-22-0162-FI
102. Serag A, Salem MA, Gong S, Wu J-L, Farag MA (2023) Decoding metabolic reprogramming in plants under pathogen attacks, a comprehensive review of emerging metabolomics technologies to maximize their applications. Metabolites 13:424. https://doi.org/10.3390/metabo13030424

Chapter 2

Investigating the Involvement of Target Protein in Plant-Pathogen Interactions Through Ectopic Expression in *Arabidopsis*

Annapurna Sahoo, Sanjoy Majumder, Rutupurna Das, Kunja Bihari Satapathy, and Gagan Kumar Panigrahi

Abstract

Exploring the involvement of specific gene(s) during pathogen invasion in plants requires explicit strategy or technique(s), which would facilitate in unraveling the functional mechanism of the gene. Here, we have primarily described the process for analyzing the target protein by using immunoblotting approach with an aim to unravel the functional role of protein of interest. Immunoblotting technique where specific antibodies are used ensures the presence or absence of the target protein and also allows confirming the turnover of the protein(s) at different time points of pathogen infection to plants. The plants expressing the target protein is confirmed by the immunoblotting approach, so that they can be used for the plant-pathogen interaction studies. Also, the interacting partners of the target protein are monitored by subsequent immunoprecipitation process with an aim to see the effect of pathogen infection on the protein-protein interaction in *Arabidopsis*. Essentially, here we focused protein monitoring techniques to investigate the involvement of target protein(s) through ectopic expression in *Arabidopsis*.

Key words *Arabidopsis*, Pathogen, Infection, Protein, Ectopic expression, Immunoblotting, Antibodies

1 Introduction

Ectopic expression is used as a powerful tool in developing plants with desired traits. In transgenic plants, ectopic expression is used to study gene function, to modify particular traits or to get desired trait(s), and also to understand gene expression regulatory mechanisms [1–4]. It helps in understanding how specific gene (s) functions in different plant tissues at different stages of plant growth, understanding various developmental mechanisms, identifying regulatory elements, and understanding gene interactions [5–7]. Various studies with transgenic *Arabidopsis* plants have shown ectopic expression of genes that are able to resist different biotic

Muthappa Senthil-Kumar and Hema Ramanna (eds.), *Plant Biotic Stress: Methods and Protocols*, Methods in Molecular Biology, vol. 2966, https://doi.org/10.1007/978-1-0716-4746-2_2,

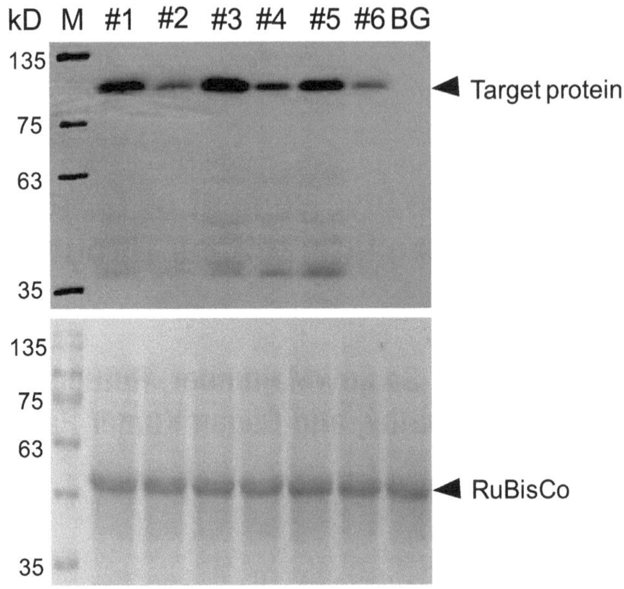

Fig. 1 Confirmation of expression of protein of interest in the transgenic *Arabidopsis* plants through western blotting using X-ray films (manual autoradiography). The transgenic *Arabidopsis* plants used for plant-pathogen interaction studies must express the target protein, and thus could be confirmed by western blotting. Several plants (#1, #2, #3, #4, #5, and #6) are tested for differential expression of the protein and can be further classified as strong expresser, medium expresser, and low expresser. *M* prestained protein ladder, *BG* background wild-type plant sample as the negative control, *RuBisCo* loading control to show equal amount of proteins been used (Ponceau S stained)

and abiotic stress conditions [8–12]. That is why ectopic expression is considered as a powerful tool that helps researchers to understand various functional aspects of the gene expression mechanism and also contributes in the development of plants with improved traits [13–16]. Understanding the target protein's function is important in studies involving plant-pathogen interactions. SDS-PAGE and western blotting are considered potent tools to study the protein expression (*see* Fig. 1) and also assist in unraveling the differential expression of proteins in pathogen-challenged *Arabidopsis* [1, 2]. Likewise, confocal laser scanning microscopy (CLSM) is also considered as one of the best platforms to see the localization of the target protein in *Arabidopsis* plants exposed to varied challenges (*see* Fig. 2). Protein studies including protein–protein interactions (PPIs) are the functional basis of any living cell. Many important cellular functions like gene regulation signal transduction and metabolic pathways are maintained through these interactions. So, understanding these PPIs is crucial for understanding the complex mechanism of plant physiology. There are many

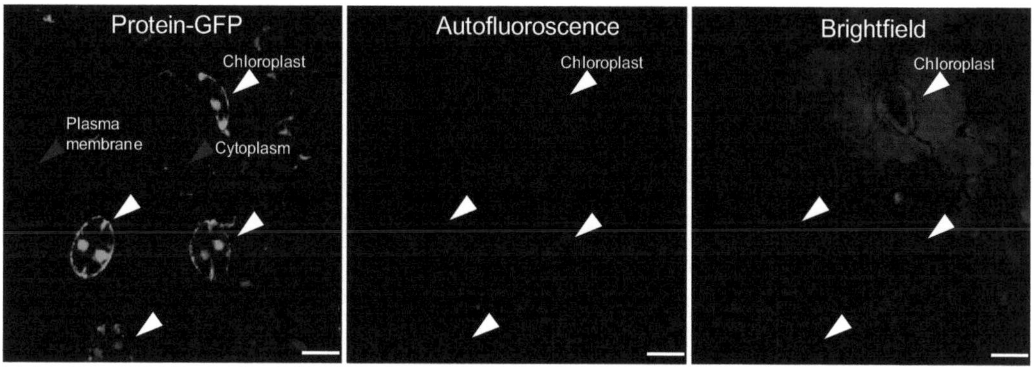

Fig. 2 Ectopic expression and localization of the target protein (post-pathogen infection) in *Arabidopsis*. The target protein fused to GFP is expressed in the *Arabidopsis* transgenic lines. Post-pathogen infection, the localization of the protein was studied using CLSM. The autofluorescence and brightfield signal shows the chloroplast (white arrowheads), whereas the true GFP signal corresponding to the cytoplasm and plasma membrane (red arrowheads) can be confirmed through the green laser channel of the confocal microscope. Scale bar: 10 μm

techniques like yeast-two hybrid (Y2H) assay, co-immunoprecipitation, CLSM, and cross-linking mass spectrometry are available, which can be used to study PPIs [17, 18].

2 Materials

2.1 Plant Materials and Pathogen Infection

1. *Arabidopsis thaliana* seeds.
2. Commercially available potting mixture soil.
3. Pots with an outer diameter of 8 cm.
4. Forceps.
5. Plant growth chamber.
6. *Pseudomonas syringae* (*Pst* DC3000) King's B medium, rifampicin, and spectinomycin.
7. 10 mM $MgSO_4$ for resuspension of bacterial cells.

2.2 Protein Isolation and SDS-PAGE

2.2.1 Sample Preparation

1. Liquid nitrogen.
2. Disposable plastics micropestles.
3. Homogenization buffer: 50 mM Tris acetate (pH 7.9), 100 mM potassium acetate, 1 mM EDTA, 1 mM DDT (dithiothreitol), 20% glycerol, protease inhibitors of choice.

2.2.2 Reagents for Preparing Precasting Gel

1. Acrylamide.
2. Bis-acrylamide (N,N′-methylenebisacrylamide).
3. Tris (2-hydroxymethyl-2-methyl-1,3-propanediol).
4. SDS (sodium dodecyl sulfate or sodium lauryl sulfate).

5. TEMED (N,N,N,N′-tetramethylene ethylenediamine).
6. Ammonium persulfate.
7. β-mercaptoethanol.
8. Glycerol.
9. Bromophenol blue.
10. Glycine.
11. Dithiothreitol (DTT).

2.2.3 Stock Solutions

1. 5 M Tris-HCl (pH 8.8), 100 mL: 18.15 g of Tris base distilled water, HCl (to adjust pH).
2. 0.5 M Tris-HCl (pH 6.8), 100 mL: 3.03 g of Tris base, distilled water, HCl (to adjust pH).
3. 10% SDS (w/v), 100 mL: 10 g of SDS, distilled water.
4. 50% glycerol (v/v), 100 mL: 50 mL of 100% glycerol, 50 mL of distilled water.
5. 1% bromophenol blue (w/v), 10 mL: 100 mg of bromophenol blue, 10 mL of distilled water.

2.2.4 Working Solutions

1. 30% acrylamide stock solution, 100 mL (Solution A): 29.2 g of acrylamide, 0.8 g of N,N′-bis-acrylamide, distilled water (make up to 100 mL) (*see* **Note 1**).
2. 4× separating gel buffer, 100 mL (Solution B): 75 mL of 2 M Tris-HCl (pH 8.8), 4 mL of 10% SDS, 21 mL of distilled water.
3. 4× stacking gel buffer, 100 mL (Solution C): 50 mL of 1 M Tris-HCl (pH 6.8), 4 mL of 10% SDS, 46 mL of distilled water.
4. 10% ammonium persulfate, 5 mL: 0.5 g of ammonium persulfate, distilled water to make up to 5 mL.
5. 1× electrophoresis buffer (pH 8.3), 1 L: 3 g of Tris, 14.4 g of glycine, 1 g of SDS, distilled water to make up to 1 L.
6. 4× sample buffer, 25 mL: 7.5 mL of 30% glycerol, 5 mL of 1 M Tris-HCl (pH 6.8), 2.5 mL of 10% SDS, 1.5 mL of β-mercaptoethanol, 0.0125 g of ServaBlue, distilled water to make up to 25 mL.
7. Coomassie brilliant blue (CBB) staining solution (400 mL): 0.5 g of CBB R-250, 200 mL of methanol, 40 mL of acetic acid, distilled water to make up to 400 mL.
8. Destaining solution (500 mL): 150 mL of methanol, 35 mL of acetic acid, distilled water to make up volume.

2.2.5 Equipments and Accessories

1. Mini gel apparatus.
2. Power supply (capacity 200 V, 500 mA).
3. Boiling water bath or 100 °C sand bath.

4. Eppendorf centrifuge.
5. Hamilton syringes or disposable.
6. Gel-loading tips for micropipette.
7. Gel dryer and high vacuum pump or water pump.
8. Small glass or plastic container.
9. Eppendorf tubes.
10. Rocking or rotary shaker.

2.3 Western Blotting

2.3.1 Reagents

1. Transfer buffer (1 L): 1.93 g of Tris base, 9 g of glycine, 200 mL of methanol, distilled water to make up to 1000 mL (*see* **Note 2**).
2. 1× TBS (1 L): 2.42 g of Tris, 8.76 g of sodium chloride, distilled water to make up to 1000 mL.
3. Wash buffer (200 mL): 200 mL of TBS, 0.1 mL of Tween-20.
4. Blocking solution (100 mL): 5 g of nonfat dry milk powder, 1× TBS to make up to 100 mL.
5. Antibodies: primary antibody and secondary antibody.
6. Luminol solution: Substrate solution and enhancer solution (mix both solution in 1:1 ratio while using).
7. Developer and fixer solutions (for manual autoradiography).

2.3.2 Equipments and Accessories

1. Electroblotting transfer apparatus.
2. Power source with capacity of 220 V.
3. Nitrocellulose paper, 0.2 or 0.45 μm pore size.
4. Whatman 3 mm filter paper.
5. Rotary shaker.
6. Magnetic stir plate.
7. Shallow tray for gel transfer.
8. Heat seal plastic bags.

2.4 Protein Interaction Studies

2.4.1 Reagents

1. Resin (New-England Biolabs).
2. Complete inhibitor.
3. Phenylmethylsulfonyl fluoride (PMSF).
4. Triton X-100.
5. Sodium dodecyl sulfate.
6. Glycerol.
7. Antibodies.
8. β-mercaptoethanol,

9. 1× TBS (1 L): 2.42 g of Tris, 8.76 g of sodium chloride, distilled water to make up to 1000 mL (final concentration: 20 mM Tris, 150 mM NaCl, pH 7.4)
10. Lysis buffer: 100 mM NaCl, 40 mM Tris-HCl (pH 7.5), 1.0 mM EDTA, pH 8.0, 10% (v/v) IGEPAL (NP-40) (*see* **Note 3**).
11. Wash buffer: 100 mM NaCl, 20 mM Tris-HCl (pH 7.5), 1 mM EDTA pH 8.0, 10 mM 2-mercaptoethanol, 0.1% IGEPAL (NP-40) [1].
12. SDS sample loading buffer.

2.4.2 Equipments and Accessories

1. Microcentrifuge for 1.5 and 2 mL standard reaction tubes.
2. Refrigerated centrifuge for 50 mL conical "Falcon"-type tubes.
3. Sonicator device.
4. Paper towels.
5. 1.5 mL microcentrifuge tubes
6. 50 mL polypropylene conical centrifuge tubes
7. Shaking incubator.
8. End-over-end rotor.
9. Spectrophotometer capable of measuring optical density at 600 nm.
10. SDS and western apparatus.

2.5 Confocal Laser Scanning Microscopy

2.5.1 Reagents

1. 1× TBS (1 L): TBS (Tris buffer saline, 20 mM Tris,150 mM sodium chloride, pH 7.4); 2.42 g of Tris, 8.76 g of sodium chloride, distilled water (make up to 1000 mL).

2.5.2 Equipments and Accessories

1. Confocal laser scanning microscopy (CLSM).
2. Scissors.
3. Glass slides.
4. Double-sided tape.
5. Immersion oil.
6. Forceps.

3 Methods

3.1 Plant Growth Conditions and Pathogen Infection

1. Sow 15 seeds (wild-type *Arabidopsis thaliana* and transgenic showing ectopic expression of target gene) in a small pot filled with soil. Keep it in 4 °C for 2 days for stratification (*see* **Note 4**).

2. Move the pots to automated growth chambers adjusted to short-day (12/12 h light/dark) or long-day (16/8 h light/dark) conditions at 22 ± 1 °C with 60–70% relative humidity (*see* **Note 4**).
3. After 7 days post-germination, transfer single plant per pot gently using forceps (*see* **Note 5**).
4. For plant-pathogen studies using *Pseudomonas syringae* pv. *tomato* DC3000 (*Pst*DC3000), prepare fresh culture of *Pst*DC3000. Use King's B medium containing rifampicin and spectinomycin as antibiotics and grow the bacterial cells overnight at 28 °C. Centrifuge (4500×*g* for 10 min) the overnight-grown culture and harvest the cells. Wash the cells three times with double-distilled water and subsequently resuspend in 10 mM $MgSO_4$ for infection [3, 5].
5. Before infecting the *Arabidopsis* plants with *Pst*DC3000, the OD of the bacterial culture must be maintained at $OD_{600} = 0.01$ [1].
6. Mark the leaves of healthy 4-week-old *Arabidopsis* plants and proceed for needleless syringe infiltration [5] (*see* **Note 6**).

3.2 Protein Isolation and SDS-PAGE

3.2.1 Protein Extraction

1. Take 0.05 g leaf sample (*see* **Note 7**).
2. Add 100 μL of 2× SDS sample buffer.
3. Macerate with small pestle, vortex it briefly, incubate at 95 °C for 5 min.
4. Centrifuge at high speed (15,000×*g*) for 3 min.
5. Take the supernatant (10–15 μL) and use it for loading during electrophoresis and discard the pellet (*see* **Note 8**).
6. Supernatant can be stored at −80 °C for future use (*see* **Note 9**).

3.2.2 Preparation of Separating Gel

1. Mix Solution A and Solution B in a test tube and then add required amount of water into this (*see* **Note 10**).
2. Add ammonium persulfate and TEMED to it and gently mix the solution by swirling or inverting (*see* Table 1).
3. Pour appropriate amount of this gel solution into the gel sandwich by using a pipette.
4. Wait for 30–60 min to allow the gel to polymerize.

3.2.3 Preparation of Stacking Gel

1. Mix Solution A and Solution C in a test tube and then add required amount of water to it (*see* Table 1).
2. Remove the water covering the separating gel plates.
3. Add ammonium persulfate and TEMED to it and gently mix the solution by swirling or inverting.

Table 1
Recipe for preparation of gel polymerization mixture for two gels

Reagents	Separating gel 10% acrylamide (10 mL)	Stacking gel 5% acrylamide (4 mL)
Acrylamide stock (30%)	3.34 mL	0.67 mL
Tris-HCl, pH 8.8	2.5 mL	–
Tris-HCl, pH 6.8	–	0.50 mL
SDS (10%)	100 μL	40 μL
APS	5.0 μL	6.0 μL
TEMED	5.0 μL	6.0 μL
Distilled water	4.05 mL	2.77 mL

4. Pour this solution on top of the separating gel.
5. Insert a comb into gel sandwich to make well.
6. Wait for 30 min to allow the gel to polymerize.
7. Take out the comb carefully.
8. Clamp the gel in the electrophoresis chamber.
9. Pour electrophoresis buffer to the inner and outer reservoir (*see* **Note 11**).

3.2.4 Loading Samples and Gel Run

1. Add 5 μL of sample (supernatant portion) to equal quantity (5 μL) of 2× loading buffer (1:1), mix well, heat at 95 °C for 5 min. Include protein marker in the first well (*see* **Note 12**).
2. Load the resulting 10 μL samples in a predetermined order into the bottom of wells.
3. Connect the electrode plugs appropriately.
4. Switch on the power supply.
5. Set voltage at 150–200 V and start the run till the dye front reaches about 5 mm from the bottom of the gel (*see* **Note 13**).
6. Switch off the power supply and remove the electrodes.
7. Take out the gel plates from the electrode assembly and place them on a paper towel. Using a spatula, pry the plates apart.
8. Mark the orientation of the gel by cutting a corner from the bottom of the gel.

3.3 Western Blotting

1. Remove the gel from the electrophoresis apparatus after completion of the SDS-PAGE and incubate with gentle shaking in western transfer buffer (1×)/distilled water for approximately 2 min to remove detergent.

2. Cut the membrane and filter paper to the dimensions of the gel. Place the PVDF membrane for at least 5 min prior to the transfer process (*see* **Note 14**).
3. Equilibrate the filter paper and fiber pads in transfer buffer for 15 min.

3.3.1 Prepare the Gel Sandwich

1. Position the cassette on a lean surface so that the gray/black side is downward. On the gray/black side of the cassette, place an equilibrated fiber pad.
2. Set a Whatman filter paper (pre-wetted sheet) on the fiber pad.
3. Put the gel on the filter paper.
4. Put the pre-wetted PVDF membrane over the gel. Place another piece of pre-wetted filter paper on the membrane followed by another pre-wetted fiber pad.
5. Remove all air bubbles (using a glass tube to gently roll air bubbles out).
6. Close the cassette firmly and lock it with the white latch. Place the gel sandwich cassette in module. Repeat the same for other cassettes if more gels are there for transfer.
7. Add the frozen Bio-ice cooling unit. Place it in the tank and completely fill the tank with 1× western transfer buffer. Add a standard small magnetic stir bar to help maintain even buffer temperature and ion distribution in the tank.
8. Plug the cable into the power supply, carefully watching the color codes. Unclamp the blot sandwich after the transfer process is completed and remove blotting papers. The membrane facing the gel could be marked (*see* **Note 15**).

3.3.2 Detecting the Protein-Specific Antibodies

1. Blocking the membrane with 5% nonfat dried milk in 10 mL TBS (1×) for 1 h with gentle rotation.
2. Incubate with primary antibody (for example, 1:5000 dilution for PAP antibody; #P1291) in 5% nonfat dried milk in 10 mL TBS (1×) solution overnight at 4 °C with gentle rotation (*see* **Note 16**).
3. Wash in 1× TBS with 0.1% Tween-20, 3 times, with 15-min intervals.
4. Incubate with secondary antibody following the manufacturer's instruction.
5. Wash in 1× TBS with 0.1% Tween-20, 3 times in every 15 min of time interval.

3.3.3 Developing

1. Drain out excess of washing buffer from the membrane by slightly touching the corner of the membrane to Kimwipe tissue.

2. Prepare the luminol mixture in a 1:1 ratio. (Add the solutions A and B provided by the manufacturer in a 1:1 ratio).
3. Incubate the luminol mixture to the membrane with continuously pipetting the luminol mixture over the membrane for a minute.
4. Then use the Gel-Doc system to visualize the bands by exposing the membranes treated with luminol starting from 1 min to several minutes depending on the nature of experiment.
5. Instead of Gel-Doc system, manual autoradiography can be done (strictly in dark room) by using 1× developing solution, 1× fixation solution, and water taken in three separate tanklike containers. Cut X-ray films as per the size of membrane and then place the X-ray film over the membrane treated with luminol, which is fixed in suitable cassette used for autoradiography.
6. Exposing time may vary from 1 min to several minutes. Take the X-ray film out, and then wash it thoroughly in the sequence starting from 1× developer solution, then 1× fixer solution and ending with water.
7. X-ray blot is ready to be observed for results (*see* Fig. 1).

3.4 Protein Interaction Studies

3.4.1 Protein Extraction from *Arabidopsis*

1. Take 0.2 g of leaf tissues in 1.5 mL microcentrifuge tubes.
2. Grind the tissues with 5 times (1000 μL to 0.2 g samples) the weight of leaf tissues with cold lysis buffer.
3. Use a sonicator to disrupt the tissues until they are lysed, which is indicated by reduced opacity. This has to be carried out on ice, and it is recommended to sonicate in intervals (5 times; 10 s pulse in 1 min). Care has to be taken not to sonicate with too high intensity.
4. Centrifuge at 4 °C, 14,000×*g* for 20 min, and take all supernatant to fresh tubes.

3.4.2 Preparation of Resin

1. In the meantime, prepare the resin. For each reaction, take 1 mL of resin (50% solution) in 15 mL Falcon tube, add 10 mL TBS buffer.
2. Centrifuge at 400×*g* for 5 min. Let the beads settle down and then discard the supernatant. Repeat this for 3 times. Finally, wash the resin with 10 mL lysis buffer, and centrifuge.

3.4.3 Protein Binding to Resin and Pull-Down

1. Take 20 μL from each sample and mix with 20 μL of sample buffer for input in immunoblotting with antibodies.
2. Mix 500 above supernatant with 5 μL (1/100 vol) of 10% IGEPAL to final 0.1% IGEPAL and mix well on ice. Make 250 μL aliquots.

3. Add 1 μL RNase A (2.5 mg/mL stock solution to final 2 μg/mL) to one of two aliquots for RNase A-resistant interaction.
4. Add each aliquot to 20 μL prewashed, settled resin, and incubate at 4 °C for 4 h by using a gyrator.
5. Centrifuge at 400×*g* for 5 min. Discard the flow-through (it may be kept if necessary). Then wash the settled resin 5 times with 1.2 mL washing buffer at 4 °C.
6. Remove all supernatant. Add 20 μL of 2× SDS sample buffer to elute the bead-bound proteins. Boil at 95 °C for 5 min and run in SDS for protein analysis.

3.5 Localization of Proteins

1. Select the plant leaves for CLSM. Cut small square discs (3 mm × 3 mm) using sharp scissors.
2. Place the leaf discs in 1× TBS solution until it is put onto the slide for microscopy.
3. Once the CLSM is ready for use, prepare the glass slides by fixing double-sided tape over it and gently place the leaf disc.
4. Select the laser according to the nature of protein tag used, and capture picture images for the given sample (*see* Fig. 2; **Note 17**).

4 Notes

1. The ingredients must be mixed thoroughly with the help of a magnetic stirrer and subsequently filtered through a Whatman No. 1 filter paper. The solution should be stored in an amber-colored bottle at 4 °C.
2. The transfer buffer must be at 4 °C for electrophoretic transfer in advance and refrigerate.
3. Add 1× complete inhibitor (Roche), 1.0 mM PMSF, 10 mM (v/v) 2-mercaptoethanol (A 0.72 mL concentrate/L) freshly just before use. Add RNase A for RNase A-resistant interaction.
4. Ecotype to be confirmed. Flower dipping method was used to produce transgenic plants. T_3 homozygote plants were screened and subsequently used in this study. Developmental and pathophysiological experiments to be done using selected set of *Arabidopsis* plants following standard methodology [4, 5].
5. Physical damage to the plants must be avoided. Periodic phenotype studies will assist in understanding the health condition of the plant.
6. Pathogen infection can be carried out at different time points, and also the leaves can be harvested at different time points to

examine the levels of the transcript(s)/protein(s) of interest [1, 2].

7. Weight of the leaf tissue may vary according to the need of the experiment. For example, freeze the leaf sample in liquid nitrogen and use a micropestle to grind the tissue to a fine powder.
8. The protein concentration in the supernatant can be measured using the Bradford assay.
9. It is always recommended to use freshly isolated protein extracts.
10. Choose the percentage of SDS-polyacrylamide gel according to protein molecular size.
11. Add Tris-glycine electrophoresis running buffer (1×) to the top and bottom reservoirs.
12. The total volume of sample may be increased based on the nature and expected outcome of experiment.
13. For a few experiments, such as when the protein of interest is a phosphorylated protein and Phos-tag analysis is to be done, the electrophoresis may be initiated at a lower voltage, like 50 V, for a longer duration.
14. Wearing gloves should be strictly followed. Handling the gels and PVDF membranes with forceps and gloved hands reduces the background.
15. Transfer conditions: (always use the Bio-ice cooling unit).

    Overnight transfer conditions: 30 V, 90 Ma.

    2 h transfer conditions: 100 V, 300 Ma.
16. Optimal antibody concentration must be determined experimentally if not recommended by the manufacturer; 1:1000 dilutions are good starting point.
17. For GFP-fused proteins, green laser can be used. Autofluorescence from chloroplasts needs to be cross-checked by using red laser channel and bright field. Immersion oil should be used for imaging at higher magnification.

Acknowledgments

Authors would like to thank the Vice Chancellor, Centurion University of Technology and Management, Odisha, for providing financial support to GKP (grant approval letter no: CUTM/VC Office/45 to GKP). Special thanks to Prof. Sang Hyon Kim, MJU, Republic of South Korea for his encouragement, motivation, and scientific discussions resulting in development of meticulous methodologies. We apologize to all colleagues whose work could not be included owing to space limitations.

References

1. Jung HW, Panigrahi GK, Jung G-Y, Lee YJ, Shin KH, Sahoo A, Choi ES, Lee E, Kim KM, Yang SH, Jeon JS, Lee SC, Kim SH (2020) PAMP-triggered immunity involves proteolytic degradation of core nonsense-mediated mRNA decay factors during early defense response. Plant Cell 32(4):1081–1101
2. Panigrahi GK, Satapathy KB (2021) *Pseudomonas syringae* pv. *syringae* infection orchestrates the fate of the *Arabidopsis* J domain containing cochaperone and decapping protein factor 5. Physiol Mol Plant Pathol 113(101598):1–9
3. Sahoo A, Satapathy KB, Panigrahi GK (2023) Ectopic expression of disease resistance protein promotes resistance against pathogen infection and drought stress in *Arabidopsis*. Physiol Mol Plant Pathol 124(101949):1–7
4. Arciga-Reyes L, Wootton L, Kieffer M, Davies B (2006) UPF1 is required for nonsense-mediated mRNA decay (NMD) and RNAi in *Arabidopsis*. Plant J 47:480–489
5. Yoo S-J, Choi HJ, Noh SW, Cecchini NM, Greenberg JT, Jung HW (2022) Genetic requirements for infection-specific responses in conferring disease resistance in *Arabidopsis*. Front Plant Sci 13:1068438
6. Sahoo A, Satapathy KB, Panigrahi GK (2023) Security check: plant immunity under temperature surveillance. J Plant Biochem Biotechnol 33:1–4
7. Vogel CM, Potthoff DB, Schäfer M et al (2021) Protective role of the *Arabidopsis* leaf microbiota against a bacterial pathogen. Nat Microbiol 6:1537–1548
8. Yuan M, Jiang Z, Bi G et al (2021) Pattern-recognition receptors are required for NLR-mediated plant immunity. Nature 592: 105–109
9. Jones JD, Dangl JL (2006) The plant immune system. Nature 444:323–329
10. Lee MW, Lu H, Jung HW, Greenberg JT (2007) A key role for the *Arabidopsis* WIN3 protein in disease resistance triggered by *Pseudomonas syringae* that secrete AvrRpt2. Mol Plant-Microbe Interact 20(10):1192–1200
11. Gloggnitzer J, Akimcheva S, Srinivasan A et al (2014) Nonsense-mediated mRNA decay modulates immune receptor levels to regulate plant antibacterial defense. Cell Host Microbe 16(3):376–390
12. Janda M, Lamparová L, Zubíková A, Burketová L, Martinec J, Krčková Z (2019) Temporary heat stress suppresses pamp-triggered immunity and resistance to bacteria in *Arabidopsis thaliana*. Mol Plant Pathol 20(7):1005–1012
13. Kim JH, Castroverde CD, Huang S, Li C et al (2022) Increasing the resilience of plant immunity to a warming climate. Nature 607(7918): 339–344
14. Zhang J, Coaker G, Zhou J, Dong X (2020) Plant immune mechanisms: from reductionistic to holistic points of view. Mol Plant 13(10): 1358–1378
15. Prasad A, Chirom O, Prasad M (2022) Shedding light on immune suppression at high temperature. Trends Microbiol 30(10):918–919
16. Macho A, Zipfel C (2014) Plant PRRs and the activation of innate immune signaling. Mol Cell 54(2):263–272
17. Green MR, Sambrook J (2012) Molecular cloning: a laboratory manual, 4th edn. Cold Spring Harbor Laboratory Press, Cold Spring Harbor
18. Kurien BT, Scofield RH (2006) Western blotting. Methods 38:283–293

Chapter 3

RNA Interference Strategy for the Development of Okra (*Abelmoschus esculentus* L. Moench) Lines Resistant to *Yellow Vein Mosaic Virus*

Vipul Ganesh Kelkar, Deepu Mathew, Shivaji Ajinath Lavale, and Feba Jacob

Abstract

RNA interference (RNAi) is a promising molecular biology approach for combating viral diseases. In this study, we present an interference strategy targeting the viral *βC1* gene to develop resistant okra lines against *yellow vein mosaic virus* (YVMV). The methodology involves several key steps, including total DNA isolation from infected plants, PCR amplification of the target gene, design of a hairpin RNAi construct harboring the *βC1* gene of the β satellite of *Begomovirus* and utilizing the pRNAi-LIC (CD3-1285) vector. Successful transformation of the construct into the highly susceptible okra cv. Salkeerthi could be achieved using an *in planta* method. Additionally, the stem loop primers designed could amplify the target small interfering RNA (siRNA) through thermal cycling. Screening of transgenic lines is performed under containment conditions, involving the controlled release of infected vectors. This chapter provides detailed insights into the methodology and procedures for performing RNAi for the development of virus-resistant okra (*Abelmoschus esculentus*).

Key words Begomovirus, Genetic modification, In planta transformation, Transgenesis protocol, Ligation-independent cloning (LIC), RNAi

1 Introduction

The yellow vein mosaic disease caused by *yellow vein mosaic virus* (YVMV, genus *Begomovirus*, family Geminiviridae) causes 60–100% crop losses in okra. This disease was first reported in India in 1924 [1], and subsequently named *yellow vein mosaic of okra* in 1940 [2]. YVMV is characterized by the yellowing of veins, chlorosis, smaller leaves and fruits, and stunted growth of the plants [3]. RNA interference is a promising approach in plants to develop the lines with resistance to viral diseases. Naturally, RNAi is operational in eukaryotes as a conserved regulatory mechanism for gene expression. This plays important roles in plant growth, development, and

Muthappa Senthil-Kumar and Hema Ramanna (eds.), *Plant Biotic Stress: Methods and Protocols*, Methods in Molecular Biology, vol. 2966, https://doi.org/10.1007/978-1-0716-4746-2_3,

viral defense. The mechanism supresses the gene expression at RNA level through dsRNA (double-stranded RNA) produced by the transription of ihpRNA (hairpin RNA) construct. The dsRNA processed by RISC (RNA-induced silencing complex) interferes with the expression of specific endogenous or viral genes [4].

The genome of monopartite begomoviruses resembles the DNA-A of bipartite begomoviruses, but they have associated satellite DNA (beta-satellite), which is about half the size of the *Begomovirus* genome components. DNA-A is responsible for multiple protein production involved in rolling circle replication, gene transcription, cell-to-cell and long-distance movement, suppression of host gene silencing, and encapsidation of the viral genome [5]. Beta-satellites have been found associated with monopartite begomoviruses in the Old World. Beta-satellites encode the βC1 protein in the complementary-sense strand, which has important roles in symptom induction and suppression of transcriptional and posttranscriptional gene silencing [6].

Success of RNAi is highly dependent on the ihpRNA cassette and method of construct designing. The ligation-independent cloning constructs enhance the efficiency of RNAi mechanism [7]. The ligation of sense and antisense strands in such vectors involve a single step with T4 DNA polymerase. Restriction enzymes are not involved during ligation, which increases possibilities of getting ihpRNA cassette with correct insertion.

Transformation in okra has been a challenging task, due to high polyphenolic and mucilage content [8–11]. Different protocols have been reported for the okra tissue culture [12–14], but were not helpful in our studies for the somatic embryogenesis for *Agrobacterium*-mediated transformation. In the recent years, *in planta*-mediated *Agrobacterium* transformation is reported, which is more promising in okra [11, 15–17].

2 Materials

2.1 DNA Isolation

1. Liquid nitrogen.
2. 2% cetrimonium bromide (CTAB).
3. 3% polyvinylpyrilidone (PVP).
4. 20 mM ethylenediaminetetraacetic acid (EDTA).
5. 100 mM Tris base or Tris (hydroxymethyl) aminomathane.
6. β-mercaptoethanol.
7. Chloroform.
8. 50 μg/mL isoamyl alcohol, RNaseA.
9. 0.3 M potassium acetate.
10. 2-isopropanol.

11. Ethanol.
12. Nuclease-free water.

2.2 Amplification of βC1 ORF

1. 200 μM dNTPs (1.0 μL).
2. 10 pM Forward primer (1.0 μL) (VβC1F: 5′-TTCTGC TTTTGAACCCCAGT-3′).
3. 10 pM Reverse primer (1.0 μL) (VβC1R: 5′-GCTCATGCAAC AACAAAGGA-3′).
4. Taq polymerase (3U/μL) (0.4 μL).
5. 1× Taq polymerase buffer (2.5 μL).
6. Nuclease-free water.

2.3 ihpRNA Vector Construction

2.3.1 Plasmid Isolation

1. Resuspension buffer A1 (containing RNase).
2. Lysis buffer A2.
3. Neutralization buffer A3.
4. Wash buffer A4 (containing ethanol).
5. Elution buffer.

2.3.2 Digestion of Plasmid Vector

1. 5.0 μg plasmid
2. 10× Buffer Tango
3. 10 U/μL *Sma*I.

2.3.3 Preparation of Sense Strand

1. Forward primer (VLIC1:5′-CGACGACAAGACCC TTTCTGCTTTTGAACCCCAGT-3′).
2. Reverse primer (VLIC2:5′-GAGGAGAAGAGCCC TGCTCATGCAACAACAAAGGA-3′).

   (Underlined sequence are adapters to ligate sense strand with pRNAi-LIC backbone, namely LIC1 and LIC2. These adapter sequence in VLIC1 primer was complementary to the 3′ region of the plasmid backbone at the first *Sma*I site. Adapter sequence in VLIC2 primer was complementary to the 5′ region of the Pdk intron.)

2.3.4 Preparation of Antisense Strand

1. Forward primer (VLIC3: 5′-CCAGCACGGAACCCTTGA GGAGAAGAGCCCT-3′).
2. Reverse primer (VLIC4: 5′-AGAGCACACGACCCTTCGAC GACAAGACCCT-3′).

   (The underlined sequence in VLIC4 primer was similar to the underlined sequence in VLIC1 primer. Similarly, the underlined sequence in VLIC3 was similar to the underlined sequence in VLIC2 primer. Rest of the nucleotides in the VLIC3 primer were complementary to the 3′ region of the Pdk intron of the second SmaI restriction site in the plasmid backbone and rest of the nucleotides in VLIC4 primers were complementary to the 5′ region of the third SmaI restriction site in the plasmid backbone.)

2.3.5 Elution of the Sense and Antisense Fragments

1. Buffer NTI.
2. Buffer NT3.
3. Molecular-grade nuclease-free water.

2.3.6 T4 DNA Polymerase Treatment of PCR Products

1. 50 ng purified PCR product
2. 5× T4 polymerase buffer
3. 100 mM dATP
4. 5 U/μL T4 DNA polymerase
5. Molecular-grade water.

2.3.7 T4 DNA Polymerase Treatment of SmaI Digested Plasmid

1. *Sma*I-digested and purified plasmid.
2. 5× T4 polymerase buffer
3. 100 mM dTTP
4. 5 U/μL T4 DNA polymerase.

2.4 Transformation of Vector into E. coli

1. TransformAid bacterial transformation kit including C-medium, T-solution (A), and T-solution (B).
2. Luria Bertani agar, Miller (Himedia): containing tryptone, yeast extract, sodium chloride, and agar.
3. Kanamycin.
4. Chloramphenicol.

2.5 Confirmation of Insert by Sequencing

1. Reverse primer (VLIC5: 5′-TCTTCTTCGTCTTACACATC-3′).
2. Reverse primer (VLIC6: 5′-AAGACCGGCAACAGGATTC-3′).

2.6 Preparation of Agrobacterium Competent Cells for Transformation

1. *Agrobacterium tumefaciens* strain GV3101.
2. Luria Bertani broth, Miller (Himedia): containing tryptone, yeast extract, and sodium chloride.
3. 25 mg/L rifampicin
4. 100 mM $MgCl_2$
5. 20 mM $CaCl_2$
6. Glycerol.

2.7 In Planta Transformation

1. MS medium.
2. 50 mg/L kanamycin.
3. 5 mg/L chloramphenicol.

4. 25 mg/L rifampicin
5. 5% sucrose.
6. 100 μM acetosyringone.
7. 1% carbendazim.
8. 0.1% mercuric chloride.
9. Ethanol.
10. 500 mg/L cefotaxime.

2.8 Confirmation of siRNA Synthesis

1. Liquid nitrogen.
2. PureLink® Plant RNA Purification Reagent (Invitrogen).
3. 0.1% DEPC treated water.
4. 1.5 M sodium chloride.
5. Chloroform.
6. 2-isopropanol.
7. 70% ethanol.
8. VStem: 5′-GTCGTATCCAGTGCAGGGTCCGAGGTATTC GCACTGGATACGACCACATC-3′.
9. VqPCR-F: 5′-GCGGCCGTAATGAAATTT-3′.
10. VqPCR-R: 5′-CCAGTGCAGGGTCCGAGGTA-3′.
11. 5× reaction buffer (4 μL) (Thermo Scientific).
12. 20 U/μL RiboLock RNase Inhibitor (1 μL).
13. 200 U/μL dNTP Mix 10 mM (2 μL).
14. RevertAid M-MuLV RT (1.0 μL).
15. Nuclease-free water.

3 Methods

3.1 DNA Isolation from YVMV-Infected Plant

DNA can be isolated from YVMV-infected plant, following standard protocol with modifications [18] (*see* **Note 1**):

1. Collect a YVMV-infected okra leaf weighing approximately 100 mg, wipe with 100% ethanol, and grind with liquid nitrogen using prechilled sterile mortar and pestle.
2. Add 1.5 mL of pre-warmed working CTAB buffer, 3% PVP, and 50 μL β-mercaptoethanol to the samples. Transfer contents to 2 mL centrifuge tubes (microtip can be cut for easy sucking of sample), and incubate the tubes at 65 °C for 45 min, with gentle inversions every 15 min.
3. Centrifuge the tubes at 10,800×*g* for 10 min at 4 °C, and transfer the supernatant to fresh 2 mL centrifuge tubes.

4. Add 400 μL chloroform: isoamyl alcohol (24:1) and centrifuge the tubes at 10,800×*g* for 10 min at 4 °C.
5. Transfer the supernatant to fresh tubes, add 2 μL RNaseA (50 μg/mL), and incubate at 37 °C for 30 min.
6. Add 300 μL sodium chloride (3 M) and 200 μL potassium acetate (0.3 M), and incubate the contents in ice for 10 min.
7. Add 400 μL chloroform: isoamyl alcohol (24:1) to the tubes and centrifuge at 10,800×*g* for 10 min at 4 °C.
8. Transfer the supernatant to fresh 1.5 mL tubes, add 500 μL chilled isopropanol, and incubate overnight at −20 °C.
9. Centrifuge the tubes at 10,800×*g* for 10 min at 4 °C, discard supernatant, and wash the pellets with 70% ethanol (100 μL).
10. Centrifuge tubes at 10,800×*g* for 3 min at 4 °C and air-dry the pellets in laminar airflow.
11. Dissolve the pellet in 100 μL sterile molecular-grade nuclease-free water, incubate at 37 °C for 10 min, and store at −20 °C.

3.2 Amplification of βC1 ORF of YVMV

1. Use VβC1F and VβC1R primers for amplification of the specific region of *βC1* ORF in the thermal cycler with initial denaturation (94 °C for 4 min), followed by 32 cycles of denaturation (94 °C for 30 s), annealing (58 °C for 30 s), extension (72 °C for 45 s), and final extension (72 °C for 7 min).
2. Analyze PCR product in agarose gel (1.6%) using horizontal gel electrophoresis unit. First well can be loaded with 4 μL of 100 bp molecular weight marker (step-up 100 bp ladder) (Fig. 1).

3.3 Sequencing of the Coding Region (Partial) of Viral βC1 ORF

1. Perform paired-end sequencing of the PCR product on Sanger platform employing the primers used in PCR.
2. Confirm that the region amplified is *βC1* ORF.

3.4 ihpRNA Vector Construction

3.4.1 Prediction of siRNA Targets in the *βC1* ORF

1. Predict siRNAs in the partial sequences of *βC1* encoding genes (partial cds) using RNAi design tool of Integrated DNA Technologies (IDT).
2. Identify the DICER substrates (siRNA target sites) in target gene using the "Custom Dicer-Substrate siRNA" (DsiRNA) tab in RNAi design tools available at Integrated DNA Technologies (IDT) webpage.
3. Set the duplex RNA GC content at 30–70% and the asymmetrical end stability base pair length of 5, along with other target parameters.

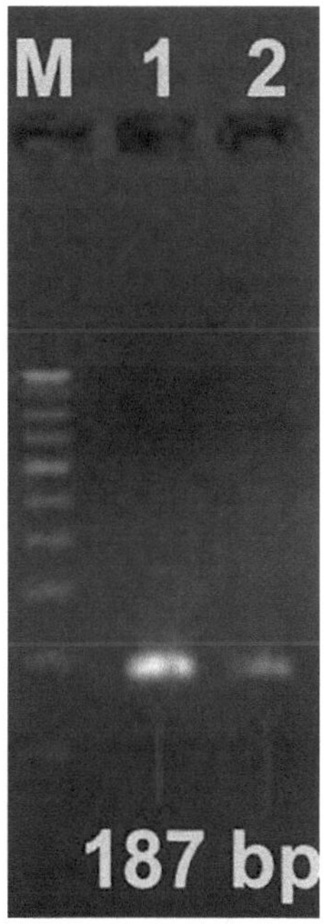

Fig. 1 Amplification of *βC1* gene of YVMV using gene-specific primer combination (VβC1)

4. For higher efficiency, select and analyze the targets identified by the software for siRNA attributes.

3.4.2 Restriction Mapping

1. Subject the partial *βC1* gene sequence to restriction mapping using the tool "Restriction Map" in "Sequence Manipulation Suite" available at www.bioinformatics.org.
2. For this, the sequence should be converted to FASTA format and submitted.
3. The output will be given as a textual map showing the positions of restriction endonuclease cut sites.

3.4.3 Viral Silencing Suppressor Prediction

1. Subject the *βC1* gene sequence to viral silencing suppressor prediction using the tool VSupPred.
2. Convert the nucleotide sequence to protein sequence in FASTA format and upload for prediction of non-viral silencing suppressor regions (NVSR).

3.4.4 Plasmid Isolation

Procurement of pRNAi-LIC vector construct and plasmid isolation (*see* **Note 2**):

1. Inoculate single colony of pRNAi-LIC CD3-1285 (not more than 10 days old) in 10 mL of LB broth containing 50 mg/L of kanamycin and incubate overnight in an orbital shaker at 120 rpm at 37 °C.

2. Add 1 mL of the overnight grown culture to 25 mL of LB broth containing 50 mg/L of kanamycin and incubate in orbital shaker at 120 rpm at 37 °C until O.D reaches to 0.3.
3. Centrifuge the culture at 13,000×*g* for 1 min, resuspend the pellet in 250 μL of Buffer A1, and mix by vortexing.
4. Add 250 μL of Buffer A2 mix by inverting the tubes 8–10 times and incubate at room temperature for 5 min.
5. Add 300 μL of Buffer A3, and mix by inverting tubes until blue color solution changes to colorless (do not vortex to avoid shearing of plasmid).
6. Centrifuge the tubes at 13,000×*g* for 5 min (repeat if clear supernatant is not found), place the Nucliospin® plasmid column in 2 mL collection tube, and add a maximum of 700 μL supernatant into column.
7. Centrifuge the tubes at 13,000×*g* for 1 min and add 600 μL of Buffer A4.
8. Dry the silica membrane by spining the tubes at 13,000×*g* for 2 min.
9. Fix the silica column into 1.5 mL microcentrifuge tubes, add 50 μL of molecular grade DNase free water in the column, and incubate for 1 min at room temperature.
10. Elute the plasmid DNA by spinning the tubes at 13,000×*g* and store at −20 °C.

3.4.5 SmaI Digestion of the Vector

1. Digest the plasmid with SmaI by incubating at 30 °C for over 3 h and then inactivate the enzyme by incubating at 65 °C for 20 min.
2. The reaction mixture contains 30.0 μL (5.0 μg) plasmid, 10.0 μL 10× Buffer Tango, and 5.0 μL SmaI (10 U/μL), with final volume made to 100 μL using sterile molecular-grade nuclease-free water (Fig. 2).

3.4.6 Preparation of Sense Strand Through PCR

1. Amplify the sense strand by using DNA of YVMV-infected plant as template for PCR and oligo pair of primers VLIC1 and VLIC2 to get PCR product-1 (sense strand) (Fig. 3).
2. Use standard PCR components and temperature profile as mentioned in Subheading 3.2.

3.4.7 Preparation of Antisense Strand Through PCR

1. Amplify the antisense strand using PCR product-1 (sense strand) as template for PCR and oligo pair of primers VLIC3 and VLIC4 to get PCR product-2 (antisense strand).
2. The LIC3 and LIC4 are adapters to ligate antisense strand with pRNAi-LIC backbone. Use standard PCR components and temperature profile as mentioned in Subheading 3.2 (Fig. 4).

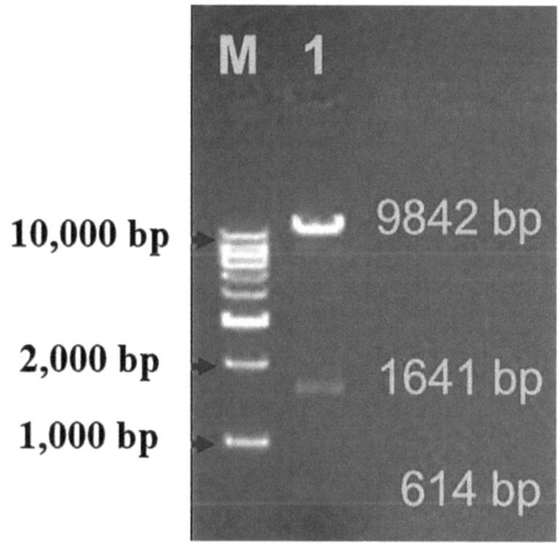

Fig. 2 SmaI-digested pRNAi-LIC vector. M—1Kb DNA ladder, vector backbone (9842 bp), Pdk intron (1641 bp), and *ccdB* gene (614 bp)

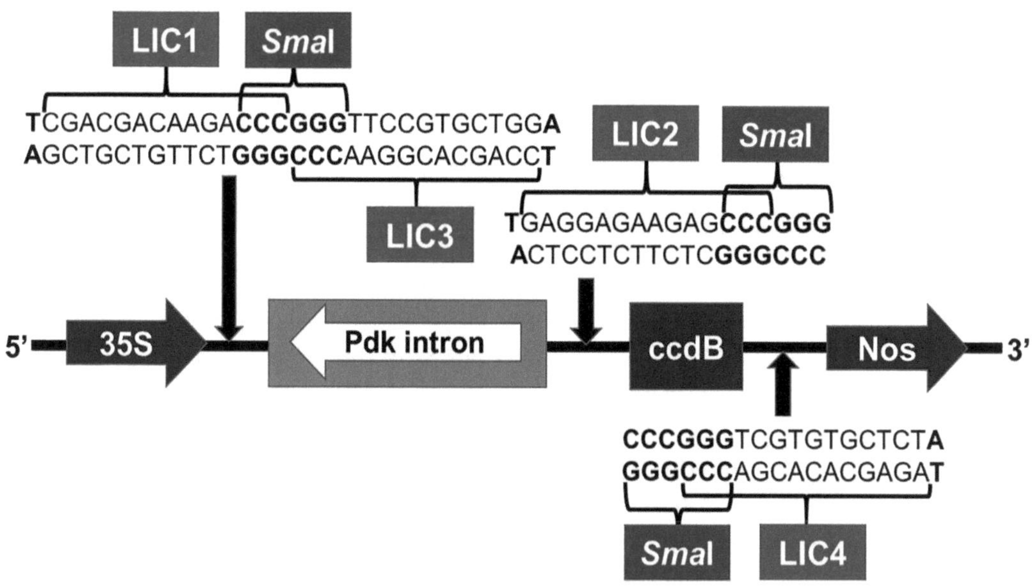

Fig. 3 Position of SmaI and LIC adapters on pRNAi-LIC (CD3-1285) vector. The target fragment of the gene of interest was PCR-amplified using gene-specific primers carrying the adaptors (VLIC1 and VLIC2) to obtain PCR Product 1, which has terminal adapter sequences LIC1 and LIC2. PCR Product 2 was amplified by using PCR Product 1 as template and primers (VLIC3 and VLIC4), and thus has terminal sequences LIC3 and LIC4. PCR products 1 and 2 were further treated with T4 DNA polymerase in the presence of dATP. Vector was digested with SmaI and treated with T4 DNA polymerase in the presence of dTTP. The treated vector and two PCR products with sticky ends were mixed and then transformed into *E. coli* cells (DH5α)

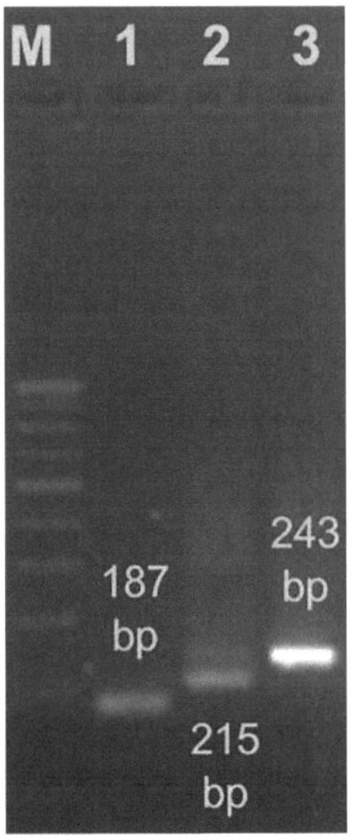

Fig. 4 PCR amplification of (*1*) *βC1* gene with primers VβC1F and VβC1R, (*2*) sense strand with primers VLIC1 and VLIC2, and (*3*) antisense strand with primers VLIC3 and VLIC4, M—StepUp 100 bp ladder

3.4.8 Elution of the Sense and Antisense Fragments and Cloning in Vector

1. Resolve the sense and antisense fragments in agarose gel (1.4%).
2. Slice out the bands with a clean scalpel with minimum amount of agarose from the gel, using a clean scalpel under UV transilluminator.
3. Transfer them to fresh pre-weighed 1.5 mL microcentrifuge tube, check weight (around 150 mg), and store.
4. Perform gel extractions from the excised bands. Commercially available extraction kits, such as Macherey-Nagel NucleoSpin® PCR clean-up gel extraction kit, may be used.
5. Add 300 μL Buffer NTI and maintain the tubes at 40 °C for 5 min to dissolve the gel slice completely.
6. Place NucleoSpin® gel and PCR clean-up silica column in 2 mL collection tube, add 700 μL sample, and centrifuge the tubes at 13,000×*g* for 30 s.
7. Discard the flow-through and place the silica column back in collection tube, add 700 μL Buffer NT3, and centrifuge the tubes at 13,000×*g* for 30 s.
8. Discard the flow-through, place silica column back in the collection tube, and dry the column by spinning the tubes at 13,000×*g* for 1 min.

9. Place silica column in fresh autoclaved 1.5 mL microcentrifuge tube, add 30 μL of molecular-grade nuclease-free water, and incubate the tubes at room temperature for 1 min.
10. Centrifuge the tubes at 13,000×*g* for 1 min, and store the eluted DNA at −20 °C.
11. Quantify the eluted PCR product using a nanovolume spectrophotometer (NanoDrop® ND-1000, NanoDrop Technologies Inc., USA).
12. To clone the PCR Product 1 and PCR Product 2 into the vector, PCR products have to be treated using T4 DNA polymerase, with dATP by incubating the tubes at 22 °C for 30 min.
13. Inactivate the reaction by incubating at 75 °C for 20 min.
14. To clone the PCR Product 1 and PCR Product 2 into the vector, SmaI-digested vector components have to be treated using T4 DNA polymerase, with dTTP by incubating the tubes at 22 °C for 30 min.
15. Inactivate the reaction by incubating at 75 °C for 20 min.
16. To clone the sense and antisense fragments into vector, incubate 10 μL (~50 ng) each of dATP-treated PCR products 1 and 2 along with 6 μL (~100 ng) of dTTP-treated SmaI digested vector backbone at 65 °C for 5 min in a thermal cycler and then at 22 °C for 15–30 min.

3.4.9 Transformation of Vector into E. coli

Prepare the competent cells of *E. coli* strain DH5α using TransformAid bacterial transformation kit. Protocol for the preparation of competent cells is given below:

1. Inoculate single colony of fresh DH5α (not more than 10 days old) in 1 mL of C-medium with no antibiotic and incubate in an orbital shaker at 37 °C at 120 rpm for 3 h.
2. Pellete out the bacterial cells by centrifugation for 1 min at 10,800×*g*, discard supernatant, resuspend the pellet in 300 μL of T-solution, and incubate on ice for 5 min. [T-solution has to be prepared fresh by mixing equal volumes of T-solution (A) and T-solution (B)].
3. Centrifuge the tubes for 1 min at 10,800×*g*, discard supernatant, resuspend the pellet in 120 μL of T-solution, and incubate on ice for 5 min.
4. Add 10 μL of ligation mixture containing 80–100 ng vector DNA to fresh microcentrifuge tubes, and incubate on ice for 2 min.
5. Add 50 μL of the prepared competent cells into the tube containing ligation mixture, mix and incubate on ice for 5 min.

6. Immediately spread 20 μL of culture on fresh LB agar plates (prepared 1 h before) containing antibiotics (kanamycin 50 mg/L and chloramphenicol 5 mg/L), and incubate overnight at 37 °C.

3.4.10 Confirmation of Insert by Colony PCR and Sequencing

1. Confirm the transformed colonies by colony PCR using same primer combination (forward—VLIC1 and reverse—VLIC2) used to prepare the PCR Product 1 (Fig. 5).
2. PCR components and temperature profile required for amplification are mentioned under Subheading 3.2.
3. Grow the positive DH5α cells carrying the cloned plasmids overnight, in LB broth containing 50 mg/L of kamamycin and 5 mg/L of chloramphenicol, for plasmid isolation (as per Subheading 3.4.4).
4. Sequence the plasmid with two reverse primers VLIC5 and VLIC6 [one for sense strand (VLIC5), which is located on intron region and second for antisense strand (VLIC6), which is located on Nos terminator of plasmid backbone in reverse direction] [7].
5. Align the sequences from the sense and antisense strands, and confirm the direction and position of the inserts.

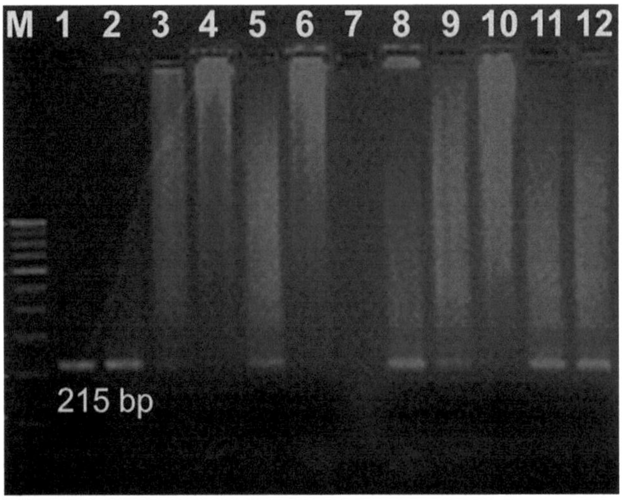

Fig. 5 Colony PCR of DH5α containing ihpRNA-βC1 construct with primers VLIC1 and VLIC2. M—StepUp 100 bp ladder; (*1*)—positive control (target gene *βC1*), (*2–12*) DH5α colonies—colonies *2*, *3*, *5*, *8*, *9*, *11*, and *12* are positive for transformation

3.5 Transfer of the Binary Vector to Agrobacterium tumefaciens

3.5.1 Preparation of Competent Cells

1. Inoculate single colony from a fresh plate of GV3101 in 50 mL LB broth containing 25 mg/L of rifampicin, and incubate overnight at 28 °C at 120 rpm.
2. Transfer 1 mL of the overnight-incubated culture of GV3101 to 50 mL of autoclaved LB broth containing rifampicin (25 mg/L), and maintain at 28 °C at 120 rpm until the OD reaches 0.3.
3. Chill the culture on ice for 10 min, and pellete out by centrifugation at 1000×*g* for 3 min.
4. Resuspend the pelette in 10 mL of $MgCl_2$ (100 mM), and incubate on ice for 1 h.
5. Centrifuge the tubes at 1000×*g* for 3 min, resuspend the pellet in 1 mL of 20 mM $CaCl_2$, and incubate tubes on ice for 30 min.
6. Make aliquots (100 μL each) by adding 20 μL glycerol, and store at −80 °C.

3.5.2 Isolation of Plasmid with ihpRNA Cassette and Transformation of A. tumefaciens

1. Isolate plasmids from *E. coli* as mentioned under Subheading 3.4.4.
2. Thaw the stock of *Agrobacterium* aliquots on ice for 10 min, and add 10 μL of plasmid (100 ng/ μL).
3. Freeze the tubes in liquid nitrogen for 2–3 min, and incubate in water bath at 37 °C for 5 min.
4. Add 1 mL LB broth with no antibiotic into the tube and incubate at 28 °C for 1 h at 160 rpm.
5. Spread 10 μL culture on LB agar plate containing 50 mg/L of kanamycin, 5 mg/L of chloramphenicol, and 25 mg/L of rifampicin, and incubate for 20 h at 28 °C.

3.5.3 Confirmation of Transformation by Colony PCR and Sanger Sequencing

1. Confirm the transformed colonies by colony PCR using the same primer combination used to prepare the PCR product 1 (VLIC1 and VLIC2).
2. Additionally, confirm using the antisense strand specific primers (VLIC3 and VLIC4) also. PCR components and temperature profile required for amplification are mentioned under Subheading 3.2 (Fig. 6).
3. To confirm the transformed colonies by Sanger sequencing, grow the positive GV3101 cells with cloned plasmids overnight in LB broth containing 50 mg/L of kanamycin, 5 mg/L of chloramphenicol, and 25 mg/L of rifampicin, and isolate the plasmids (as per Subheading 3.4.4).
4. Sequence the sense and antisense strands using two reverse primers VLIC5 and VLIC6 (Subheading 2.5).
5. Align the sequences from the sense and antisense strands to confirm the direction and position of the inserts.

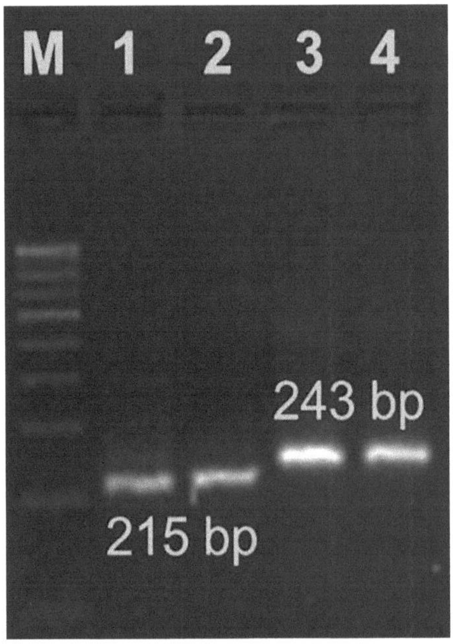

Fig. 6 Colony PCR of *Agrobacterium tumefaciens* GV3101 cells containing ihpRNA-βC1 constrct. M—StepUp 100 bp ladder; (*1–2*)—colonies amplified with sense strand-specific primers (VLIC1 and VLIC2, 215 bp); (*3–4*)—colonies amplified with antisense strand-specific primers (VLIC3 and VLIC4, 243 bp)

3.6 In Planta Transformation of Okra

3.6.1 Preparation of Agrobacterium Suspension for Cocultivation

1. Inoculate single colony of *Agrobacterium* strain GV3101 carrying the binary vector pRNAi-LIC in 10 mL LB broth containing 50 mg/L of kanamycin, 5 mg/L of chloramphenicol, and 25 mg/L of rifampicin, and incubate overnight in an orbital shaker, at 28 °C and 160 rpm.
2. Transfer 1 mL of the overnight-incubated culture of GV3101 to 50 mL LB broth containing 50 mg/L of kanamycin, 5 mg/L of chloramphenicol, and 25 mg/L of rifampicin, and incubate in an orbital shaker at 28 °C and 160 rpm until OD reaches 0.3.
3. Centrifuge the culture at 2700×*g* for 5 min, and resuspend the pellet in 100 mL MS liquid medium containing 5% sucrose and 100 μM acetosyringone.

3.6.2 Preparation of Plant Material and Cocultivation with Agrobacterium

1. Sterilize okra seeds with 1% carbendazim for 10 min, followed by 0.1% mercuric chloride for 10 min and 70% ethanol for 1 min. After every treatment, rinse the seeds 2 to 3 times with autoclaved distilled water.
2. Soak the sterilized seeds overnight in autoclaved distilled water, and then incubate on moist tissue paper for 4 h.
3. Select well-germinated seeds for transformation.

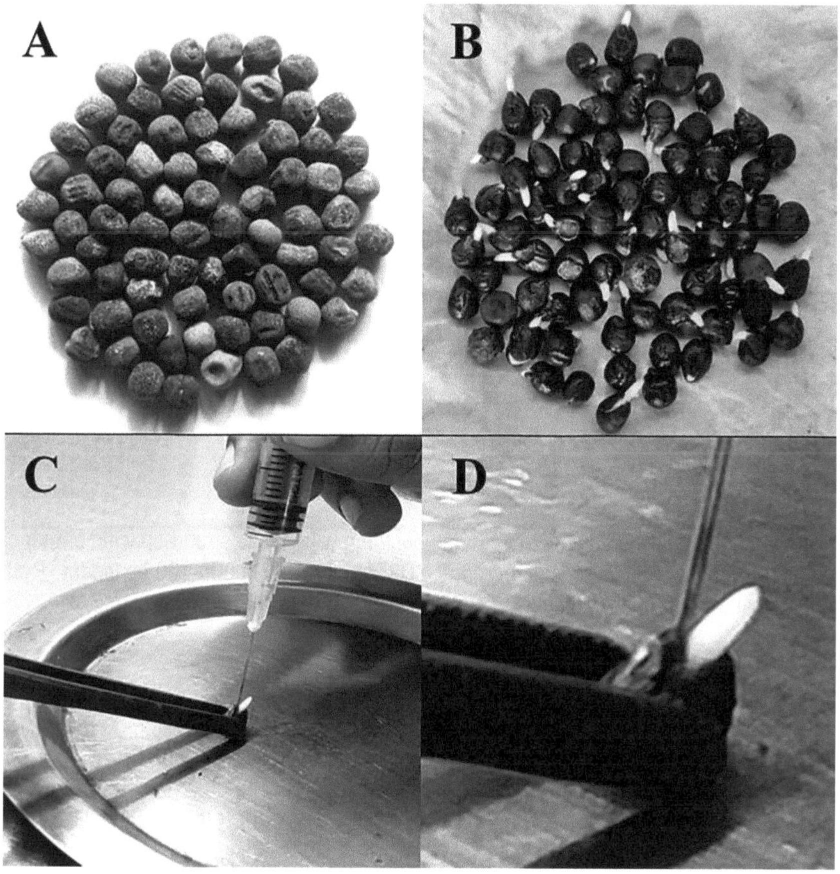

Fig. 7 *In planta* transformation in okra cv. Salkeerthi. (**a**) Seeds; (**b**) sterilized, overnight-soaked seeds germinated on sterilized moist tissue paper; (**c**) plumule prickled with syringe; (**d**) magnified image of plumule pricking

4. With a sterile syringe needle, make injury to the germinated seeds near the plumule (Fig. 7).
5. Cocultivate the prickled seeds with *Agrobacterium* culture (MS liquid medium containing 5% sucrose and 100 μM acetosyringone), in an incubator shaker for 30 min at 28 °C and 100 rpm.
6. Wash the cocultivated seeds once with 500 mg/L of cefotaxime solution, and sow in a paper cup containing soil and FYM (1:1).

3.7 Confirmation of Transformed Plants

1. Isolate DNA from 15-day-old plants (as given under Subheading 3.1)
2. Quantify using nanovolume spectrophotometer, and check the quality by agarose gel electrophoresis.

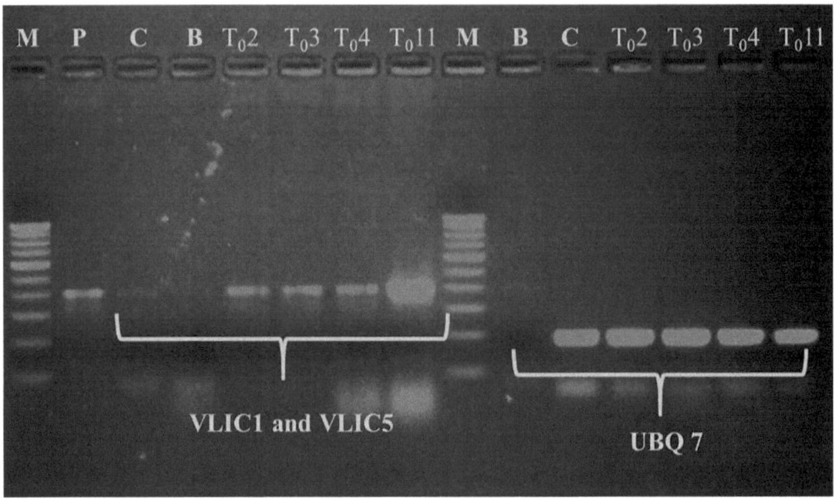

Fig. 8 Confirmation of transformed plants by PCR amplification with sense strand-specific primer combination (400 bp) and a housekeeping gene-specific primer (UBQ7, 187 bp). *M* StepUp 100 bp ladder, *P* plasmid DNA, *C* control (non-transformed) plant, *b* blank, *T02*, *T03*, *T04*, and *T011* transformed plants

3. Perform PCR using sense strand-specific primers VLIC1 and VLIC5 at an annealing temperature of 58 °C.
4. Amplify the housekeeping gene ubiquitin using the UBQ7 primer combination (F 5′-CTCCACCTCGTCCTCCGTC TG-3′ and R 5′-CGCCGCACTCCGCATTAAGG-3′; Annealing temperature 58 °C).
5. PCR components and temperature profile required for amplification are mentioned under Subheading 3.2 (Fig. 8).

3.8 Confirmation of siRNA Synthesis

3.8.1 RNA Isolation

Isolate total RNA from the transgenic lines using Purelink® Plant RNA Purification Reagent (Invitrogen, India) [19]. Procedure for RNA isolation is given below:

1. Grind 100 mg of emerging leaves with liquid nitrogen using DEPC-treated mortar and pestle.
2. Add 1 mL of the plant RNA purification reagent and transfer the homogenate to a 1.5 mL DEPC treated tube.
3. Homogenize the sample by vortexing for 2 min, and incubate the tubes horizontally at room temperature (21 °C) for 5 min.
4. Centrifuge the tubes at 15,600×*g* for 2 min at room temperature, and transfer the supernatant to a 2 mL centrifuge tube.
5. Add 100 μL of 1.5 M sodium chloride to the centrifuge tube, and homogenize by inverting the tube gently. Thereafter, add 300 μL of chloroform and mix the contents by inverting eight times.

6. Centrifuge the tubes at 15,600×*g* for 10 min at 4 °C, and transfer the supernatant to a new tube.
7. Add 500 μL of 2-isopropanol, gently invert and incubate at room temperature for 10 min.
8. Centrifuge the tubes at 15,600×*g* for 10 min at 4 °C, add 1 mL of 75% ethanol to wash the pellet, and thereafter centrifuge the tubes at 15,600×*g* for 1 min at room temperature.
9. Discard the supernatant and air-dry the pellet for 5 min.
10. Dissolve the pellet in 70 μL of autoclaved DEPC water or nuclease-free water and store at −80 °C.

3.8.2 Primer Designing for βC1 Gene-Specific siRNA

The siRNA in *βC1* gene as revealed by IDT software (Subheading 3.4.1) can be used for designing stem loop primers. Reported conserved stem-loop sequence and reverse primers have to be selected as suggested by Kramer [20].

3.8.3 Synthesis of First Strand of cDNA

1. To synthesize the complementary DNA (cDNA) using the small RNA isolated, add 500 ng RNA (10 μL) (DNase free), 15 pmol of stem loop primer specific for siRNA sequences in *βC1* gene sequence (1 μL) (VStem), 5× reaction buffer (4 μL), 20 U/μL RiboLock RNase Inhibitor (1 μL), 10 mM dNTP Mix (2 μL), and 200 U/μL RevertAid M-MuLV RT (1.0 μL) in an RNase-free PCR tube and make up volume to 20 μL with nuclease-free water.
2. Incubate for 30 min at 42 °C, followed by 95 °C for 2 min to stop the reaction.
3. Store the cDNA at −20 °C until use.

3.8.4 PCR with siRNA-Specific Primers

1. Prepare a standard PCR mix containing 100 ng of template cDNA, 200 μM dNTPs, 10 pM each of siRNA-specific forward and reverse primer (VqPCR-F and VqPCR-R), 1 unit of Taq DNA polymerase and 1× Taq polymerase buffer, and molecular-grade nuclease-free water to make volume up to 20.0 μL.
2. Amplify the siNA through thermal cycling.
3. Analyze the PCR product electrophoretically in 2% agarose gel along with 100 bp DNA marker.

3.9 Screening of Transgenic Lines Under Containment Facility

1. Maintain 20-day-old transgenic plants along with control and YVMV-infected plants for 1 week inside an insect cage within the containment facility.
2. Release virus-carrying whitefly vectors inside the insect cage, permit the whiteflies to infect the plants, and observe plants for disease response.
3. Consider the plants free from symptoms even after 30 days of infestation as resistant (Fig. 9).

Fig. 9 Transformed plants (after 30 days of infection) showing resistance to YVMV, maintained within containment along with infected vectors

4 Notes

1. **DNA Isolation from YVMV-Infected Plant**
   DNA can be isolated from YVMV-infected plant using CTAB method [19] with modifications.
2. **Plasmid isolation**
   Procure the backbone of the construct pRNAi-LIC (Genbank GQ870263, CD3-1285, 12 kb), a derivative of pCAMBIA2300 from ABRC (Arabidopsis Biological Resource Centre, The Ohio State University), and isolate the plasmid using plasmid DNA purification kit.

Acknowledgments

This work was supported by doctoral research grant from the Kerala Agricultural University, India. We thank Arabidopsis Biological Resource Centre, The Ohio State University, for providing plasmid.

References

1. Kulkarni CS (1924) Mosaic and other related diseases of crops in the Bombay Presidency. Poona Agric Coll Mag, Pune, 16p
2. Uppal BN, Varma PM, Capoor SP (1940) Yellow vein mosaic of okra. Curr Sci 9:227–228
3. Venkataravanappa V, Reddy CNL, Jalali S, Reddy MK (2012) Molecular characterization of distinct bipartite Begomovirus infecting bhendi (*Abelmoschus esculentus* L.) in India. Virus Genes 44:522–535

4. Castel SE, Martienssen RA (2013) RNA interference in the nucleus: roles for small RNAs in transcription, epigenetics and beyond. Nat Rev Genet 14:100–112
5. Lazarowitz SG (1992) Geminiviruses: genome and structure and gene function. Rev Plant Sci 11:327–349
6. Akhtar S, Akmal M, Khan J (2017) Resistance to cotton leaf curl disease in transgenic tobacco expressing *βC1* gene derived intron hairpin RNA. Indian J Biotechnol 16:56–62
7. Xu G, Sui N, Tang Y, Xie K, Lai Y, Liu Y (2010) One-step, zero-background ligation independent cloning intron-containing hairpin RNA constructs for RNAi in plants. New Phytol 187:240–250
8. Ganesan M, Chandrasekar R, Ranjitha BD, Jayabalan N (2007) Somatic embryogenesis and plant regeneration of *Abelmoschus esculentus* through suspension culture. Biol Plant 51(3):414–420
9. Kabir AH, Sarker KK, Sharmin SA, Islam MS, Alam MF (2008) Callus induction and plantlet regeneration in *Abelmoschus esculentus* (L.) Moench. J Agric Technol 4(1):193–204
10. Anisuzzaman M, Kabir AH, Sarker KK, Jarin S, Alam MF (2010) Micropropagation of *Abelmoschus esculentus* L. (Moench.) for disease free plantlets through meristem culture. Arch Phytopathol 43:460–466
11. Narendran M, Deole SG, Harkude S, Shirale D, Nanote A, Bihani P, Parimi S, Char BR, Zehr UB (2013) Efficient genetic transformation of okra (*Abelmoschus esculentus* (L.) Moench) and generation of insect-resistant transgenic plants expressing the *Cry1Ac* gene. Plant Cell Rep 32:1191–1198
12. Mallela RR, Vutukuri PR, Kanuri M (2009) *In-vitro* plant regeneration and genetic transformation of okra (*Abelmoschus esculentus* L. Moench). Fruit Veg Cereal Sci Biotechnol 3(1):1–6
13. Dhande GA, Patil VM, Raut RV, Rajput JC, Ingle AG (2012) Regeneration of okra (*Abelmoschus esculentus* L.) via apical shoot culture system. Afr J Biotechnol 11(86):15226–15230
14. Irshad M, He B, Liu S, Mitra S, Debnath B, Li M, Rizwan HF, Qiu D (2017) In vitro regeneration of *Abelmoschus esculentus* L. cv. Wufu: influence of anti-browning additives on phenolic secretion and callus formation frequency in explants. Hortic Environ Biotechnol 58(5):503–513
15. Manickavasagam M, Subramanyam K, Ishwarya R, Elayaraja D, Ganapathi A (2015) Assessment of factors influencing the tissue culture-independent *Agrobacterium*-mediated in planta genetic transformation of okra (*Abelmoschus esculentus* (L.) Moench). Plant Cell Tissue Organ Cult 123:309–320
16. Menon R, Sarao NK, Pathak M (2018) *In planta Agrobacterium*-mediated genetic transformation in okra (*Abelmoschus esculentus* (L.) Moench). Appl Biol Res 20(3):221–228
17. Anandan R, Tohidfar M, Kumar PS, Kamaraj A, Vijayakumar N (2019) Transgenic okra plants expressing a *Cry3a* gene for fruit borer (*Helicoverpa armigera*) resistance-an in planta transformation approach. J Pharmacogn Phytochem 8(3):79–83
18. Ghosh R, Paul S, Ghosh SK, Roy A (2009) An improved method of DNA isolation suitable for PCR-based detection of *Begomoviruses* from jute and other mucilaginous plants. J Virol Methods 159:34–39
19. Sreedevi JK (2019) Cloning and characterisation of *myo-inositol phosphate synthase* gene coding for phytates in *Dolichos lablab* L. M.Sc. (Ag.) thesis. Kerala Agricultural University, Thrissur, 112p
20. Kramer MF (2011) Stem-loop RT-qPCR for miRNAs. Curr Protoc Mol Biol Chapter 15: Unit 15.10

Chapter 4

Spot Inoculation of *Begomovirus* by Single Whitefly (*Bemisia tabaci*) Transmission in Mungbean Using Microcage

Shikha Dixit, Pandian Rajendran, Revanasiddappa, Bhimeshwari Sahu, J. Sridhar, Ashish Marathe, and Palaiyur Nanjappan Sivalingam

Abstract

The *Begomovirus* genus is a prominent group of viruses that are transmitted by the whitefly vector, *Bemisia tabaci*, and have been known to cause significant yield reductions in several economically important crops worldwide. Among these viruses, the mungbean yellow mosaic India virus is particularly problematic, causing substantial yield losses in mungbean and proving difficult to control. Therefore, developing new techniques to study the vector-virus systems involving begomoviruses is a research priority for mungbean and other crops affected by these viruses. In this chapter, we present a technique suitable for enhancing the acquisition and inoculation of begomoviruses using a single whitefly. By confining the whitefly in an environment that closely mimics its natural habitat, we are able to increase the transmission efficiency of the virus. This technique can be easily and efficiently replicated in other host-begomovirus systems for various studies.

Key words *Begomovirus*, Whitefly, *Bemisia tabaci*, Mungbean, Microcage, Spot inoculation, Virus transmission

1 Introduction

Whitefly (*Bemisia tabaci*) is a global pest that attacks important crops. Its piercing and sucking mouthparts damage plants and cause problems like honeydew and sooty mold. Whiteflies also transmit plant viruses, posing a threat to food crops [1]. There are 1556 known species of whitefly, and several genera of plant viruses can be transmitted by them, including *Begomovirus*, *Carlavirus*, *Crinivirus*, *Ipomovirus*, and *Torradovirus* [2]. Among those viruses, begomovirus is one of the major concerns for many food crops including mungbean [2]. Begomoviruses are transmitted in a persistent and circulative manner that does not replicate within the whitefly vector [3]. The virus is acquired from an infected plant by

Muthappa Senthil-Kumar and Hema Ramanna (eds.), *Plant Biotic Stress: Methods and Protocols*, Methods in Molecular Biology, vol. 2966, https://doi.org/10.1007/978-1-0716-4746-2_4,

the vector (whitefly) and it is dispensed to a healthy plant which is called transmission of the virus. The successful introduction of viable virus particles into the plant system, for its further replication and disease establishment, are critical steps in virus-vector and host-plant interaction [4].

In past research, whitefly has been used to transmit the virus to healthy plant under controlled conditions to study the factors influencing the complicated outcome of virus-vector relations. Such factors includebiological parameters, such as the development, fecundity, fertility, and mortality rates of the vector along with the successful acquisition of the virus particles from the infected tissue to dispense it into a healthy plant for further replication [2, 3]. In addition, the development of new approaches to management, such as the evaluation of new chemical compounds [2, 5], new cultural approaches, or the selection and development of resistant cultivars, requires the use of whiteflies for virus transmission [6]. The use of whitefly transmission of plant viruses for the selection and development of resistant cultivars in breeding programs is particularly challenging [4]. Effective selection and screening for resistance employ large numbers of plants and there is a need for inoculation of 100% plants to find the few genotypes, which possess resistance genes [7]. These studies use very large numbers of viruliferous whiteflies, often several times per year [8]. To perform such studies, various types of cages were designed and employed to study parameters related to whitefly by covering the whole plant or single leaf or by restricting whitefly within a small area. The whole-plant cage method used to cover one or many plants is easy but difficult for handling during acquisition and inoculation. It is also difficult to monitor the presence and feeding of whitefly on plants, particularly on the non-preferred hosts of whitefly. This method has the chance of whitefly escapes during handling and is not suitable for localized inoculation of the virus. On several occasions, the covering of the whole plant or leaf with plastic cages/falcon tubes was not conducive to suitable growth and development of the whitefly [9]. Based on the above facts, we hypothesize that higher acquisition of begomovirus by whitefly can be achieved by keeping whitefly in a confined area with an environment as close as natural and thereby enhancing inoculation and transmission efficiency. In past research, people have also used clip cage technique in various studies, which require external support to hold the cage along with the leaf on both surfaces. This creates a higher chance of leaf damage, it also interferes with the light passage and changes the microclimate inside the cage. However, these problems are reduced to a greater extent in the microcage developed in this study and the whiteflies are allowed in a near-natural environment. To overcome the operational difficulty faced in performing the virus-vector studies using large-sized cages (macrocage) and other techniques. We have developed and standardized a more efficient, accurate, and easy technique for virus inoculation in mungbean by whitefly. This technique involved the construction of a small-sized cage using a 1.5 mL microcentrifuge tube, which was cut and the end cylindrical portion was removed, and one end was closed using a fine muslin cloth to design a "microcage," which can be used to acquire 1–6 whitefly (detailed procedure of microcage preparation is mentioned in the Methods) [10]. This cage when pasted on the targeted area of the host leaf can lead to accurate and localized virus transmission. The microcage method of virus transmission has many advantages over the conventional method: (i) Microcage can be easily constructed with available material in a common laboratory; (ii) the cage is potentially suitable for whitefly studies, particularly for a screening program for resistance against the vector and *Begomovirus*, molecular studies concerning *Begomovirus* of whitefly-induced host gene expression, proteomics etc.; (iii) this technique can be used for the transmission of begomoviruses in a small area (as low as 1.57 cm^2 leaf areas); (iv) this method provides better environmental conditions like aeration and light to whitefly as close as to that of natural conditions, which ensure higher whitefly survival; and (v) the microcage developed is lesser in weight, size (height 1.6 cm and dia 1 cm), and volume (1.256 cm^3) is well suited to plants with a narrow leaf of a minimum width of 1 cm. In brief, this technique involves collection of the whitefly from the field and rearing. This technique presented higher virus acquisition and transmission to healthy plants with no leaf damage and higher recovery of infected tissues as compared to the conventional macrocage techniques used for virus-vector studies. Apart from that, this technique is easy to perform and cost-effective, which ensures higher adaptability and usage under different conditions as per the researcher's requirements.

To shed some light on the technical aspect, we observed that the attachment and retention of the cage with leaf surface depend on cage weight, base material, and adhesive. The ring-shaped muslin cloth used as a base material was better supported to retain the microcage with leaf surface and with high retention at least 72 h [10]. When tested on other crop species like brinjal and tomato better whitefly survival, high cage retention without gaps, and no leaf damage were observed, suggesting the wider usage and applicability of this cage [10, 13]. The whiteflies prefer to stay on the abaxial surface of the leaf, with the light on the upper surface and better aerations makes it conducive for whitefly feeding. The acquisition and inoculation efficiency of begomoviruses was higher while using the microcage [8]. The probable reasons may be, uniform inoculation pressure and other favorable microenvironmental factors, which ultimately enhanced the acquisition and inoculation of *Begomovirus* [9]. This method may be beneficial in screening crop cultivars against begomoviruses over the field inoculation method as numerous plants escape inoculation by whiteflies [11, 12] and suitable for screening wild species/accessions as they differ in whitefly preference by the presence of trichome, antibiosis, etc. Interestingly, this technique can be easily replicated in other crop species with ease and a similar level of efficiency since the feeding pattern of whitefly was observed to be unaffected but controlled and efficient [9, 11].

2 Materials

2.1 Whitefly Collection and Rearing

1. Rectangular sponge.
2. Plastic box with ventilation holes (Dimensions120 × 100 × 15 mm).
3. 4-in. pot.
4. Soilrite: soil (50:50).
5. Brinjal, tomato, cotton, or any gourd family plant with 15-day-old seedlings.
6. Glasshouse (27 ± 2 °C and 60–70% RH) with preferably natural light (*see* **Note 1**).
7. Insect-proof cage—(60 cm × 60 cm × 90 cm) 60-mesh size nylon net.
8. Classic insect handheld aspirator collection tube length (25 × 75 mm), collecting pipe—25 cm length.

2.2 Microcage Preparation

1. Microcentrifuge tubes (1.5 mL) height—1.6 cm, diameter—1 cm, volume—1.256 cm^3.
2. Fine muslin cloth (approximately 40 mesh).
3. Knife.
4. Scissor.
5. Synthetic resin adhesive—Fevicol or any synthetic-grade glue.
6. Candle.

2.3 Acquisition of Begomovirus *by Non-viruliferous Whiteflies Using Microcage*

1. 8-in. pot
2. Soilrite: soil (50:50).
3. Plastic tray (85 mm × 453 mm × 478 mm).
4. Agro-inoculated plant/virus-infected plant.
5. Polythene cage sized to fit the pot placed inside the cage.
6. Classic insect handheld aspirator, collection tube length (25 × 75 mm), collecting pipe—25 cm length.

7. Polypropylene 1 mL micropipette tips (white color)—71 mm length hole diameter: 8.5 mm/0.32 in.
8. Muslin cloth.
9. Wheat flour—3 g.
10. 200 mL glass beaker
11. 20 mL water
12. Both-end flat spatula—20.32 cm length (8 in.).
13. Mixer with heater or oven.
14. Collected non-viruliferous whitefly (in 1 mL pipette tip).
15. Insect-proof cage—(60 cm × 60 cm × 90 cm) 60-mesh size nylon net.
16. Microcage from Subheading 2.2.

2.4 Transmission of the Virus Through Whitefly to Healthy Mungbean Plants

1. Healthy mungbean plant.
2. Collected viruliferous whitefly (in 1 mL pipette tip).
3. Insect-proof cage—(60 cm × 60 cm × 90 cm) 60-mesh size nylon net.
4. Material required for the construction and assembly of the microcage on the healthy plants remain the same as mentioned in the above list (Subheadings 2.2 and 2.3).

2.5 DNA Isolation and PCR Confirmation

1. DNA extraction reagent: CTAB buffer: 100 mM Tris HCl, pH 8.0; 20 mM EDTA, 1.4 M NaCl, 2% CTAB, 0.02% β-mercaptoethanol.
2. Ethanol (absolute and 70%).
3. Chloroform: isoamyl alcohol (24:1).
4. RNAse A (10 mg/mL).
5. Centrifuge, water bath, Ice tub, microcentrifuge tube, micropipette.
6. Micro-pestle, PCR machine, and incubator.
7. Vortex mixer and mini spin.
8. PCR reagents: 10× PCR buffer, 25 mM MgCl2, 10 mM dNTP, 10 pmole of each primer [primers—F-GTATTTGCAK CAWGTTCAAGA (nucleotide coordinate—225-246): R-AGGDGTCATTAGCTTAGC (nucleotide coordinate—1185-1198)] specific to MYMIV (*see* **Note 6**), *Taq* DNA polymerase, and nuclease-free water.
9. Gel electrophoresis unit including agarose gel (1.0%).

3 Methods

3.1 Rearing of Non-viruliferous Whiteflies

1. Collect whiteflies from the field condition, using a handheld aspirator. Approach the whiteflies slowly and carefully, and gently aspirate them into the device. Avoid applying too much suction that could cause the whiteflies to be damaged.
2. Release the whiteflies into the ventilated plastic box with a leaflet plugged into a wet sponge. Transport the collected flies to the glasshouse quickly and keep them at room temperature.
3. Healthy host plants (*Nicotiana tabacuum* and brinjal) should be planted in 4-in. diameter pots containing Soilrite:soil mix (50:50), and the collected whiteflies should be released on these host plants in an insect-proof cage [(60 cm × 60 cm × 90 cm) 60-mesh size nylon net].
4. Allow whitefly to feed and lay eggs, since begomovirus does not exhibit transovarial transmission, therefore the eggs are free from any begomovirus infestation.
5. Collect whitefly eggs along with the leaf and attach them to the leaf of healthy host plants in a separate cage in a separate glasshouse to avoid cross infection (temperature 27 ± 2 °C and 60–70% RH).
6. The hatched nymphs will feed on healthy plants provided inside the cage, and therefore grow up into a virus-free adult population of whiteflies. From this population, a male and a female should be selected for further development of pure culture (*see* **Note 2**).
7. Rear these whiteflies in a separate cage to maintain a pure culture of non-viruliferous whiteflies, which can be used for transmission studies.

3.2 Construction of a Microcage

1. Cut the regular 1.5 mL microcentrifuge tube close to the 0.5 mL mark, remove the cap, and make a hollow cylinder of 1 cm (Fig. 1a–c).
2. Cut a ring and a disk of muslin cloth and paste on the cut end of the tube using an adhesive, and allow to dry for 15 min (Fig. 1d, e).
3. After the cloth is properly stick to the tube, the cage can be used for the operational purposes.

3.3 Acquisition of the Begomovirus by Non-viruliferous Whitefly Using Microcage

1. Place the begomovirus-infected plants or 30–35 days agro-inoculated mungbean (fifth to sixth trifoliate leaf stage) in an insect-proof cage (*see* **Note 3**).

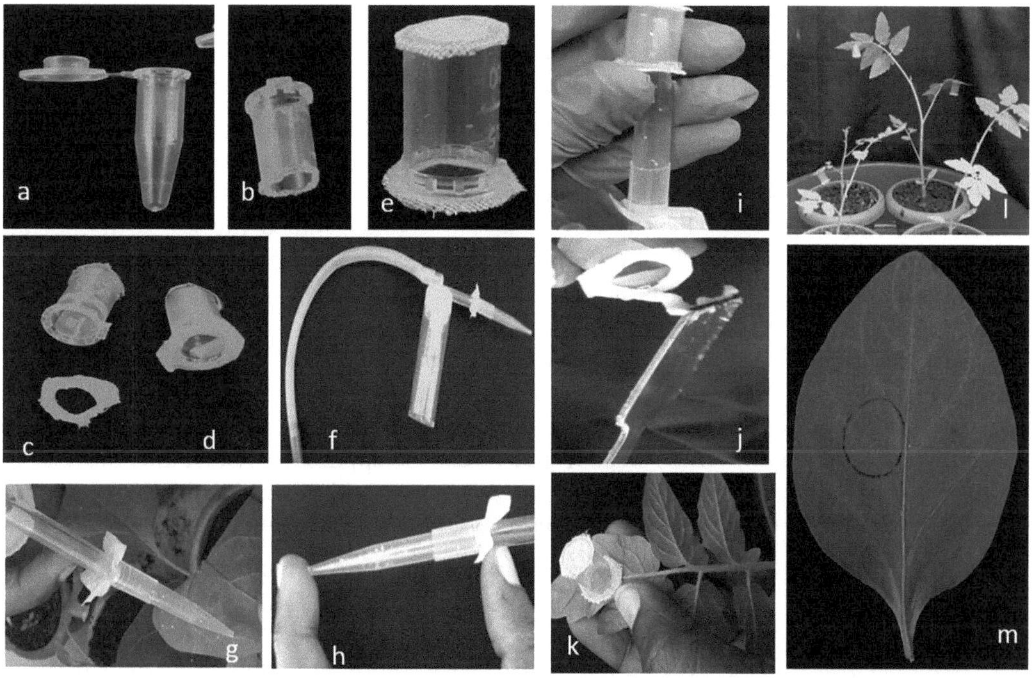

Fig. 1 Steps for the preparation and use of the microcage developed in this study. (**a**) A 1.5-mL microcentrifuge tube was used for the preparation of the microcage. (**b**) The microcentrifuge tube was cut from the 0.5 mL mark at the bottom. (**c**) A ring of muslin cloth was used to cover the rim of the tube. (**d**) The ring was then attached to the cut tube. (**e**) A small piece of muslin, cut to the size of the tube, was pasted on the other end of the tube using Fevicol. (**f**) For collection tube preparation, 1 mm of the tip end of a 1-mL micropipette tip was cut, and the other end was covered with muslin. The tip was then securely fitted onto the suction tube of an aspirator. (**g**) Whiteflies were collected by sucking them using this modified aspirator. (**h**) After collecting the whiteflies, the tip end was closed with one finger, and the other end of the microtip was closed with muslin using the thumb. (**i**) To release the whitefly into the microcage, hold the microcage with the muslin-covered side facing up, and with the tip end of the microtip pointing up, remove the finger and place the tip up into the microcage, allowing the whitefly to move into the microcage. (**j**) Wheat glue was applied to the rim of the microcage with the whiteflies attached. (**k**) The microcage was then immediately attached to the abaxial leaf surface. (**l**) After the cage has attached, return the leaf with the microcage to its natural position. (**m**) The area of the leaf in contact with the microcage was marked for sample collection after inoculation for 24 h

2. Fix a fine-tipped 1 mL microtip to the classic standard handheld aspirator suction tube (Fig. 1f), cut the microtip using scissor to increase the aperture size, and cover the other end with muslin cloth and fix firmly to the aspirator's suction tube so that the whiteflies remain in the microtip rather than being suck into the collection bottle (Fig 1g–h).
3. Collect non-viruliferous whiteflies into microtip of the aspirator by gentle sucking, then close with a finger to prevent whitefly escape (Fig. 1i).

4. Hold the microcage with the muslin cloth end facing up, and insert cut-end microtip into open end of microcage, then release the whiteflies gradually.
5. Before turning the healthy host plant's leaf surface upside down, apply the wheat gum to the cage base (Fig. 1j). Attach the microcage containing whiteflies to the leaf's lower side (Fig. 1k). After a few seconds, return the leaf to its normal position for proper whitefly feeding for the next 24 h (Fig. 1l) (*see* **Note 4**).

3.4 Spot Inoculation by Viruliferous Whitefly

1. After 24 h, collect a single whitefly using handheld aspirator by removing the muslin cloth attached to the top of the microcage. The same releasing procedure as described in Subheading 3.3, **steps 4** and **5**, can be applied to release the viruliferous whitefly into the microcage and place the microcage on the mungbean leaf.
2. Attach the microcage containing a single virulent whitefly to a healthy mungbean plant leaf for a 24-h inoculation period.

3.5 DNA Extraction and PCR Confirmation

1. After 24 h (IAP), mark the place of microcage-attached leaf surface by circular ring to indicate the location of the whitefly feeding area (Fig. 1m).
2. Collect leaf disc (1.57 cm^2) samples from the mungbean leaf for DNA extraction, and kill whitefly mechanically (*see* **Note 5**).
3. Use a micro-pestle and liquid nitrogen to powder the leaf disc samples in a centrifuge tube (1.5 mL), to extract genomic DNA using CTAB method and dissolve in 30μL of sterile distilled water or 1× TE buffer.
4. Measure the quality and quantity of DNA using a spectrophotometer and verify the DNA integrity by performing a gel check on 1.0% agarose gel and store samples in a deep freezer for future use.
5. Prepare the PCR mixture containing 0.5μg of genomic DNA, 2.5μL of 10× PCR buffer, 1.5μL of 25 mM MgCl2, 0.5μL of 10 mM dNTP, 10 pmole of each primer, and 0.5 unit of *Taq* DNA polymerase. The reaction volume should be set to 25μL in a thermal cycler (thermal cycler profile—*see* **Note 6**). After the run, the tubes can be stored at 4 °C.
6. To prepare the agarose gel, take 1% agarose and prepare it in 1× Tris acetate EDTA buffer (TAE) buffer. Add 0.05μg/mL of ethidium bromide to the agarose gel and mix properly. To load the PCR samples onto the agarose gel, add 6× loading dye to the DNA samples and PCR samples. Load the samples into the agarose gel.
7. Confirm all inoculated plants for the presence of MYMIV with PCR. The expected product size is approximately 970 bp.

4 Notes

1. This method is best used within the temperature range of 27–33 °C and relative humidity (RH) of 70–80%. Temperatures above 33 °C can cause the wheat adhesive to dry out, leading to leaf damage during cage removal. Similarly, under high humidity (above 80% RH), the wheat adhesive becomes hygroscopic, causing the cage to fall off the leaf.
2. The pure culture of the whitefly genetic group (Asia II-7) is perpetuated on eggplant (VNR 212 brinjal) in an insect-proof cage (60 cm × 60 cm × 90 cm) with a 60-mesh size nylon net on three sides and glass (on the top and front door) under glasshouse conditions (temperature 27 ± 2 °C and 60–70% RH). It is preferable to grow whitefly cultures under natural light. The whiteflies are confirmed to be of Asia II-7 type using PCR confirmation, as mentioned in (10).
3. The source of begomovirus can be symptomatic plants infected in the field, or we can grow mungbean agro-inoculated plants by multiplying the agrobacterium culture containing viral DNA insert and allowing the transformation of injured freshly germinated mungbean seeds. This will result in virus-infected agro-inoculated plants. A detailed protocol for agro-inoculation is provided in (8).
4. Two different adhesives, synthetic resin adhesive and wheat flour gum, can be tested individually to attach the microcage on the abaxial leaf surface of the eggplant. (i) For synthetic resin adhesive, commercially available Fevicol can be used directly on the base of the cage, while (ii) wheat flour gum can be prepared using locally available wheat flour by adding and mixing 3 g of wheat flour in 20 mL of water in a 200 mL beaker. The mixture can be boiled in a microwave oven for 2.5 min with intermittent mixing of the content once every approximately 40 s using a spatula. Once the mixture becomes sticky, heating can be stopped and allowed to cool in the room. It can then be applied on the cage base after releasing the whitefly into the microcage.
5. Carefully mark the upper surface of the leaf corresponding to the attachment point of the cage with a fine marker before removing the cage from the leaf. Collect the marked area, or the leaf disc (i.e., feeding spot), and store it at −80 °C for total DNA extraction. Maintain the inoculated plants under an insect-proof polythene cage for symptom development. Grind each leaf disc sample individually with liquid nitrogen using a micro-pestle in a 1.5 mL microcentrifuge tube for DNA isolation. Extract the DNA using the CTAB method of DNA extraction as per (14). Check the quality of the DNA by 0.8%

agarose gel electrophoresis and quantify it using the Nanodrop ND 1000 UV/VIS spectrophotometer. Carry out the PCR with 100 ng of total DNA extracted from the leaf disc using MYMIV-specific primers.

6. Use the following thermal cycler profile: initial denaturation at 94 °C for 5 min, denaturation of 94 °C for 1 min, annealing temperature of 55 °C for 2 min, and the synthesis at 72 °C for 3 min for 30 cycles. Keep the final extension at 72 °C for 10 min. After the run, store the tubes at 4 °C. Prepare 1% agarose gel pre-stained with ethidium bromide (0.05μg/mL) in 1× Tris acetate EDTA buffer (TAE). Mix the samples to 6× loading dye and check on agarose gel. The expected product size is approximately 973 bp.

Acknowledgments

The authors sincerely thank ICAR-National Agricultural Science Fund, New Delhi (F. No. NASF/ABP-7021/2018-19/252) for providing funding assistance.

References

1. Muñiz M, Nombela G (2001) Bemisia tabaci: a new clip-cage for biological studies. Eur Whitefly Stud Netw 90(1):618–624. https://core.ac.uk/download/pdf/36023466.pdf
2. Kumar V, Kular JS, Kumar R, Sidhu SS, Chhuneja PK (2020) Integrated whitefly Bemisia tabaci (Gennadius) management in Bt-cotton in North India: an agroecosystem-wide community-based approach. Curr Sci 119(4): 618–624
3. Afreen B, Baghel G, Khan AA, Naqvi QA (2020) Virus-vector relationship of yellow vein disease of calendula and the whitefly. J Plant Prot Sci 2(2):60–65
4. Mishra GP, Dikshit HK, Ramesh SV, Tripathi K, Kumar RR, Aski M et al (2020) Yellow mosaic disease (YMD) of Mungbean (Vigna radiata (L.) Wilczek): current status and management opportunities. Front Plant Sci 11:918. https://doi.org/10.3389/fpls.2020.00918
5. Sahu B, Dokka N, Mahajan MM, Sharma KC, Singh HK, Marathe A et al (2021) Begomoviruses affecting pulse and vegetable crops are unevenly distributed in distinct agroecological zones of eastern India. J Phytopathol 169(4): 209–228
6. He C, Liu S, Liang J, Zeng Y, Wang S, Wu Q et al (2020) Genome-wide identification and analysis of nuclear receptors genes for lethal screening against Bemisia tabaci Q. Pest Manag Sci 76(6):2040–2048
7. Fiallo-Olivé E, Pan LL, Liu SS, Navas-Castillo J (2020) Transmission of begomoviruses and other whitefly-borne viruses: dependence on the vector species. Phytopathology 110(1): 10–17
8. Sivalingam PN, Dokka N, Mahajan MM, Sahu B, Marathe A, Kaushal P et al (2022) Achieving maximum efficiency of Mungbean yellow mosaic India virus infection in mungbean by agro inoculation. 3 Biotech 12(1):29. https://doi.org/10.1007/s13205-021-03088-w
9. Mware B, Narla R, Amata R, Olubayo F, Songa J, Kyamanyua S et al (2009) Efficiency of cassava brown streak virus transmission by two whitefly species in coastal Kenya. J Gen Mol Virol 1(4):40–45
10. Sivalingam PN, Sahu B, Mahajan MM, Sridhar J, Dokka N, Marathe A et al (2022) Enhanced transmission efficiency of begomoviruses by a single whitefly (*Bemisia tabaci*) using a micro cage technique. J Gen Plant Pathol 88(6):378–382. https://doi.org/10.1007/s10327-022-01086-1
11. Vidavsky F, Leviatov S, Milo J, Rabinowitch HD, Kedar N, Czosnek H (1998) Response

of tolerant breeding lines of tomato, Lycopersicon esculentum, originating from three different sources (L. peruvianum, L. pimpinellifolium and L. chilense) to early controlled inoculation by tomato yellow leaf curl virus (TYLCV). Plant Breed 117(2): 165–169

12. Bellotti AC, Arias B (2001) Host plant resistance to whiteflies with emphasis on cassava as a case study. Crop Prot 20(9):813–823
13. Naveen NC, Chaubey R, Kumar D, Rebijith KB, Rajagopal R, Subrahmanyam B et al (2017) Insecticide resistance status in the whitefly, Bemisia tabaci genetic groups Asia-I, Asia-II-1 and Asia-II-7 on the Indian subcontinent. Sci Rep 7:40634. https://doi.org/10.1038/srep40634
14. Rouhibakhsh A, Priya J, Periasamy M, Haq QMI, Malathi VG (2008) An improved DNA isolation method and PCR protocol for efficient detection of multicomponent of begomovirus in legumes. J Virol Methods 147(1): 37–42

Chapter 5

One-Step RT-LAMP: A Method for Closed-Tube Colorimetric Detection of RNA Viruses in Plants

Smita Nair and K. Midhuna Madhu

Abstract

A good virus indexing platform is essential for early detection of plant viruses and for providing virus free planting materials. Majority of the plant viruses are RNA viruses and the available detection methods include ELISA and RT-PCR. The ELISA is time consuming, and RT-PCR needs post PCR sample handling, predisposing to sample cross contamination. The LAMP (loop-mediated isothermal amplification) is an auto-cycling strand displacement isothermal DNA synthesis catalyzed by a strand displacing DNA polymerase like the *Bst* DNA polymerase. The RT-LAMP (reverse transcription LAMP) is a specific, highly sensitive, and relatively fast diagnostic method for RNA viruses. In one-step RT-LAMP, the total RNA extracted from symptomatic samples can be directly amplified under isothermal conditions by incorporating a reverse transcriptase and an RNase inhibitor in the LAMP assay. The inclusion of hydroxy naphthol blue dye in the RT-LAMP reaction mixture facilitates closed-tube visual detection of positive samples by a color change from violet to sky blue. Molecular fidelity of the LAMP amplicons can be assessed through restriction profiling.

Key words RT-LAMP, RNA viruses, *Bst* DNA polymerase, Colorimetric detection, Hydroxy naphthol blue

1 Introduction

The rapid and accurate detection of RNA viruses in plants is of utmost importance for disease management and crop protection. The traditional methods for virus detection often involve time-consuming and labour-intensive procedures, making them less suitable for rapid and on-site diagnostics. However, advancements in molecular biology techniques have led to the development of innovative methods that overcome these limitations. Isothermal amplification techniques such as helicase-dependent amplification (HDA), nucleic acid sequence-based amplification (NASBA), loop-mediated isothermal amplification (LAMP), and recombinase polymerase amplification (RPA) allow nucleic acid amplification at a single temperature, facilitating on-site disease diagnosis in a

Muthappa Senthil-Kumar and Hema Ramanna (eds.), *Plant Biotic Stress: Methods and Protocols*, Methods in Molecular Biology, vol. 2966, https://doi.org/10.1007/978-1-0716-4746-2_5,

low-resource environment [1, 2]. One such technique is the one-step reverse transcription loop-mediated isothermal amplification (RT-LAMP), a one-step detection method for RNA viruses in plants.

1.1 LAMP: An Isothermal Amplification Technique

The LAMP is an auto-cycling, strand displacement DNA synthesis catalyzed by a strand displacing DNA polymerase using four specially designed primers, a pair of inner primers and a pair of outer primers, which bind to six distinct regions on a target producing 10^9 copies of a target in less than an hour. The inner primers have regions binding to both sense and antisense sequences of the target DNA, and by the interaction of the forward and reverse internal primers binding at each end of the target region, an intermediate structure with a stem-loop at both end is formed, which acts as the template for further rounds of isothermal amplification [1]. The isothermal amplification is typically carried out at 60–65 °C by a DNA polymerase with strand displacing property like the *Bst* DNA polymerase. Additionally, loop primers can be included in the LAMP reaction to increase the rate of amplification [3]. Due to its high specificity and sensitivity, LAMP has become a popular tool for pathogen detection.

The LAMP products consist of a mixture of stem-loop DNAs of different stem lengths and alternately inverted repeats of the target sequence forming cauliflower like structures with multiple loops [1]. In agarose gel electrophoresis, the amplicons have a ladder like pattern. We can utilize colorimetric indicators, such as pH-sensitive dyes or metal ion indicators, to visualize positive amplification without the need for gel electrophoresis or additional processing steps. A colorimetric detection method for LAMP positives has been developed by adding a metal ion indicator, hydroxy naphthol blue (HNB), to the reaction mixture prior to amplification and by monitoring the color change from purple to sky blue in positive reactions in response to change in the Mg^{2+} ion concentration [4]. The molecular identification of LAMP amplicons can be made by restriction analysis using a restriction enzyme, which has recognition sites between the LAMP internal primer binding region [1, 5].

1.2 One-Step RT-LAMP: Combining Reverse Transcription and LAMP for RNA Virus Detection in Plants

The one-step RT-LAMP combines the principles of reverse transcription and LAMP to detect the presence of RNA viruses [6, 7]. Using a heat-stable reverse transcriptase, RT-LAMP can directly detect the RNA virus from the total RNA purified from a symptomatic plant sample [8]. In one-step RT-LAMP, the template is the total RNA isolated from diseased plant sample. In addition to the components of a LAMP reaction, the one-step RT-LAMP reaction includes M-MLV reverse transcriptase and RNase inhibitor. The reverse transcription and isothermal amplification take place at around 65 °C. The RT-LAMP reaction can be set on a dry bath or a water bath. The inclusion of HNB dye in the RT-LAMP reaction mixture prior to amplification enables closed-tube visual detection of positives.

One-step RT-LAMP has been successfully applied for the detection of economically important viruses like the banana bract mosaic virus [7], tomato yellow leaf curl virus [9], cucumber green mottle mosaic virus [6], wheat yellow mosaic virus [10], etc. The one-step RT-LAMP method has demonstrated exceptional sensitivity and specificity in the detection of RNA viruses in plants [7]. Its high specificity is attributed to the use of multiple primers that target different regions within the viral genome. This multi-primer approach enhances the specificity of amplification, reducing the chances of false positives caused by nonspecific primer binding and amplification. Moreover, RT-LAMP eliminates the need for sophisticated equipment and complex sample preparation steps, making it accessible even to resource-limited settings [11]. The closed-tube detection system eliminates the need for additional reagents and post-amplification processing steps, thereby reducing the overall cost of the assay, and eliminating the possibility of sample cross contamination. The rapid amplification kinetics of RT-LAMP, typically detecting within 30 min to an hour, make it an ideal tool for time-sensitive applications such as phytosanitary inspections, quarantine operations, and field surveillance. The combination of high sensitivity, specificity, and efficiency makes one-step RT-LAMP a promising technique for the rapid and reliable detection of RNA viruses in plant samples, facilitating effective disease management and safeguarding agricultural productivity. This chapter aims to serve as a comprehensive guide for researchers and molecular plant pathologists interested in implementing one-step RT-LAMP, a powerful tool for RNA virus detection in plant samples.

2 Materials

2.1 Isolation of Total RNA

1. Plant sample (100 mg of leaf tissue is preferable).
2. DEPC-treated water.
3. RNeasy plant mini kit from Qiagen or any other source.
4. Liquid nitrogen.
5. DEPC-treated mortar and pestle.
6. DEPC treated plastic wares.
7. Microcentrifuge.

2.2 RT-LAMP

1. RT-LAMP primers—forward internal primer, backward internal primer, forward external primer, backward external primer, forward-loop primer, and backward-loop primer.

Table 1
Components of the primer mixture for setting RT-LAMP reaction

Primer	Concentration (μM)
Forward internal primer (FIP)	20
Backward internal primer (BIP)	20
Forward external primer (F3)	5
Backward external primer (B3)	5
Forward-loop primer (LF)	10
Backward-loop primer (LB)	10

Prepare a primer mixture containing all the six RT-LAMP primers according to Table 1 (*see* **Note 1**).

2. RNA sample.
3. M-MLV reverse transcriptase.
4. RNase inhibitor.
5. *Bst* DNA polymerase enzyme.
6. 10× ThermoPol buffer, with 2 mM $MgSO_4$
7. Betaine.
8. dNTPs.
9. $MgSO_4$.
10. HNB dye.
11. Molecular biology-grade water.

    Now prepare the RT-LAMP reaction mixture according to Table 2 (*see* **Note 2**).
12. Dry bath.

2.3 Agarose Gel Electrophoresis

1. Agarose.
2. Ethidium bromide (10 mg/mL stock).
3. 50× TAE buffer, pH 8.0 (242.2 g of Tris buffer, 100 mL 0.5 M EDTA, and 57.1 mL glacial acetic acid made up to 1 L)
4. 6× gel loading dye
5. 1 Kb DNA ladder
6. Horizontal gel electrophoresis unit with power pack.

2.4 Restriction Analysis

1. Restriction enzyme.
2. 10× buffer for restriction enzyme.
3. Dry bath.

Table 2
Components of one step RT-LAMP reaction mixture with HNB-based colorimetric detection

Reagents	Final concentration
RT-LAMP primers	F3/B3—0.2 μM, FIP/BIP—0.8 μM, LF/LB—0.4 μM (1 μL from the primer mixture prepared as per Table 1)
Total RNA	2 μg
Reverse transcriptase (200 U/μL)	100 U
RNase inhibitor (40 U/μL)	20 U
Bst polymerase (8 U/μL)	8 U
10× ThermoPol buffer with 2 mM $MgSO_4$	1×
Betaine	1 M
dNTPs	1.6 mM each
$MgSO_4$	4 mM
HNB	120 μM
Molecular biology grade water	To make up the volume to 25 μL

3 Methods

3.1 Isolation of Total RNA

1. Ensure all procedures are carried out at room temperature.
2. Begin by grinding 100 mg of leaf tissue using liquid nitrogen in a prechilled mortar and pestle, until a fine powder is obtained.
3. Transfer the powdered sample into a 1.5 mL microcentrifuge tube containing the extraction buffer provided with the kit.
4. Follow the RNA isolation protocol provided by the manufacturer, using the reagents supplied in the kit.
5. Once the RNA isolation is completed, store the RNA sample at −80 °C until use (*see* **Note 3**).

3.2 Designing RT-LAMP Primers

1. Design the RT-LAMP primers using the software PrimerExplorer version 5.0 (primerexplorer.jp/lampv5e/index.html) with default parameters (*see* **Note 4**).
2. Select a target region of viral genome for designing the RT-LAMP primers (*see* **Note 5**).
3. Retrieve the sequence from NCBI database.
4. For regular primers (external and internal primers), open "primerexplorer.jp/lampv5e/index.html", click on the browse button.
5. Choose and upload the target sequence file.

6. Choose the corresponding parameter set.
7. Click on the primer design button. The regular primers will be designed.
8. After getting the regular primer sets, save the file (Primer Information File).
9. Then, for loop primers, click on the browse button and open the primer information file for the selected regular primers.
10. Click on the primer design button.
11. Save the loop primer file. We now have a pair of RT-LAMP external primers (F3 and B3), internal primers (FIP and BIP), and loop primers (LF and LB) (*see* **Note 6**).

3.3 One-Step RT-LAMP Reaction

1. One-step RT-LAMP is performed using total RNA as the template.
2. Prepare the RT-LAMP primer mixture containing all the six primers as per Table 1.
3. Prepare the RT-LAMP reaction mixture as per the Table 2 (*see* **Note 7**).
4. Include a positive control, negative control, and a no-template control (NTC) along with each assay.
5. Prepare the positive control using the total RNA isolated from a virus confirmed plant.
6. Prepare the negative control using the total RNA isolated from a healthy plant (you can take samples from tissue culture-raised virus-indexed plants, if available).
7. Prepare NTC without the template RNA, and instead, add molecular biology-grade water for check (*see* **Note 8**).
8. Set the reaction for 60 min at 65 °C on a dry bath/water bath. The time required for amplification can vary depending on the initial template concentration.

3.4 Analysis of RT-LAMP Amplicons

3.4.1 Agarose Gel Electrophoresis

1. Make 1× TAE buffer, and to 100 mL 1× TAE buffer, add 2 g of agarose to prepare 2.0% agarose gel. Heat the mixture in a microwave oven till a clear solution is obtained (*see* **Note 9**).
2. Allow the solution to cool down to room temperature.
3. Add 3.0 μL ethidium bromide to the solution and mix well.
4. Place the gel casting tray on the caster by closing both the sides and wipe the tray with 70% ethanol.
5. Pour the warm gel with ethidium bromide into the casting tray with the comb avoiding air bubbles and allow the gel to solidify for a few minutes.
6. Gently remove the comb after solidification of the gel, and place the gel along with tray in the electrophoresis tank

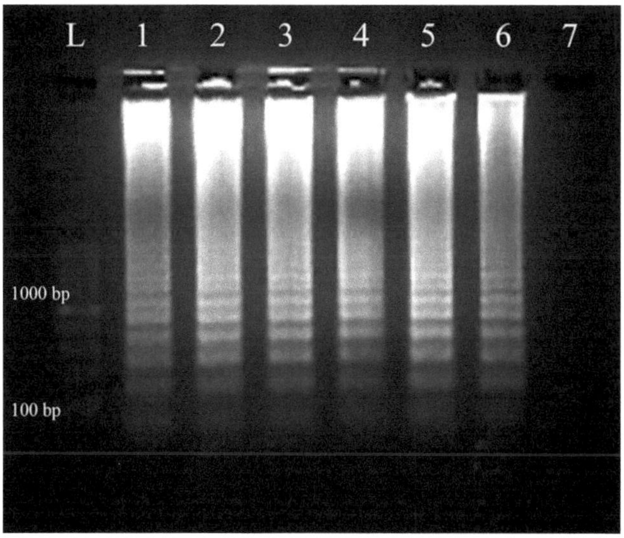

Fig. 1 Detection of RT-LAMP amplicons by agarose gel electrophoresis. Lanes L—1 Kb plus DNA ladder; *1–6*—RT-LAMP positive samples; *7*—no template control. The positive RT-LAMP amplicons produce a ladderlike pattern on the agarose gel, while the negative samples show no amplification/band on the agarose gel. (This forms a part of the study by Kovileri et al. [7] and the original publication may be referred for further information)

containing 1× TAE buffer with wells placed at the cathode (negative electrode) end.

7. Load DNA ladder (4.0 μL) on to the first well after mixing it with gel loading dye.
8. Load the samples (RT-LAMP products) to be analyzed into the wells after mixing with gel loading dye.
9. Run the electrophoresis at 70 V.
10. Finally, visualize the bands using gel a documentation unit.
11. Observe the positive amplicons showing ladder like bands on 2% agarose gel representing the stem-loop DNA with inverted repeats of the target sequence, whereas healthy sample and NTC shows no amplification (Fig. 1; *see* **Note 10**).

3.4.2 Colorimetric Detection

1. Use hydroxy naphthol blue dye for colorimetric detection of RT-LAMP amplicons.
2. Add 120 μM HNB dye to the RT-LAMP reaction mixture before the start of the reaction, causing the mixture to turn violet.
3. After the RT-LAMP reaction, a positive result, indicating the presence of the target RNA, is identified as a visible color change in the reaction tube from violet to sky blue. This color

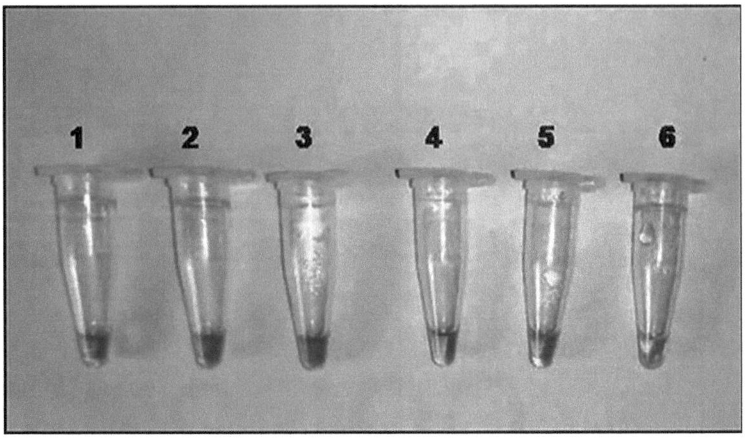

Fig. 2 Closed-tube colorimetric analysis of RT-LAMP products using HNB dye. *1*—no template control; *2* and *3*—RT-LAMP negative samples; *4–6*—RT-LAMP positive samples. The positive samples show sky blue color after the RT-LAMP reaction, while NTC and negative samples remain violet. (This forms a part of the study by Kovileri et al. [7] and the original publication may be referred for further information)

change is attributed to the inclusion of HNB dye in the reaction (*see* Fig. 2).

4. Ensure negative samples retain violet color after the RT-LAMP reaction (*see* **Note 11**).

3.4.3 Restriction Profiling of RT-LAMP Product

Perform restriction digestion of RT-LAMP product with a specific enzyme to demonstrate the fidelity of the RT-LAMP assay.

Identification of Suitable Restriction Enzyme

1. After obtaining the amplified DNA product, choose an enzyme for digestion.
2. Identify a restriction enzyme having internal cut site in the F2-B2 (internal primers) flanking region.
3. Open http://nc2.neb.com/NEBcutter2/ (*see* **Note 12**).
4. Paste/enter the sequence (sequence between F2 and B2).
5. Set the preference and give the project name.
6. Click on submit.
7. Then you can select the enzyme of your choice to digest (*see* **Note 13**).

Restriction Digestion of RT-LAMP Product

1. Digest the RT-LAMP product using the chosen restriction enzyme.
2. Set the restriction digestion reaction in a 0.2 mL PCR tube as given in Table 3.

Table 3
Components of restriction digestion reaction for the analysis of RT-LAMP products

Components	Volume (μL)
RT-LAMP product	10
Restriction enzyme (10 U/μL)	2
Buffer (10×)	5
Molecular biology-grade water	33
Total volume	50

3. Incubate the reaction at 37 °C for 3 h, and terminate the reaction with heat inactivation at 65 °C for 20 min (*see* **Note 14**).
4. Analyze the restricted product on 2.5% agarose gel stained with ethidium bromide along with a 50 bp DNA ladder.
5. Depending on the position and number of restriction sites for the selected enzyme present in the F2-B2 region, analyze the agarose gel to observe differently sized bands.

4 Notes

1. It is better to make a primer mixture of 50 μL as stock, and 1 μL primer from this can be used for each RT-LAMP reaction. First, reconstitute all the synthesized primers to 100 μM concentration. To prepare 50 μL primer stock, take 10 μL each of FIP and BIP, 5 μL each of LF and LB, and 2.5 μL each of F3 and B3, and make up the volume to 50 μL with molecular biology-grade water.
2. Optimize the concentration of the reagents in the RT-LAMP reaction by varying the concentration of each component. Different concentrations of $MgSO_4$ (4–8 mM), dNTP (1.2–1.6 mM each), betaine (0.8–1 M), and *Bst* polymerase (4–8 U) could be tried for optimisation.
3. Isolation of total RNA from symptomatic plants can be carried out using RNeasy plant mini kit (Qiagen) or any other protocol/kit suitable for plant RNA isolation. All the materials for RNA isolation such as centrifuge tubes, microtips, mortar, and pestle must be treated with 0.1% DEPC overnight, and then double-autoclaved. Double-autoclaved DEPC water should be used for the preparation of TAE buffer and 75% ethanol. Work bench, micropipettes, and gel electrophoresis unit should be

wiped with 75% ethanol and RNase ZAP (Invitrogen) or any RNase control reagent before use.

4. PrimerExplorer is a web-based LAMP primer designing software developed by Eiken Chemical Co., Ltd. Other software like NEB LAMP Primer Design Tool, LAMP Designer (Optigene), etc. can be alternately used for designing LAMP/RT-LAMP primers.
5. Select any region of the virus genome such as coat protein gene, replicase gene, and movement protein gene for picking RT-LAMP primers.
6. Synthesize the primers after validating the designed primers with BLASTN analysis. The RT-LAMP reaction can be carried out with external and internal primers alone. The loop primers are used to increase the sensitivity of the assay. Instead of two-loop primers, we can use a single-loop primer also.
7. Thaw the reagents on ice before preparing the reaction mixture. Maintain a separate sterile area for setting up the RT-LAMP reaction, ideally a UV enabled PCR workstation. Also, maintain a separate space for sample preparation. Never take the RNA sample or amplified product inside the PCR workstation. All these measures will help reduce sample cross contamination and false positives. Closed tube system reduces the chances of cross contamination to a great extent by eliminating the process of post-amplification sample handling.
8. It is always recommended to include a positive and a negative control in each assay. Positive control ensures that the RT-LAMP reaction is set properly, and all the reagents are working. The negative control is included for monitoring contamination and false positives in the assay.
9. The 1× TAE buffer can be prepared from 50× stock, pH 8.0 (242.2 g of Tris buffer, 100 mL of 0.5 M EDTA, and 57.1 mL glacial acetic acid made up to 1 L). Take 20 mL of 50× TAE buffer and make up the volume to 1000 mL with distilled water to get 1× TAE buffer.
10. It is always ideal to use a closed-tube detection system for RT-LAMP. Since agarose gel electrophoresis-based detection needs post-amplification sample handling, it is not recommended for routine diagnostic assays. However, it may be used during initial optimization for visualisation of positive amplicons.
11. The concentration of $MgSO_4$ and dNTPs in the reaction need to be optimized to get a clear color change. The RT-LAMP reaction mixture with 120 μM HNB dye, 1.6 mM dNTPs, and 6 mM $MgSO_4$ has violet colour prior to amplification. The HNB is a colorimetric titration indicator for the Mg^{2+} ions,

and it changes its color depending on pH and Mg^{2+} ion concentration. As the reaction proceeds, the concentration of Mg^{2+} ions decrease due to the formation of magnesium pyrophosphate leading to a color change to sky blue indicating a positive reaction. If desired color change is not observed, try to optimize the concentration of $MgSO_4$. You can try a range of $MgSO_4$ concentration from 4 to 8 mM. Note that the ThermoPol buffer also contains 2 mM $MgSO_4$. The dNTPs at 1.2 to 1.6 mM each can be tried for optimizing the color change. Also, note that the agarose gel-based detection of RT-LAMP positives is possible even with a lower concentration of dNTPs.

12. The NEB cutter or any virtual restriction digestion software can be used to identify the suitable restriction enzyme for your RT-LAMP product.
13. Preferably, select an enzyme that has one or two restriction sites within the DNA product (F2-B2 region). The purpose of selecting such an enzyme is to achieve two or three distinct bands up on digestion.
14. The incubation temperature depends on the restriction enzyme used. Most of the restriction enzymes have optimal activity at 37 °C. Follow the manufacturer's instructions while setting and running the restriction digestion reaction.

Acknowledgments

The authors acknowledge the facilities at the Centre for Plant Biotechnology and Molecular Biology, College of Agriculture, Kerala Agricultural University, Thrissur, Kerala.

References

1. Notomi T, Okayama H, Masubuchi H, Yonekawa T, Watanabe K, Amino N, Hase T (2000) Loop-mediated isothermal amplification of DNA. Nucleic Acids Res 28:63
2. Rubio L, Galipienso L, Ferriol I (2020) Detection of plant viruses and disease management: relevance of genetic diversity and evolution. Front Plant Sci 11:1092. https://doi.org/10.3389/fpls.2020.01092
3. Nagamine K, Hase T, Notomi T (2002) Accelerated reaction by loop-mediated isothermal amplification using loop primers. Mol Cell Probes 16(3):223–229. https://doi.org/10.1006/mcpr.2002.0415
4. Goto M, Honda E, Ogura A, Nomoto A, Hanaki KI (2009) Colorimetric detection of loop-mediated isothermal amplification reaction by using hydroxy naphthol blue. BioTechniques 46(3):167–172
5. Nair S, Manimekalai R, Gangaraj P, Hegde V (2016) Loop mediated isothermal amplification (LAMP) assay for detection of coconut root wilt disease and arecanut yellow leaf disease phytoplasma. World J Microbiol Biotechnol 32:108
6. Balic A, Savić V, Vucurović A, Kovačević V, Krstić B (2018) One-step reverse transcription loop-mediated isothermal amplification (RT-LAMP) for rapid detection of cucumber green mottle mosaic virus. J Virol Methods 252:62–66

7. Kovileri MM, Nair S, Loius V (2023) One-step reverse transcription-loop mediated isothermal amplification (RT-LAMP) for closed-tube colorimetric detection of Banana bract mosaic virus in Banana (Musa spp.). 3 Biotech 13(5): 131
8. Przewodowska A, Zacharzewska B, Chołuj J et al (2015) A onestep, real-time reverse transcription loop mediated isothermal amplification assay to detect potato virus Y. Am J Potato Res 92:303–311. https://doi.org/10.1007/s12230-015-9430-3
9. Pandey B, Khan MA, Prasad KV (2019) Development of a one-step reverse transcription loop-mediated isothermal amplification assay for rapid detection of tomato yellow leaf curl virus. Arch Virol 164(7):1861–1867
10. Zhang Z, Liu X, Li D, Yu J, Han C (2011) Rapid detection of wheat yellow mosaic virus by reverse transcription loop mediated isothermal amplification. Virol J 8:550
11. Yan L, Zhou J, Zheng Y, Gamson AS, Roembke BT, Nakayama S, Sintim HO (2020) Isothermal amplified detection of DNA and RNA. Mol BioSyst 16(11): 2302–2317

Chapter 6

Detection of Sesame Phyllody Disease and Its Transmitting Vectors

K. B. Durga Bhavani, E. Ashwini, V. Dinesh Kumar, T. Boopathi, and P. Duraimurugan

Abstract

Sesame (*Sesamum indicum* L.) is a herbaceous annual plant belonging to the family Pedaliacae. It is cultivated for its edible seed oil, which is an excellent source of balanced fatty acids. A sizable amount of the world's edible oil is obtained from sesame. Phyllody is the most destructive disease of sesame and is caused by phytoplasma, a cell wall-less bacteria that is a phloem-harboring obligate pathogen. One of the distinctive features of this disease is the transformation of the floral organs into green leafy structures, which results in a loss of up to 80% of the seed output. Phytoplasmas are obligate intracellular parasites and are often present in very small numbers. As phytoplasmas are non-culturable, with unpredictable distribution, low concentration, seasonal variation, and enzyme-inhibitory plant polysaccharides and polyphenolic substances, they are frequently challenging to detect and identify. To lessen the economic loss due to phytoplasma infections, rapid, sensitive, accurate, and early detection is essential and might provide better scope to manage. Detection of phytoplasma through conventional methods such as microscopy, and protein-based assays is not sensitive and requires more expertise and resources. Traditional PCR methods offer some advantages and overcome the limitations of conventional methods to a great extent, but real-time polymerase chain reaction (qPCR) offers a number of benefits over normal PCR. It is a high-throughput, quick, sensitive, and reliable detection method. Hence, qPCR is crucial for the detection of phytoplasma as well as the analysis of host-pathogen interactions and epidemiological investigations. This chapter describes the procedures used in the qPCR assay for the identification of phytoplasma in both plants and insect vectors. To provide a holistic idea, the other methods have also been dealt with briefly.

Key words Phyllody, Phytoplasma, Detection, qPCR, Vectors, Sesame

1 Introduction

1.1 Sesame Production and Utility

Sesame (*Sesamum indicum* L.) is a diploid, annual plant of the Pedaliaceae family that is often grown in tropical and subtropical regions of the world, particularly in Africa and Asia [1]. Sesame seeds contain 32.8–62.7% vegetable oils rich in monounsaturated essential fatty acids (~85%), mainly oleic and linoleic acid [2]. Sesame oils are rich in lipid-soluble lignans (mainly sesamol, sesamin,

Muthappa Senthil-Kumar and Hema Ramanna (eds.), *Plant Biotic Stress: Methods and Protocols*, Methods in Molecular Biology, vol. 2966, https://doi.org/10.1007/978-1-0716-4746-2_6,

and sesamolin), which protect it from oxidative-induced rancidity, prolonging its shelf life [3]. Sesame seeds are an excellent source of proteins (14.1–29.5%), carbohydrates (4.5–20.5%), fiber (2.7–6.7%), vitamins (E—α-tocopherol and B1—thiamine), minerals, and phytosterols [4]. The global cultivated acreage in 2023 was ~12.82 million hectares, producing ~64.5 million tons of sesame seeds that supplied ~12% of the gross world consumption of vegetable oils [5]. India ranks first in world with 19.47 lakh ha area and 8.66 lakh tons production. The average yield of sesame (413 kg/ha) in India is low as compared to that of other countries in the world (535 kg/ha) [6].

1.2 Sesame Phyllody

Among diseases afflicting sesame, phyllody, a phytoplasma-related disease is considered one of the main hindrances in many areas, especially in Asia and Africa, causing significant losses in sesame yield quantity and quality. In India, as one of the top ten sesame-producing countries globally, the incidence of sesame phyllody disease reaches from 10% to 100% in some croplands (Fig. 1). Akhtar et al. [7] and Salehi et al. [8] reported that sesame phyllody has been a destructive disease and may cause 80% seed yield loss.

Phytoplasmas are intracellular plant pathogens originated from a single lineage derived from gram-positive bacteria and belong to the class mollicutes. They usually colonize the nutrient-rich phloem tissue and appear as very small (400–900 nm diameter) spherical to oval or elongated cells in sieve elements (SEs), with a tiny (680–1600 kb) genome size [9]. Phytoplasma is transmitted among plants via diseased plant materials in vegetatively propagated

Fig. 1 Phyllody, a phytoplasma-related disease that damages sesame crops

Fig. 2 Leafhopper (**a** nymphs; **b** adult), vector transmitting phyllody disease in sesame

crops, phloem-feeding insects (mainly leafhoppers Fig. 2, planthoppers, and psyllids), and through dodder [10].

1.3 Pathogen Detection

Detection of pathogen is defined as drawing conclusion from the disease sample that a particular organism is associated with the disease. The best detection method should be reliable, specific, sensitive and accurate, less time-consuming, easy to handle, and it should be cost-effective. Detailed description of the individual detection methods has been well-documented in this chapter [11].

DNA-based methods have been used for detection of phyllody. Based on nucleic acid-based methods such as RFLP, virtual RFLP, and 16S rRNA gene sequence analysis, 32 16Sr groups of phytoplasma strains have been identified to date [11, 12]. The genomes of four *Candidatus Phytoplasma* strains, including Ca. *P. asteris* strains OY-M and AY-WB, Ca. *P. australiense*, and Ca. *P. mali*, have all been fully sequenced thus far [13].

Due to the nonavailability of culturing on artificial media under in vitro conditions, the phytoplasma's detection and identification methods and ecological studies are still restricted [14]. Traditionally, phytoplasma detection was based on biological properties and microscopic techniques (light microscopy and transmission electron microscopy). However, these techniques need experienced and skilled laboratory staff, and also suffer from precise identification issues leading to incorrect interpretation of results and incorrect disease diagnosis, affecting ultimately the treatment of the disease [15].

1.4 Detection Methods Used earlier for Phytoplasma Identification

1.4.1 Symptomatology and Light Microscope

Symptomatology is one of the major criteria for preliminary diagnosis of phytoplasma diseases. Symptoms basically include the transformation of floral parts into leaflike structure and thus appearance of a cluster of leaves in place of the inflorescence (phyllody). Various types of symptoms for phyllody disease to be associated with phyllody infected sesame plants, namely flowery phyllody, virescence, intense leaf and flower bud proliferation,

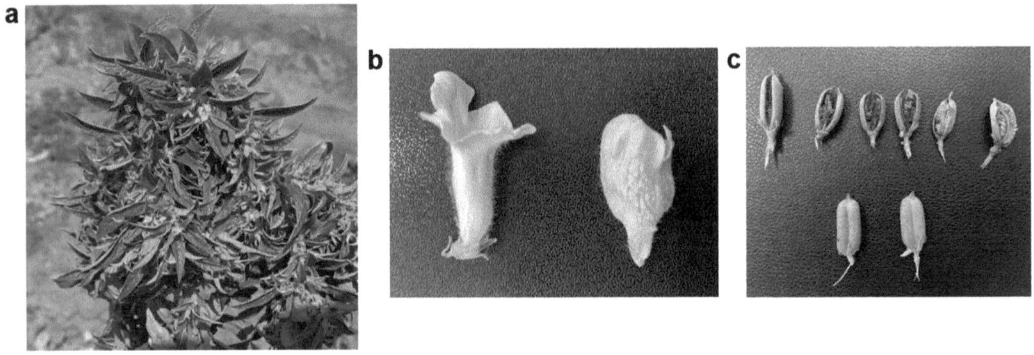

Fig. 3 Sesame plants showing different types of phyllody symptoms such as (**a**) little leaf symptom, (**b**) flower virescence (healthy and infected flowers), and (**c**) seed capsule cracking (healthy and infected capsules)

ovivipary condition on sesame plant, and cracking of seed capsules (Fig. 3). However, "virescence" and "phyllody" have registered more often compared to the floral "proliferation." Of this major class of symptoms, "phyllody" refers to the leafy structures development in flowers instead of corolla; "virescence" refers to corolla changes from pink to green or white to green. Subsequently, cracking of seed capsule, dark exudates formation on floral and foliage parts, seeds germinating within capsules, and yellowing occasionally associated with the disease. Actually depending on the time of infection, the inflorescence if already initiated and progressed to produce flowers and siliques will start showing the symptoms in the later flower buds. However, the duration between the exact time of infection and manifestation of the disease varies between a few days to weeks.

Two different methods can be used to identify phytoplasma in the infected tissue by means of light microscopy. The stain most commonly employed for phytoplasma identification is Dienes' reagent, which eventually produces a blue color [16] and 4′-6-diamidino-2-phenylindole (DAPI), a fluorochrome that binds to DNA. These methods are easy, rapid, reasonably priced approaches, and can be employed as a first step in identifying phytoplasmas in plants that have high levels of the pathogen but lack specificity for the various phytoplasmas [17]. However, it is constrained when the phytoplasma population in the infected host plant's phloem tissue is relatively low.

1.4.2 Transmission Electron Microscopy (TEM)

For more than 30 years, TEM was one of the most effective and dependable ways to find phytoplasma [18]. This method has made it possible to investigate as well as make diagnoses. Electron microscopy (EM) techniques, however, are time-consuming, expensive, still require specialized staff, and are not appropriate for high-throughput applications [19].

1.4.3 Confocal Laser Scanning Microscope

Phytoplasmas in living plant tissues can be found using a confocal laser scanning microscope with vital dyes—DiOC7(3) and SYTO are both 3,3′-diheptyloxacarbocyanine iodides. During multiphoton confocal laser scanning, SYTO 13 can identify both living and dead phytoplasmas, while DiOC7(3), a potentiometric dye, can be used to see alive and active phytoplasmas.

1.4.4 Fluorescence In Situ Hybridization (FISH) Technique

The in situ detection of phytoplasmas in both host plant tissues and insect vectors is made possible using nonradioactive oligonucleotide probes. As hybridization probes in FISH, primer sequences can be used [20–22]. FISH could be used to specifically identify phytoplasma taxonomic categories, subgroups, and/or strains, as well as to examine the patterns of phytoplasma colonization.

1.4.5 ELISA

It is a serological method employed for phytoplasma detection in the early 1980s. Immunodominant membrane proteins (IMPs), SecA membrane proteins, and entire phytoplasma cells isolated from infected plants have all been targets of antibodies [23]. Unfortunately, due to inadequate specificity and sensitivity in producing particular high-titre, polyclonal antisera, serological approaches have not been widely used in phytoplasma detection and identification. Moreover, due to cross-reactivity with plant host proteins and nonspecific background reactivity, the creation of specific antibodies to unculturable phytoplasmas is challenging, laborios, and time-consuming [24].

1.4.6 Dot and Southern Blot Hybridization

Nucleic acid-based approaches for phytoplasma detection appear to be more sensitive, specific, and reliable than the ELISA method. While some probes allowed for the detection of closely related phytoplasma strains, others allowed for the detection of a wide variety of phytoplasmas [25–27]. However, difficulties encountered in generating specific probes and insufficient sensitivity of the probes is the major problem of phytoplasma detection by DNA hybridization assays [28].

1.4.7 PCR Amplification of Phytoplasma-Specific DNA

In the recent years, much effort has been devoted to the development of methods for detecting and identifying plant pathogens based on DNA/RNA probe technology and PCR amplification of nucleic acid sequences, such as polymerase chain reaction (PCR) and sequence analysis have been used to characterize and differentiate phytoplasma strains by amplifying the highly conserved gene coding in the dispersed regions of the 16S rRNA [29]. All prokaryotes include the highly conserved 16S rRNA gene, which has been used to detect and categorize unknown phytoplasmas. PCR offers a sensitive, focused, rapid, and affordable method for the identification of phytoplasmas. When it comes to identifying phytoplasmas in plant and insect hosts, PCR has significantly outperformed compared to microscopic, serological, and hybridization approaches in terms of sensitivity and specificity.

Phytoplasma group-specific primers have been designed for ribosomal RNA genes [30]. The nested-PCR assay is designed to overcome the low concentration of phytoplasma in infected samples through increasing both sensitivity and specificity. However, improvements in fluorescence technology enabled the creation of qPCR, which has a number of benefits over traditional PCR [31].

1.4.8 Loop-Mediated Isothermal Amplification Assay (LAMP)

LAMP is a single tube method for DNA amplification that uses an incubation temperature range of 60 °C to 65 °C. LAMP is a straightforward, quick [there is no time lost because of using a robust (heat cycling)] approach. DNA may be amplified with a high degree of specificity and sensitivity. LAMP assay has been developed for detection of sesame phyllody in Vitenam [32] though it has not been popularly used. In this method, the LAMP assays were carried out using a real-time fluorometer with amplification plots and annealing curves visualized directly.

1.4.9 Real-Time PCR

The real-time PCR technique has gained widespread popularity in diagnostic and research labs. Real-time flourogenic PCR has various benefits, including quick analysis times and provide reliable results. Compared to conventional or nested PCR for phytoplasma detection, the latter offers superior sample processing, accuracy, sensitivity, high automation capability, high-throughput capacity, and the absence of post-PCR analysis such as gel electrophoresis. Furthermore, a real-time PCR system fitted with various measuring channels can accomplish the simultaneous multiplex detection of various DNA targets in the same reaction. Real-time PCR has been used to identify and measure a wide variety of phytoplasmas from various groups. Real-time PCR methods based on SYBR green and TaqMan have been utilized more often and are presently available for the detection of phytoplasmas, both generally and specifically [33, 34].

Real-time polymerase chain reaction (qPCR) offers several advantages over conventional PCR. It is a fast, sensitive, and reliable detection technique amenable to high throughput. Two fluorescent chemistries, intercalating dyes or hybridization probes [35], are employed in qPCRs. Intercalating dyes are relatively less expensive than TaqMan® hybridization probes, but the latter chemistry is the most commonly used for phytoplasma detection. PCR may be designed for universal detection of phytoplasma, group- or subgroup-specific detection, or for simultaneous detection of up to three or four phytoplasmas (multiplexing) [36]. For relative or absolute quantification of phytoplasma in insect vectors and host plants, qPCR can be performed. Hence, qPCR is crucial for the detection of phytoplasma as well as the analysis of host-pathogen interactions and epidemiological investigations.

The procedures used in qPCR assays for phytoplasma identification and quantification are described in this chapter, with a particular emphasis on the more adopted method that uses TaqMan® probes.

2 Materials

2.1 DNA Isolation

1. mL of CTAB buffer: 2.0 g CTAB (hexadecyl trimethylammonium bromide), 10.0 mL of 1 M Tris EDTA pH 8.0, [4.0 mL of 0.5 M EDTA pH 8.0 (ethylene diamine tetra acetic acid disodium salt)], 28.0 mL of 5 M NaCl, 40.0 mL of H_2O, 1 g of PVP 40 [polyvinyl pyrrolidone (vinylpyrrolidone homopolymer) Mw 40,000], and maintain the pH to 5.0 with HCl and make up to 100 mL with H_2O.
2. Microfuge tubes.
3. Mortar and pestle.
4. Liquid nitrogen.
5. Microfuge.
6. Absolute ethanol (ice cold) 70% ethanol (ice cold).
7. Ammonium acetate (7.5 M) at 55 °C water bath.
8. Chloroform: Iso amyl alcohol (24:1).
9. Tris EDTA (1 M) pH 8.0: Dissolve 121.1 g of Tris EDTA base in 800 mL of H_2O. Adjust pH to 8.0 by adding 42 mL of concentrated HCl. Allow the solution to cool to room temperature before making the final adjustments to the pH. Adjust the volume to 1 L with H_2O. Sterilize using an autoclave.
10. 1% agarose gel: 1 g agarose dissolved (by boiling in a microwave) in 100 mL TBE.
11. 6× loading buffer
12. 5× TBE solution: For 10× TBE buffer (100 mL), 54 g of Tris base, 27.5 g of boric acid, 20 mL of 0.5 M EDTA (pH 8.0), and make up the volume to 1 L with water. To make a 0.5× working solution, do a 1:10 dilution of the concentrated stock. 1% agarose gel 1 g agarose dissolved in 100 mL TBE
13. Agarose gel electrophoresis system.
14. Ethidium bromide solution.
15. Water (sterile).

2.2 Samples and Controls

1. DNA to be tested is isolated from the infected host plants (sesame) and transmitting vectors (leafhoppers) as positive control (*see* **Note 1**).

2. Extracted DNA from the healthy sesame plants is used as negative control (*see* **Note 2**).
3. Sterile deionized water is used instead of DNA template as no template control (NTC) (*see* **Note 1**) in the PCRs.

2.3 Quantitative PCR

1. qPCR Taq polymerase Master Mix or SYBR Green Master Mix (*see* **Note 3**).
2. Primers and probes (*see* **Note 4** and Table 1 for universal detection of phytoplasma from sesame and vectors).
3. Sterile DNase- free water (*see* **Note 5**).

2.4 qPCR Thermocycler and Accessories

1. 96 or 384-well hard quantitative PCR reaction plates.
2. Optical adhesive plate covers.
3. Real-time thermocycler.
4. Software process of plate setup, data collection, and analysis of real-time PCR results.
5. Autoclaved tubes, pipettes, and aerosol-barrier pipette tips.
6. UV chamber.
7. Micro centrifuge and plate centrifuge.
8. UV-Vis spectrophotometer for the quantification and purify of DNA template.

3 Methods

1. All the procedures are carried out at room temperature unless otherwise specified.

3.1 DNA Extraction

1. Make a fine paste of 200 mg of plant tissue and 500 μL of CTAB buffer.
2. To a microfuge tube, transfer the CTAB/plant extract mixture.
3. The CTAB/plant extract mixture should be incubated in a recirculating water bath for approximately 15 min at 55°C.
4. Spin the CTAB/plant extract mixture at 12,000×*g* for 5 min after incubation to remove any remaining cell debris. Supernatant should be transferred to clean microfuge tubes.
5. Add 250 μL of chloroform: iso amyl alcohol (24:1) to each tube and invert the mixture many times to combine. After mixing, spin the tubes for 1 min at 13,000 rpm.
6. Transfer the aqueous portion of the solution (contains the DNA) to a fresh microfuge tube.
7. To each tube, add 50 μL of 7.5 M ammonium acetate followed by 500 μL of ice-cold absolute ethanol.

Table 1
List of primers and probes designed for some internal control genes of sesame and insect vectors

Host	Target gene	Forward primer 5′–3′	Reverse primer 5′–3′	Probe 5′–3′	Size	References
Sesame	18SrDNA	CGCGGAAGTTTGAGGCAATA	CTGTCGGCCAAGGCTATAGACT	HEX-TAGATGTTCTGGGCCGCACGCG-TAMRA	103 bp	[22]
Leafhopper species	18SrDNA	MqFw-AACGGCTACCACATCCAAGG	MqRV-GCCTCGGATGAGTCCCG	Mq Probe AGGCAGCAGGCACGCAAATTACCC	100 bp	[23]
Hyalesthes obsoletus	18SrDNA	TCTGTCTGCTGCGTGAACAT	ACTGAGCTCGCCCAGGTT	TR--TAACAAACTCCGGAAGAACATAACTCA	128 bp	[24]
Macrosteles quadrilinatus	CP6	GGGCAAGAAGGGCAAG	AGGCTCCAGATACACTAGGTC	EvaGreen	90 bp	[25]

8. To precipitate the DNA, slowly invert the tubes multiple times. In general, it is possible to observe that DNA precipitates out of solution. After adding the ethanol to help the DNA precipitate, the tubes can also be left at −20 °C for 1 h.
9. After precipitation, the DNA can be removed with a pipette by slowly spinning or rotating a pipette tip in a cold solution. DNA that has precipitated is evident as a thick, clear precipitate that adheres to the pipette tip. Transfer the precipitate into a microfuge tube filled with 500 μL of ice-cold 70% ethanol, and then slowly invert the tube to wash the DNA.
10. Alternatively, the precipitate may be separated by spinning the tube for 1 min at 13,000 rpm to form a pellet. Take away the supernatant, then add two changes of ice-cold 70% ethanol to wash the DNA pellet.
11. After the wash, centrifuge the DNA for 1 min at 13,000×*g* to form a pellet. Eliminate all of the supernatant, and then let the DNA pellet dry (approximately 15 min). Avoid overdrying the DNA because it will be difficult to redissolve.
12. Depending on how much DNA is isolated, a different amount of sterile, DNase-free water (about 50–400 μL) may be required to dissolve it. To get rid of any RNA in the preparation, add 10 mg/mL of RNase A to the water before dissolving the DNA (10 μL RNase A in 10 mL H_2O).
13. The DNA is then resuspended, heated at 65 °C for 20 min to kill any potential DNAses, and then stored at 4 °C.
14. The integrity of the DNA will be revealed by agarose gel electrophoresis, and the concentration and cleanliness will be revealed by spectrophotometry. DNA quality assurance.
15. Melt 1 g of agarose in 100 mL of 0.5× TBE buffer in a microwave for about 2 min to get an agarose solution at 1% concentration. After a few minutes as of cooling, add 2.5 μL of ethidium bromide and stir to combine.
16. Using the provided tray and comb, cast a gel. Let the gel set for at least 20 min on a flat surface at room temperature.
17. Load the following into separate wells—10 μL 1 kb ladder, 5 μL sample +5 μL water +2 μL 6× loading buffer.
18. Run the gel for 30 min at 100 V—expose the gel to UV light and photograph (demonstration).
19. Verify the quality of the DNA. A well-resolved high-molecular-weight band shows that the DNA is of good quality, whereas a smeared band suggests that the DNA is degrading.

3.2 Preparation of Plate

1. Arrange the samples using the plate layout manually or according to the qPCR software.
2. Assign two or three wells for each non-template control (NTC), negative control (NC), and positive control.
3. Each plate should contain at least 12 wells for duplicates and 18 wells for triplicates for standard curves.

3.3 Sample Preparation [37]

1. All reagents should be taken out of the freezer and allowed to defrost on ice away from light. All thawed tubes are subjected to a brief vortexing, and then quickly spun briefly (~5 s).
2. Create two distinct reaction mixtures: one for the internal control gene and the other for the phytoplasma primers (host primers). The total number of reactions determines the master mix's final volume. Use the qPCR machine software (e.g., Biorad CFX manager 3.1 -> Tools -> Master-mixCalculator) to calculate the volumes of each of the components of the master mix. When preparing the master mix, the total number of samples to be considered may be increased by10% to factor in the pipetting errors (*see* **Note 6**).
3. Fill the wells with appropriate 18 μL of the master mix (*see* **Note 7**).
4. For each reaction, use 300 and 100 nM of each primer and probe, respectively (*see* **Note 8**).
5. Add 2 μL of extracted DNA (*see* **Note 9**), cover the qPCR with an optical adhesive cover, and then centrifuge for 30 s at 2000×*g* to collect the DNA and the mix at the bottom of the plate wells (*see* **Note 10**).

3.4 Perform qPCR Run

1. Turn on the thermocycler.
2. Load the plate into the thermocycler.
3. Run the thermocycler's software.
4. Choose the cycling procedure (standard methodology to be optimized depending on your working conditions), import the plate data, and then select Start Run. The cycling protocol should be 40 cycles of 3 s denaturation at 95 °C and 30 s annealing and elongation at 60 °C (see **Note 11**).

3.5 Analysis of qPCR Results

1. Save the run's results after the reaction is finished, and then take off the plate.
2. Every sample's quantification cycle (Cq) values will be calculated automatically by the software, although one can choose to slightly alter the threshold line. Ideally, the NTC and NC samples would receive 0 or 38 Cq [37].

4 Notes

1. A positive control (PC) and a non-template control (NTC) should be used in every qPCR run. NTC control uses purified deionized water as template instead of sample DNA. The positive control is a sample that contains the target phytoplasma DNA.
2. In every qPCR assay, a healthy sample of the same host (plant or vector) should be included to check cross-reactions of the primers with the DNA from sample material (false positives).
3. Master Mix can be used from any supplier; it is advised to compare products from at least two suppliers to check for reproducibility.
4. The purchased lyophilized primers are to be suspended to 100 μM in sterile DNase water. Dilute each stock till 10 μM in aliquots, and store at −20 °C. Cover the tubes containing probes to protect from light exposure.
5. Use sterile DNase-free water in all recipes and protocol steps.
6. While setting up qPCR reactions, exercise the greatest caution possible. The creation of the reaction mixtures and the incorporation of DNA must be done in a UV chamber to prevent contamination.
7. Keep in mind that every sample and standard is used twice. Depending on the master mix employed, reactions can be reduced down to 10 μL.
8. Equimolar concentrations of primers (100, 200, 300, and 400 nM) and probes (100, 200, and 300 nM) can be used to screen for the highest relative fluorescence unit (RFU), the lowest quantitation cycle (Cq), and the highest reaction efficiency in order to find the best primer/probe combination for the target gene. A standard curve can be used to calculate the reaction efficiency for each primer/probe set.
9. In sterile, DNase-free water, the template DNA taken for each reaction should be diluted to less than 100 ng. Thoroughly combine the DNA tubes and their dilutions, and then spin the tubes for a brief period of time to collect the DNA solution at the bottom.
10. Use the sides to handle the plate and the optical adhesive cover. Run the plate immediately after spinning; if not, leave it on ice in the dark for no more than a few hours.
11. Using the thermocycler software: (1) adjust the reaction volume and the samples location on the plate; (2) assign the correct filters of the probes used.

References

1. Kabak B, Dobson AD (2017) Mycotoxins in spices and herbs—an update. Crit Rev Food Sci Nutr 57:18–34
2. Uzun B, Arslan C, Furat S (2008) Variation in fatty acid composition, oil content and oil yield in a germplasm collection of sesame (*Sesamum indicum* L.). J Am Oil Chem Soc 85:1135–1142
3. Rizki H, Nabloussi A, Kzaiber F, Jbilou M, Latrache H, Hanine H (2018) Comparative assessment of bioactive components, antioxidant effects from 15 cultivars of sesame (*Sesamum indicum* L.) for different crop years. Int J Sci Eng Res 9:462–468
4. Nzikou JM, Matos L, Kalou GB, Ndangui CB, Pambou-Tobi NPG, Kimbonguila A, Silou T, Linder M, Desobry S (2009) Chemical composition on the seeds and oil of sesame (*Sesamum indicum* L.) grown in Congo-Brazzaville. Adv J Food Sci Technol 1:6–11
5. Food and Agriculture Organization of the United Nations (FAO) (2020) FAOSTAT statistics. United Nations, Rome
6. Priyanka Evangilin N, Ramana Murthy B, Mohan Naidu G, Aparna B (2020) Statistical model for forecasting area, production and productivity of sesame crop (Sesamum indicum L.) in Andhra Pradesh, India. Int J Curr Microbiol Appl Sci 9(7):1156–1166
7. Akhtar KP, Sarwar G, Sarwar N, Elahi MT (2013) Field evaluation of sesame germplasm against sesame phyllody disease. Pak J Bot 45: 1085–1090
8. Salehi M, Esmailzadeh-Hosseini SA, Salehi E, Bertaccini A (2017) Genetic diversity and vector transmission of phytoplasmas associated with sesame phyllody in Iran. Folia Microbiol 62:99–109
9. Doi Y, Teranaka M, Yora K, Asuyama H (1967) Mycoplasma or P.L.T. group-like microorganisms found in the phloem elements of plants infected with mulberry dwarf, potato witches'-broom, aster yellows or paulownia witches'-broom. Ann Phytopathol Soc Jpn 33:259–266
10. Bosco D, Tedeschi R (2013) Insect vector transmission assays. Methods Mol Biol 938: 73–85
11. Nejat N, Vadamalai G, Davis RE, Harrison NA, Sijam K, Dickinson M, Abdullah SNA, Zhao Y (2013) 'Candidatus Phytoplasma malaysianum', a novel taxon associated with virescence and phyllody of Madagascar periwinkle (Catharanthus roseus). Int J Syst Evol Microbiol 63: 540–548
12. Zhao Y, Wei W, Lee I-M, Shao J, Suo X, Davis RE (2009) Construction of an interactive online phytoplasma classification tool, iPhyClassifier, and its application in analysis of the peach X-disease phytoplasma group (16SrIII). Int J Syst Evol Microbiol 59:2582–2593
13. Oshima K, Kakizawa S, Nishigawa H, Jung H-Y, Wei W, Suzuki S, Arashida R, Nakata D, Miyata S, Ugaki M, Namba S (2004) Reductive evolution suggested from the complete genome sequence of a plant-pathogenic phytoplasma. Nat Genet 36:27–29
14. Devonshire BJ (2013) Visualization of phytoplasmas using electron microscopy. In: Dickinson M, Hodgetts J (eds) Phytoplasma, Methods in molecular biology (methods and protocols), vol 938. Humana Press, Totowa
15. Ahrens U, Seemüller RE (1992) Detection of DNA of plant pathogenic mycoplasma-like organisms by a polymerase chain reaction that amplifies a sequence of the 16S rRNA gene. Phytopathology 82:828–832
16. Dienes L (1940). L Type of Growth in Gonococcus Cultures. Proc Soc Exp Biol N Y 44:468
17. Deeley J, Stevens WA, Fox RTV (1979) Use of Dienes stain to detect plantdiseases induced by MLO. Phytopathology 69:1169–1171
18. Poghosyan A, Lebsky V, Arce-montoya M, Landa L (2004) Possible phytoplasma disease in papaya (*Carica papaya* L.) from Baja California Sur: diagnosis by scanning electron microscopy. J Phutopathol 152:376380
19. Sinclair BJ, Worland MR, Wharton DA (1999) Ice nucleation and freezing tolerance in New Zealand alpine and lowland weta, Hemideina spp. (Orthoptera; Stenopelmatidae). Physiol Entomol 24:56–63
20. Lherminier J, Bonfiglioli RG, Daire X, Symons RH, Boudon-Padieu E (1999) Oligodeoxynucleotides as probes for *in situ* hybridization with transmission electron microscopy to specifically localize phytoplasma in plant cells. Mol Cell Probes 13:41–47
21. Webb MA, Arnott HJ (1983) Inside plant crystals: a study of the noncrystalline core in druses of Vitis vinifera endosperm. Scan Electron Microsc 4:1759–1770
22. Bulgari D, Casati P, Brusetti L, Quaglino F, Brasca M, Daffonchio D, Bianco PA (2009) Endophytic bacterial diversity in grapevine (*Vitis vinifera* L.) leaves described by 16S rRNA gene sequence analysis and length heterogeneity-PCR. J Microbiol 47:393–401

23. Chen TA, Jiang XF (1988) Monoclonal antibodies against the maize bushy stunt agent. Can J Microbiol 34:6–11
24. Engvall E (1972) Enzyme-linked immunosorbent assay, Elisa. J Immunol 109(1):129–135
25. Bertaccini A, Davis RE, Lee I-M, Cont M, Dally EL, Douglas SM (1990) Detection of chrysanthemum yellows mycoplasmalike organism by dot hybridization and southern blot analysis. Plant Dis 74:4043
26. Bonnet F, Saillard C, Kollar A, Seemüller E, Bove JM (1990) Detection and differentiation of the mycoplasmalike organism associated with apple proliferation disease using cloned DNA probes. Mol Plant Microbe Interact 3: 438–443
27. Davis MJ, Tsai JH, Cox RL, McDaniel LL, Harrison NA (1988) Cloning of chromosomal and extrachromosomal DNA of the mycoplasma-like organism that causes maize bushy stunt disease. Mol Plant Microbe Interact 1:295–302
28. Seemüller E, Kirkpatrick BC (1996) Detection of phytoplasma infections in plants. In: Tully JG, Razin S (eds) Molecular and diagnostic procedures in mycoplasmology, vol II. Academic Press, San Diego, pp 291–311
29. El-Banna OM, Toima NI, Youssef SA, Shalaby A (2015) Molecular and electron microscope evidence for an association of phytoplasma with citrus witches broom disease. Int J Sci Eng Res 6:127–133
30. Bertaccini A, Martini M (1999) Ribosomal and nonribosomal primers for sensitive detection and identification of phytoplasmas. Petria 9: 89–92
31. Gundersen DE, Lee IM, Rehner SA, Davis RE, Kingsbury DT (1994) Phylogeny of mycoplasma like organisms (phytoplasmas): a basis for their classification. J Bacteriol 176:5244–5254
32. Notomi T, Okayama H, Masubuchi H, Yonekawa T, Watanabe K, Amino N, Hase T (2000) Loop-mediated isothermal amplification of DNA. Nucleic Acids Res 28:E63
33. Galetto L, Marzachí C (2010) Real-time PCR diagnosis and quantification of phytoplasmas. In: Weintraub PG, Jones P (eds) Phytoplasmas genomes, plant hosts and vectors. CABI Publishers, Boston, pp 1–19
34. Hodgetts J, Boonham N, Mumford R, Dickinson M (2009) Panel of 23S rRNA gene-based real-time PCR assays for improved universal and group-specific detection of phytoplasmas. Appl Environ Microbiol 75:2945–2950
35. Pfaffl MW (2012) Quantification strategies in real-time polymerase chain reaction. In: Filion M (ed) Quantitative real-time PCR. Applied microbiology. Horizon Scientific Press, Norfolk
36. .Abou-Jawdah Y, Aknadibossian V, Jawhari M, Tawidian P, Abrahamian P (2019) Real-time PCR protocol for phytoplasma detection and quantification. Methods Mol Biol 1875:117–130
37. Abou-Jawdah Y (ed) (1875) Phytoplasmas: methods and protocols, methods in molecular biology, vol 2019. Springer Nature, New York. https://doi.org/10.1007/978-1-4939-8837-2_9. © Springer Science+Business Media, LLC, part of Springer Nature

Chapter 7

A Method for Isolation and Transfection of *Arabidopsis* Protoplast for Sucrose Feeding Assay

Anjali Anjali and Muthappa Senthil-Kumar

Abstract

Protoplasts serve as a powerful system to study various plant physiological processes and to understand crucial signaling pathways within the plant system. The isolation of *Arabidopsis* protoplasts is a well-established technique and being utilized for wide range of assays. The method described herein includes the precise cutting of *Arabidopsis* leaves using extraction buffer containing cellulase and macerozyme. Further we present a polyethylene glycol [PEG]-mediated protoplast transformation method. Here, we have elucidated the methodology for isolation of protoplast for the AtSWEET-mediated sucrose uptake assay in control, and *Pseudomonas syringe pv tomato DC3000 (Pst DC3000)* treated leaves by utilizing GC/MS analysis. Our approach includes certain modifications to the previously published protoplast isolation technique, streamlining the process and providing a more accessible alternative to the highly specialized Xenopus oocyte uptake assay.

Key words *Arabidopsis thaliana*, Membrane protein, Plant-pathogen interaction, Protoplast, *Pseudomonas syringae* pv tomato , Sugar transport

1 Introduction

The technique of protoplast isolation has been established around 45 years ago [1]. The transient gene expression in plant protoplast is a crucial advancement in the field of plant molecular biology research. The transient gene expression study in *Arabidopsis* protoplast is well-established and widely being used [2–4]. Though various systems for transient gene expression have been employed depending upon the methods of DNA introduction into the plant system encompassing *Agrobacterium* infiltration of leaf epidermal cells [5, 6], biolistic bombardment of tissues or cultured cells [7, 8], and the transformation of protoplasts utilizing polyethylene glycol (PEG) or electroporation [4, 9, 10]. Out of all the techniques, protoplast transient gene expression system has gained significant popularity. This technique eliminates the need to maintain material in a sterile environment and has no requirement for generation of

Muthappa Senthil-Kumar and Hema Ramanna (eds.), *Plant Biotic Stress: Methods and Protocols*, Methods in Molecular Biology, vol. 2966, https://doi.org/10.1007/978-1-0716-4746-2_7,

any suspension culture. Along with that it is a less time-consuming, convenient, and flexible for different assays.

In contrast to any other cell culture, protoplasts exhibit a remarkable ability to retain numerous physiological responses and cellular activities similar to each other despite the various treatments. It is extensively utilized across a range of applications such as the regulation of transcription and translation, gene expression analysis, cell wall synthesis, cell division, differentiation during regeneration, photosynthesis activity, studying the plant signaling mechanisms and its regulation, RNA interference (RNAi), exploring the cell functional genomics studies, [4] as well as the study of protein-protein interaction via bimolecular fluorescence complementation (BiFC).

Apart from protoplast system, Xenopus oocyte system has been the widely used tool for studying the expression and characterization of various proteins. The oocyte system is well-established and has been widely used for studying the channel proteins, plant membrane transporters, and protein-protein interaction. This system has been successfully utilized for the expression of membrane protein from the yeast, bacteria, and plants [11, 12]. Though the oocyte system is well-established, this technique has several drawbacks. The high maintenance requirements of cell culture and the need for specialized expertise render it less feasible in many aspects. Thus, an alternate assay to this is demonstrated in this chapter.

Bacterial pathogen infected the host plants for nutrients and multiplication. *Pseudomonas syringe* pv tomato DC3000 is a host pathogen of the *Arabidopsis*. After infection, *Pst* DC3000 resides and multiply in the apoplast of the host plant and utilize the sugars as their carbon source. Therefore, bacterial pathogens are reported to manipulate the apoplastic sugar profiling of their host plants [13, 14]. In this document, we present a refined methodology for assessing sucrose uptake activity in mock and *Pst* DC3000-treated *Arabidopsis* protoplasts, incorporating specific modifications to previously published methods over the Xenopus oocyte system, thereby enhancing feasibility.

2 Materials

2.1 For Plant Growth

1. Soil mix (agropeat and soilrite 3:1).
2. Plastic pots.
3. Hoagland solution (mix of solutions A, B, and C).
   Solution A (1 M)
   - 1 M KH_2PO_4 (1 mL/l)
   - 1 M KNO_3 (1 mL/l)

- 1 M $Ca(NO_3)_2$ (5 mL/l)
- 1 M $MgSO_4$ (2 mL/l).

  Solution B (1 mL/l)
- 1 M H_3BO_3
- 1 M $MnCl_2.4H_2O$
- 1 M $ZnSO_4.7H_2O$
- 1 M $CuSO_4.5H_2O$
- 1 M H_2MoO_4.

  Solution C (1 mL/l)
- 1 M Fe-EDTA-Na_2EDTA
- 1 M $FeSO_4$.7H2O.

Chemicals and Instruments

1. Potassium chloride.
2. $CaCl_2$.
3. $MgCl_2.6H_20$.
4. NaCl.
5. MES.
6. Mannitol.
7. BSA.
8. Cellulase R-10.
9. Macerozyme R-10.
10. PEG-4000.
11. Mannitol.
12. Sucrose.
13. 0.22 μm sterile filter
14. Miracloth.
15. 90 mm petri plate
16. 50 mL Falcon tube
17. 1.5 mL MCTs
18. Haemocytometer.
19. Glass slide.
20. Glass cover slide.
21. Vacuum concentrator.
22. Bench-top centrifuge.
23. Tomy centrifuge.

2.2 Protoplast Isolation Buffer

20 mM KCL, 20 mM MES, 0.6 M mannitol, 10 mM $CaCl_2$, 0.1% BSA, and double-distilled H_2O (*see* **Note 1**).

2.2.1 Pretreatment Buffer (10 Ml)

2.2.2 Enzyme Buffer I

1% cellulase R10, 0.5% macerozyme R 10, 20 mM KCl, 20 mM MES, 0.6 M mannitol. Then heat the solution at 55 °C for 10 min in water bath; let it cool down to RT, then add 10 mM $CaCl_2$, 0.1% BSA, and double-distilled H_2O (*see* **Note 2**).

2.2.3 Enzyme Buffer II

1.2% cellulase R-10, 0.4% macerozyme R-10, 20 mM KCl, 20 mM MES, 0.6 M mannitol. Then heat at solution at 55 °C for 10 min in water bath, let it cool down to RT then add 10 mM $CaCl_2$, 0.1% BSA, and double-distilled H_2O (*see* **Note 2**).

2.2.4 W5 Buffer

5 mM KCl, 125 mM $CaCl_2$, 154 mM NaCl, 2 mM MES, and double-distilled H_2O (*see* **Note 3**).

2.2.5 WI Solution

4 mM MES (pH 5.7), 0.5 M mannitol, and 20 mM KCl.

2.2.6 PEG-Calcium Transfection Buffer

20–40% (w/v) PEG4000, 0.2 M mannitol, 100 mM $CaCl_2$ (*see* **Note 4**).

2.2.7 Incubation Buffer for Sucrose Feeding Assay

0.2 M mannitol, 20 mM MES, 20 mM KCl, 0.2 M $CaCl_2$ with or without 20 mM sucrose.

2.2.8 For Culturing Pst DC3000

King's B (KB)medium (liquid).

3 Methods

3.1 Plant Germination and Growth

1. Sow the seeds of Arabidopsis wild-type lines in a soil mix of agropeat and soilrite in a ratio of 3:1.
2. Transfer the seed-sown pots in cold room (4 °C) in the dark for stratification for 2 days.
3. Transfer the pots in growth chamber under 8 h light (light intensity).
4. 200 μE m − 2 s − 1)/16 h dark at 20 °C and 75% relative humidity.
5. Irrigate the plants after 2 weeks of germination with water and Hoagland nutrient solution.

3.2 Bacterial Inoculum Preparation and Pathogen Inoculation via Syringe Infiltration Method

1. Culture the pathogenic bacterial strains of *Pst* DC3000 in King's B (KB) medium (liquid) containing 50 μg/mL of rifampicin, and grown at 28 °C with continuous shaking at 150 rpm.
2. Grow the culture to obtain an optical density of 0.4 at 600 nm (OD600 = 0.4).
3. Collect the bacterial cells by centrifugation at 4270× *g* for 10 min at RT.
4. Bacterial cells were further washed thrice with sterile miliQ water.
5. Resuspend the bacterial cells in sterile water to obtain the desired concentration (*see* **Note 5**).
6. For inoculating the pathogen, 32-day-old plants were used.
7. Pathogen suspension was syringe-infiltrated at the abaxial surface of the leaves using a needless syringe.

3.3 Protoplast Isolation

1. Harvest healthy leaves from 3–4-week-old Col-0 pathogen-infected and noninfected plants (*see* **Note 6**).
2. Place the leaves (15–20 leaves) in a petri plate filled with pretreatment buffer, and with the help of a fresh sharp razor blade cut 0.5–1 mm leaf strips leaving the midrib (*see* **Note 7**).
3. Transfer the leaf strips quickly and gently into the prepared enzyme solution (10–20 leaves in 5–10 mL in a glass beaker). Submerge the leaf strips completely into the buffer (*see* **Notes 8** and **9**).
4. Vacuum (~100 mbar) infiltrate the leaf strips for 10–15 min in a desiccator jar.
5. With the help of a flat tip dropper, discard the pretreatment buffer gently without disturbing the leaf strips.
6. Carefully add 5 mL of enzyme buffer I to the leaf strips and keep on the shaker at 40 rpm at RT for 1 h (*see* **Note 10**).
7. Carefully transfer the digested leaf strips along with buffer to a 50 mL Falcon, centrifuge the sample at 200 g for 5 min at RT.
8. Discard the supernatant without disturbing the partially digested strips.
9. Add 5 mL of enzyme buffer II to the sample, and fix it to the shaker at a speed of 40 rpm for 1 h at RT.
10. Add 5 mL of W5 buffer to the protoplast mixture to cease the enzymatic reaction.
11. To filter the protoplast, cut 4 × 4 cm miracloth, fold the cloth into half to create two layers, fold it in half again to have fourfold miracloth-filtering setup. Rinse it with W5 buffer. Place it inside the glass funnel for filtering the protoplast mixture.

12. Carefully transfer the protoplast mixture to the prepared strainer using flat end dropper for filtering the protoplast (*see* **Note 11**).
13. Collect the protoplast by centrifuge at 200 g for 5 min at RT.
14. Discard the supernatant gently with the help of blunt tip dropper without disturbing the protoplast pellet.
15. Resuspend the protoplast in 2 mL of W5 buffer by gentle mixing using blunt tip dropper.
16. Transfer the protoplast into a fresh 1.5 mL MCT, and place it on the ice for 15–20 min. This helps in sedimenting the protoplast.
17. Discard the supernatant gently and resuspend the protoplast into 300 μL of W5 buffer.
18. Resuspend of the protoplast to a final volume of 300 μL W5 buffer yields 1×10^9 cells/mL. This concentration can be further diluted as per the downstream experiments (Fig 1).

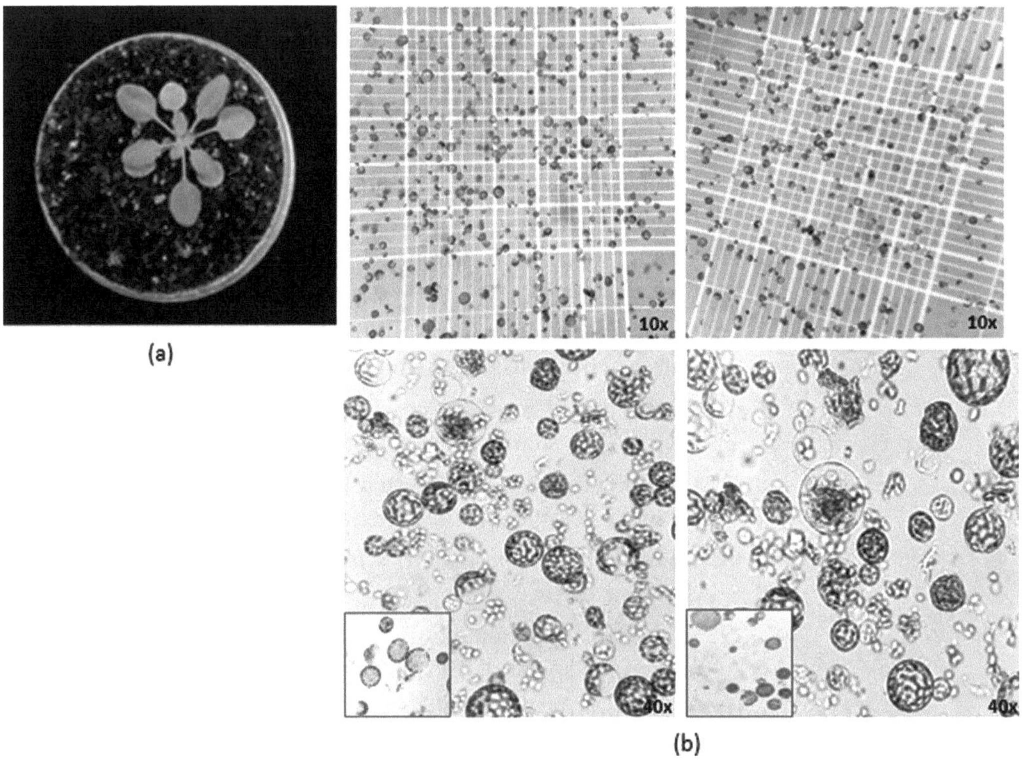

Fig. 1 Protoplast yield and viability of mock and treated samples. (**a**) Col-0 plant used for protoplast isolation. (**b**) Protoplast under compound microscope at 10× magnification: (upper left panel) mock, and *Pst* DC3000 treated (Upper right panel); 40× zoomed image of control leave protoplasts—number of viable protoplasts is 1.26×10^6/mL (lower left panel), and 40× zoomed of *Pst* DC3000 treated leaves protoplasts—number of viable protoplasts are 2.1×10^5/mL (lower right panel)

3.4 Trypan Blue Staining for the Quality Check of the Protoplast

1. Transfer 20 μL protoplast to a fresh PCR tube, and add 2 μL of 0.4% trypan blue stock solution.
2. Take 10 μL of this protoplast onto a hemocytometer, and examine the cell under a compound microscope. Count the number of blue-stained protoplast and the number of total protoplasts.
3. Use the formula for calculating the viability percentage of the protoplast.

$$\%\text{viable protoplast} = [1.00 - (\text{number of blue protoplasts}/\text{number of total protoplasts})] \times 100$$

3.5 PEG-Calcium-Mediated Protoplast Transfection (See Note 12)

1. Take 10–15 μg of plasmid DNA (for plasmid size 5–10 kb) to a fresh MCT.
2. Add 100 μL–150 μL of protoplast of concentration 2×10^4 protoplast.
3. Add 100 μL of PEG solution to the protoplast, and mix them by gently tapping the bottom of the MCT.
4. Incubate the mixture for 10–15 min at RT (*see* **Note 13**).
5. To stop the transfection process, dilute the transfection mixture by adding 400–450 μL of W5 solution, and mix well by gently inverting the tube for 4–6 times.
6. Centrifuge the solution at 200 g for 2 min at RT. Carefully discard the supernatant.

3.6 Sucrose Feeding Assay in Protoplast

1. Resuspend the protoplast in 400 μL of incubation solution containing 20 mM sucrose.
2. Centrifuge the incubated protoplast at 200 g for 2 min at RT.
3. Wash the pelleted protoplast with incubation buffer (without sucrose) by gentle pipetting, and again centrifuge at 200 g for 2 min at RT.
4. Repeat **step 3** twice.
5. Directly proceed for the GC/MS sucrose quantification as per the methodology provided by [15] or store at −20 °C for further use.

4 Notes

1. Pretreatment buffer should be prepared freshly.
2. Sterilize the buffer through 0.22 μm filter in a laminar hood.
3. W5 buffer should be prepared freshly.

4. PEG 4000 solution can be prepared an hour prior for complete dissolution. PEG solution should not be autoclaved. Freshly prepared PEG solution is recommended.
5. For infecting the leaves via syringe inoculation method here, we used 0.0005 OD_{600} of *Pst* DC3000 pathogen.
6. The selection of right-aged plant and healthy leaves is a crucial factor for the protoplast isolation.
7. Avoid pressing the leaves or causing any damage to the tissue. Make sure that freshly cut leaf strips are submerged in the pretreatment buffer and should not get dried.
8. Leaf strips should be transferred quickly but gently into the buffer. Avoid using sharp-pointed forceps. If needed, flat forcep can be used for dipping the leaf strips.
9. To avoid extraction buffer exposure to light, cover the beaker containing leaf strips with aluminium foil.
10. Maximum number of protoplasts are expected to be released after 1 h of the enzymatic digestion. To check the amount of protoplast, few drops of the digested sample can be checked under the microscope. Proceed for a second round of digestion if needed.
11. Protoplast are very fragile; use of pipette is not recommended. Avoid shaking of the mixture while filtering.
12. The methodology is similar to the one provided by [10] with minor modifications.
13. To optimize transfection timing, we performed a series of experiments, varying the time point of transfection.

Acknowledgments

The SWEET transporter project at MS-K's lab was funded by NIPGR core funding. Anjali acknowledges NIPGR-SRF fellowship. We acknowledge NIPGR plant growth facility for plant growth and space (Fig. 1).

References

1. EC, COCKING (1960) A method for the isolation of plant protoplasts and vacuoles. Nature 187:927–929
2. Damm B, Schmidt R, Willmitzer L (1989) Efficient transformation of Arabidopsis thaliana using direct gene transfer to protoplasts. Mol Gen Genet MGG 217:6–12
3. Abel S, Theologis A (1994) Transient transformation of Arabidopsis leaf protoplasts: a versatile experimental system to study gene expression. Plant J 5(3):421–427
4. Sheen J (2001) Signal transduction in maize and Arabidopsis mesophyll protoplasts. Plant Physiol 127(4):1466–1475
5. Sparkes IA, Runions J, Kearns A, Hawes C (2006) Rapid, transient expression of fluorescent fusion proteins in tobacco plants and

generation of stably transformed plants. Nat Protoc 1(4):2019–2025

6. Manavella PA, Chan RL (2009) Transient transformation of sunflower leaf discs via an agrobacterium-mediated method: applications for gene expression and silencing studies. Nat Protoc 4(11):1699–1707
7. Ueki S, Lacroix B, Krichevsky A, Lazarowitz SG, Citovsky V (2009) Functional transient genetic transformation of Arabidopsis leaves by biolistic bombardment. Nat Protoc 4(1): 71–77
8. Wang H, Jiang L (2011) Transient expression and analysis of fluorescent reporter proteins in plant pollen tubes. Nat Protoc 6(4):419–426
9. Miao Y, Jiang L (2007) Transient expression of fluorescent fusion proteins in protoplasts of suspension cultured cells. Nat Protoc 2(10): 2348–2353
10. Yoo SD, Cho YH, Sheen J (2007) Arabidopsis mesophyll protoplasts: a versatile cell system for transient gene expression analysis. Nat Protoc 2(7):1565–1572
11. Meyer S, Lauterbach C, Niedermeier M, Barth I, Sjolund RD, Sauer N (2004) Wounding enhances expression of AtSUC3, a sucrose transporter from Arabidopsis sieve elements and sink tissues. Plant Physiol 134(2):684–693
12. Stadler R, Sauer N (1996) The Arabidopsis thaliana AtSUC2 gene is specifically expressed in companion cells. Botanica Acta 109(4): 299–306
13. Fatima U, Senthil-Kumar M (2015) Plant and pathogen nutrient acquisition strategies. Front Plant Sci 6:750
14. Wang K, Senthil-Kumar M, Ryu CM, Kang L, Mysore KS (2012) Phytosterols play a key role in plant innate immunity against bacterial pathogens by regulating nutrient efflux into the apoplast. Plant Physiol 158:1789–1802. https://doi.org/10.1104/pp.111.189217
15. Abelenda JA, Bergonzi S, Oortwijn M, Sonnewald S, Du M, Visser RG, Sonnewald U, Bachem CW (2019) Source-sink regulation is mediated by interaction of an FT homolog with a SWEET protein in potato. Curr Biol 29(7):1178–1186

Chapter 8

Twitching Motility Dynamics in the Wilt-Causing Phytopathogen *Ralstonia solanacearum*

Shuvam Bhuyan, Lukapriya Dutta, Shuhada Begum, Shubhra Jyoti Giri, Monika Jain, Manabendra Mandal, and Suvendra Kumar Ray

Abstract

Ralstonia solanacearum, a gram-negative bacterium, is known to cause devastating wilting disease in many plant species. The bacterium's pathogenesis relies on various virulence determinants, including twitching motility, facilitated by type IV pili appendages. Twitching motility is a density-dependent mechanism enabling bacterial movement on solid surfaces, crucial for biofilm formation and virulence. Understanding this motility is of utmost importance in studying the pathogen's behavior. In this chapter, we present a simple and reproducible method to assess twitching motility in *R. solanacearum.* The method involves microliter spotting of a diluted bacterial suspension onto a solid agar growth medium, followed by microscopic observation of the early-stage microcolonies. Twitching motility becomes evident as finger-like projections, which merge and progressively spread in all directions, forming multilayered structures. The examination also reveals cell heterogeneity within colonies and the influence of bacterial concentration on microcolony shape, potentially attributed to quorum sensing. Our simplified protocol for assessing twitching motility in microcolonies using microliter spotting offers significant advantages over existing methods described in the literature, making it an effective tool for investigating twitching motility in bacterial pathogenesis. A detailed step-by-step description of the protocol is provided in this chapter, facilitating its adoption and application in future research.

Key words *Ralstonia solanacearum*, Twitching motility, Motility dynamics, Microcolony morphology, Cell density

1 Introduction

Plants play an indispensable role in sustaining life, bestowing us with food, oxygen, and a wealth of resources. Additionally, they form the very foundation of agro-economics, propelling key industries like agriculture, horticulture, and forestry. Not only do plants meet our fundamental nutritional requirements, but they also generate employment prospects, bolster livelihoods, and contribute to economic growth through trade and exports. However, amid their immense significance, the *Ralstonia solanacearum* species

Muthappa Senthil-Kumar and Hema Ramanna (eds.), *Plant Biotic Stress: Methods and Protocols*, Methods in Molecular Biology, vol. 2966, https://doi.org/10.1007/978-1-0716-4746-2_8,

Fig. 1 Comparison between healthy and *R. solanacearum*-infected tomato plants inside the greenhouse. (**a**) A healthy tomato plant (control). (**b**) *R. solanacearum*-inoculated tomato plant by soil drenching at 22 days post-inoculation

complex emerges as a formidable adversary, inflicting systemic devastation upon 504 plant species through its pernicious wilting disease ([15, 16]; *see* Fig. 1). *R. solanacearum* is a gram-negative, rod-shaped bacterium found in soil that infiltrates its host plants by penetrating through the root system, subsequently invading the xylem vessels and spreading throughout the stem, ultimately inflicting severe wilting disease in them [1–3]. To ensure successful pathogenesis, this phytopathogen employs multiple virulence determinants among which twitching motility serves as one [4].

Twitching motility is a type of bacterial movement mechanism that enables *R. solanacearum* to traverse solid surfaces. While this form of motility is independent of the flagellum but it involves the extension, attachment, and retraction of thin filamentous appendages called type IV pili (TFP). These pili interact with the surrounding environment, allowing the bacterium to crawl or "twitch" along surfaces such as plant tissues or other substrates. Twitching motility aids in the colonization and spread of the bacterium within host plants, contributing to its pathogenicity [5]. This is further substantiated by studies showing that mutants of *R. solanacearum* lacking functional twitching motility also exhibit a deficiency in virulence [6–8]. As such it becomes indispensable to assess the presence of twitching motility in such pathogenic bacterium.

The current literature on evaluating twitching motility encompasses various methods, primarily categorized into surface and subsurface approaches. In subsurface assays, cells are introduced underneath the agar through stabbing and left to grow at the agar/petri dish interface overnight, resulting in the development of a conspicuous halo indicating interstitial colony expansion, which serves as an observable sign of twitching motility. Improvements to this subsurface method include staining adherent bacteria with crystal violet to assess their twitching motility, following the removal of nonadherent ones by washing [9], and stab inoculation of bacterial cells onto an agar slab placed on a microscope slide, facilitating microscopic observation [10]. In the surface twitching motility assays too, bacterial colonies that are grown on the surface of the agar growth medium can be evaluated for twitching motility both microscopically and macroscopically. While the grown bacteria on the agar surface can be directly visualized under a microscope to assess twitching motility, bacteria grown on cellophane sheets placed on agar, besides facilitating microscopic observation also allow easy harvesting of actively twitching cells [11]. For macroscopic assessment, the cells can be grown directly on a soft agar growth medium supplemented with triphenyltetrazolium chloride [12].

In this chapter, we describe a surface method assisted by phase-contrast microscopy for assaying twitching motility in *R. solanacearum* [13, 14, 16, 17]. The use of time-lapse imaging in conjunction with phase-contrast microscopy to study bacterial motility dynamics reveals that cells of *R. solanacearum* spread exponentially in all directions in layers via twitching motility to produce protruding, flat, transparent, multilayered film-like colony edges. On the other hand, strains of *E. coli* lack functional TFP and are therefore devoid of twitching motility, resulting in smooth colony edges that do not actively spread across the surface of the growth medium ([13, 14]; *see* Fig. 2). In this surface method, we spot a few microliters of diluted bacterial cell suspension onto the agar-growth medium surface to observe twitching motility of *R. solanacearum* microcolonies [13, 14, 16–19]. Observation of twitching motility in individual microcolonies imparts certain advantages over the observation of the bulk of cells in terms of motility dynamics.

Upon spotting, some bacterial cells tend to arrange themselves at the periphery of the spotted area, that is, at the wet-dry interface of the microliter spot. These peripheral cells grow in numbers to become microcolonies, which in turn grow in size, merge together, and demonstrate more robust twitching motility than the central cells, suggesting heterogeneity. It is these peripheral cells where it is clearly observable that layers of bacteria twitch on top of each other in all directions, resulting in the formation of a multilayered film-like appearance (*see* Fig. 3). Whereas the microcolonies at the center

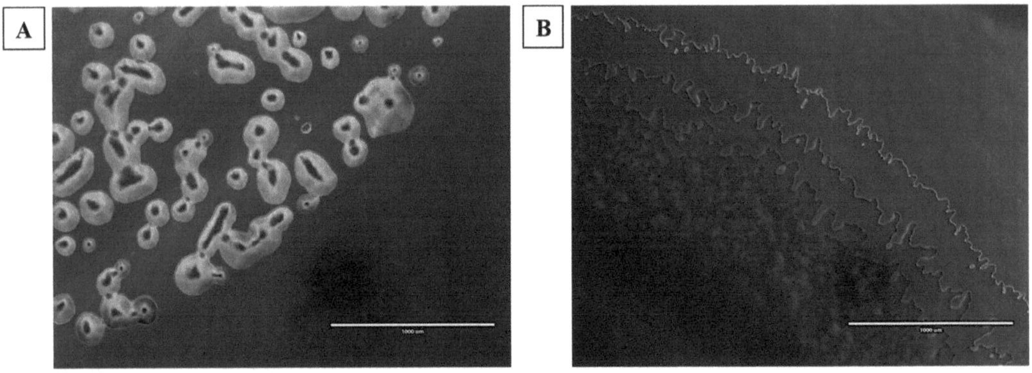

Fig. 2 Phase-contrast photomicrographs of *E. coli* and *R. solanacearum* microcolonies at 40× magnification. (**a**) *E. coli* at 11 h of incubation exhibiting smooth colony edges due to the absence of twitching motility. (**b**) *R. solanacearum* at 28 h of incubation exhibiting finger-like projections of flat, multilayered film-like colony edges due to the presence of twitching motility

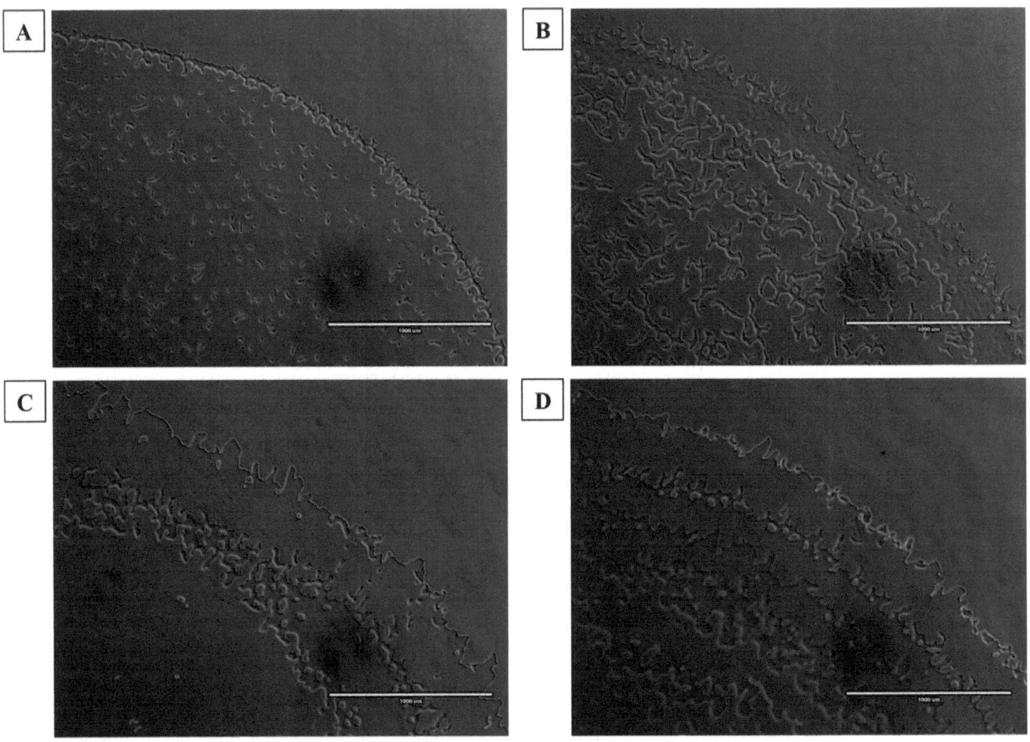

Fig. 3 Twitching motility of 10^5 CFU/mL of *R. solanacearum* cells. (**a**) Bacterial cells arrange themselves at the periphery of the spotted area in a ring-like structure that grows in numbers to become microcolonies, which in turn grow in size, merge and initiate the first layer of twitching motility at 12 h of incubation. (**b**) Twitching motility at 17 h of incubation. (**c**) A second layer of bacteria twitching on top of the first layer at 20 h of incubation. (**d**) Microcolonies at 23 h of incubation twitch in all directions across solid surfaces in multilayered film-like layers

of the spotted area at higher and lower bacterial cell concentrations reveal the influence of microcolonies on each other with respect to their shape due to bacterial sensing of their adjacent counterparts in a population [14]. Using time-lapse photography, the displacement of bacterial cells from the periphery of the spotted area due to twitching motility can be plotted on a scatter plot, and its rate of twitching motility may be calculated [13, 14].

2 Materials

2.1 Bacterial Growth

1. BG growth medium consisting of 1% peptone, 0.1% yeast extract, and 0.1% casamino acids supplemented with 5 g/L glucose is used for *R. solanacearum* growth. For preparing solid BG medium, 1.5% agar is added to it.
2. Sterile culture vials and inoculation loop.
3. Sterile 1000 μL micropipettes and micropipette tips.
4. Laboratory film.
5. Laminar airflow cabinet and incubator.

2.2 Twitching Motility Assay

1. BG-agar medium consisting of 1% peptone, 0.1% yeast extract, 0.1% casamino acids, and 1.5% agar but devoid of glucose (*see* **Notes 1** and **2**).
2. Sterile 1.5 mL microfuge tubes.
3. Sterile 5 mL of distilled water.
4. Sterile 1000 μL and 10 μL micropipettes and micropipette tips.
5. Laboratory film.
6. Laminar airflow cabinet, incubator, and phase-contrast microscope.

2.3 Preparation of Time-Lapse Twitching Motility Video

1. Free video editing applications such as iMovie (https://www.apple.com/in/imovie/) for MacOS users or the built-in Video Editor found within the Microsoft Photos Legacy application for Windows OS users. In this chapter, we discuss time-lapse video preparation in the iMovie application (*see* Subheading 3.3).
2. Microsoft PowerPoint.

3 Methods

3.1 Preparation of Diluted Bacterial Suspension

1. Take cells of *R. solanacearum* from a glycerol stock.
2. Streak the cells onto petri plates containing BG-agar medium supplemented with glucose.

3. Seal the streaked-out petri plates with laboratory film.
4. Incubate the sealed petri plates at 28 °C for 24 h.
5. Inoculate *R. solanacearum* cells into a culture vial containing 10 mL BG medium supplemented with glucose.
6. Incubate the culture vial with the inoculated cells overnight (16 ± 1 h) at 28 °C.
7. Pipette 1 mL of the saturated culture into a sterile 1.5 mL microfuge tube.
8. Centrifuge the microfuge tube at 1500 g for 15 min at 4 °C.
9. Carefully decant the supernatant from the tube.
10. Resuspend the pellet in 1 mL of sterile distilled water.
11. Serially dilute the bacterial suspension 10^4 times with sterile distilled water.

3.2 Twitching Motility Assay

1. Prepare a petri plate containing BG-agar medium without glucose (*see* **Notes 1** and **2**).
2. Air-dry the prepared petri plate for at least 1 h in a sterile environment, such as a laminar airflow cabinet (*see* **Note 3**).
3. Mark the bottom of the petri plate with circles of approximately 1 cm diameter using a permanent marker (*see* **Note 4**).
4. Using a micropipette, spot 5 μL of the diluted bacterial suspension at the center of the marked circle on the BG-agar petri plate without glucose.
5. After spotting, allow the petri plate to remain open, and let the spotted area air-dry undisturbed (*see* **Note 5**) in the laminar airflow cabinet (refer Fig. 4).
6. Once the spotted area is completely dry, cover the petri plate with its lid and seal it using laboratory film.
7. Incubate the sealed petri plate at 28 °C for 10 h.
8. Using an inverted phase-contrast microscope at 40× magnification with the 4/10 PH condenser annulus, take photomicrographs of the grown microcolonies at the periphery of the spotted area. Capture images in JPEG format hourly (or every 5 min to prepare a time-lapse twitching motility video, *see* Subheading 3.3).
9. Observe the characteristic irregular, flat, multilayered colony edges formed due to twitching motility in *R. solanacearum*, which are clearly visible from 19 to 22 h of incubation.

3.3 Preparation of Time-Lapse Twitching Motility Video

1. Click photomicrographs of the grown microcolonies at the periphery of the spotted area at regular intervals of 5 min from 10 h of incubation until the desired time. Capture the images in JPEG format using an inverted phase-contrast

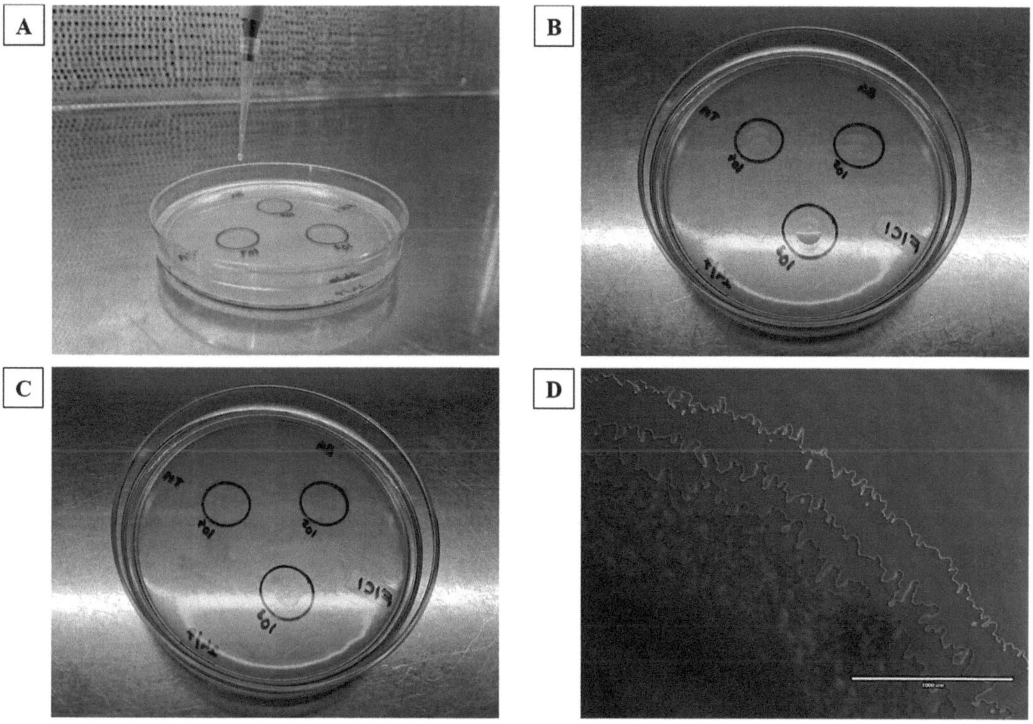

Fig. 4 Twitching motility assay of *R. solanacearum*. (**a**) Spotting of 5 μL bacterial inoculum on BG-agar medium. (**b**) Air-drying of spotted plates. (**c**) Complete drying of the spotted area. (**d**) Twitching motility of *R. solanacearum* observed under a phase-contrast microscope at 40× magnification after 28 h of incubation

microscope at 40× magnification with the 4/10 PH condenser annulus (*see* **Notes 6–8**).

2. Transfer the captured photographs from the microscope to a separate folder on a laptop.
3. Open the iMovie application.
4. Click on the "Projects" tab and select "Create New."
5. Choose "Movie" from the drop-down list to open the video editing interface of the application.
6. Drag and drop the selected photomicrographs from their stored folder directly into the video editing area located in the bottom half of the application's interface.
7. Reselect all the photomicrographs in the video editing area, and click on the "Reset All" button situated at the top right-hand corner of the application's interface.
8. The default duration of each image is set to 4.0 s. Under the "ⓘ" (clip information) button, reset the duration of each image to a range of 0.3–0.5 s.
9. Optionally, add introductory and ending video segments to the twitching motility video from the "Titles" and "Backgrounds"

tabs (*see* **Note 9**). Also, under the "Audio & Video" tab, select audio from the list of freely available "Sound Effects" to include in the video.

10. To save the finished video, click on the "File" button located in the menu bar.
11. From the drop-down list that appears, select "Share" and then choose "File...".
12. A dialog box with the name "File" will appear. In this dialog box, select the desired resolution, quality, and compression settings for the video from the provided drop-down lists corresponding to each parameter.
13. Click on the "Next..." button to proceed.
14. In the "Save As" field of the dialog box, choose a file name for the video.
15. Click on the "Save" button to save the video with the chosen settings and file name.

4 Notes

1. Supplementing BG-agar growth medium with glucose affects the formation of distinctive finger-like projections of microcolonies exhibiting twitching motility.
2. TFP is employed by bacteria to twitch across solid surfaces. If the concentration of agar in the growth medium is reduced below 1.5%, the distinctive finger-like projections of microcolonies resulting from twitching motility may not be visible.
3. To accelerate the drying of the spotted area, petri plates with BG-agar growth medium prepared for performing twitching motility assay can be left to air-dry for an extended duration of 1 h.
4. Marking petri plates with circles helps locate the spotted bacterial cells under a phase-contrast microscope easily especially in low dilutions at early stages of incubation.
5. Some bacterial cells tend to arrange themselves at the periphery of the spotted area where twitching motility is more prominently observed. Any changes to this arrangement may alter twitching motility in certain aspects with regard to the rate or the appearance of layers. Therefore, it is to be ensured that the droplet of bacterial suspension spotted on the BG-agar growth medium retains a circular shape throughout the entire drying process. It is advisable not to disturb the petri plates during the drying period until they are completely dry.

6. Fluctuations in room temperature while performing time-lapse photography cause the accumulation of water droplets beneath the petri plate lid, resulting in light obstruction. To counteract this, periodic adjustments to the brightness may be necessary.
7. When performing time-lapse photography, if bacteria are exposed to elevated room temperature for an extended period, rather than their optimal growth temperature, the distinct irregular finger-like projection characteristic of twitching motility may not be observable.
8. To create a twitching motility video for *R. solanacearum*, a time-lapse photography interval of 5 min is adequate. However, to achieve a more seamless transition of photomicrographs in the video, it may be beneficial to reduce the time interval between capturing two consecutive photomicrographs.
9. For adding customized title styles to the video, PowerPoint slides with the required text may be saved as transparent images that may be superimposed on the photomicrographs in the iMovie application by simple drag and drop in the video editing area over the required photomicrographs. To do this, the slides are first resized as per the dimensions of the photomicrographs by clicking on "Page Setup" from the drop-down list of "Slide Size" under the "Design" button. The required size and orientation of the slides are chosen and clicked on the "OK" button. Then the required slides with text are selected, and "File" from the menu bar is selected followed by selecting "Export" from the drop-down list. A dialog box appears where the file format is selected as PNG from the drop-down list. As per the requirement, then select either "Save Every Slide" or "Save Current Slide Only." An export folder name is chosen in the "Export as" field, and then the "Export" button is clicked. A pop-up message appears showing the directory of the saved slides as PNG images. The "Ok" button is clicked.

Acknowledgments

Shuvam Bhuyan is thankful for the JRF/SRF fellowship from the UGC-NFSC, GoI, New Delhi. **Lukapriya Dutta** is thankful to DST, GoI, New Delhi for the INSPIRE-JRF fellowship. **Shubhra Jyoti Giri and Shuhada Begum** are thankful for the JRF fellowship from the DBT, GoI, New Delhi grant (BT/PR41637/NER/95/1753/2021) awarded to **Suvendra Kumar Ray**. **Monika Jain** is thankful for the JRF fellowship from the CSIR, GoI, New Delhi.

References

1. Genin S (2010) Molecular traits controlling host range and adaptation to plants in Ralstonia solanacearum. New Phytol 187(4): 920–928. https://doi.org/10.1111/j.1469-8137.2010.03397.x
2. Schell MA (2000) Control of virulence and pathogenicity genes of Ralstonia solanacearum by an elaborate sensory network. Annu Rev Phytopathol 38(1):263–292. https://doi.org/10.1146/annurev.phyto.38.1.263
3. Naik B, Kumar V, Rizwanuddin S et al (2023) Genomics, proteomics, and metabolomics approaches to improve abiotic stress tolerance in tomato plant. Int J Mol Sci 24(3):3025. https://doi.org/10.3390/ijms24033025
4. Genin S, Denny TP (2012) Pathogenomics of the Ralstonia solanacearum species complex. Annu Rev Phytopathol 50:67–89. https://doi.org/10.1146/annurev-phyto-081211-173000
5. Corral J, Sebastià P, Coll NS et al (2020) Twitching and swimming motility play a role in Ralstonia solanacearum pathogenicity. mSphere 5(2). https://doi.org/10.1128/msphere.00740-19
6. Kang Y, Liu H, Genin S et al (2002) Ralstonia solanacearum requires type 4 pili to adhere to multiple surfaces and for natural transformation and virulence. Mol Microbiol 46(2): 427–437. https://doi.org/10.1046/j.1365-2958.2002.03187.x
7. Ray SK, Kumar R, Peeters N et al (2015) rpoN1, but not rpoN2, is required for twitching motility, natural competence, growth on nitrate, and virulence of Ralstonia solanacearum. Front Microbiol 6:229. https://doi.org/10.3389/fmicb.2015.00229
8. Liu H, Kang Y, Genin S et al (2001) Twitching motility of Ralstonia solanacearum requires a type IV pilus system. Microbiology 147(12): 3215–3229. https://doi.org/10.1099/00221287-147-12-3215
9. Harding CM, Tracy EN, Carruthers MD et al (2013) Acinetobacter baumannii strain M2 produces type IV pili which play a role in natural transformation and twitching motility but not surface-associated motility. MBio 4(4): 10–1128. https://doi.org/10.1128/mbio.00360-13
10. Glessner A, Smith RS, Iglewski BH et al (1999) Roles of Pseudomonas aeruginosa las and rhl quorum-sensing systems in control of twitching motility. J Bacteriol 181(5):1623–1629. https://doi.org/10.1128/jb.181.5.1623-1629.1999
11. Xu B, Wozniak DJ (2015) Development of a novel method for analyzing Pseudomonas aeruginosa twitching motility and its application to define the AmrZ regulon. PLoS One 10(8): e0136426. https://doi.org/10.1371/journal.pone.0136426
12. Leong CG, Bloomfield RA, Boyd CA et al (2017) The role of core and accessory type IV pilus genes in natural transformation and twitching motility in the bacterium Acinetobacter baylyi. PLoS One 12(8):e0182139. https://doi.org/10.1371/journal.pone.0182139
13. Bhuyan S, Yadav M, Giri SJ et al (2023) Microliter spotting and micro-colony observation: a rapid and simple approach for counting bacterial colony forming units. J Microbiol Methods 207:106707. https://doi.org/10.1016/j.mimet.2023.106707
14. Bhuyan S, Dutta L, Begum S et al (2024) A study on twitching motility dynamics in Ralstonia solanacearum microcolonies by live imaging. J Basic Microbiol 64(1):42–49. https://doi.org/10.1002/jobm.202300272
15. Lowe-Power TM, Avalos J, Bai Y et al (2020) A meta-analysis of the known global distribution and host range of the Ralstonia species complex. bioRxiv 2020–07. https://doi.org/10.1101/2020.07.13.189936
16. Bhuyan S, Jain M, Boruah L et al (2025) Tissue-dependent nature of plant susceptibility: a comparative pathogenicity study of the systemic phytopathogen Ralstonia pseudosolanacearum in eggplant and tomato seedlings through root and leaf. J Plant Pathol 1–14. https://doi.org/10.1007/s42161-025-01959-2
17. Bhuyan S, Boruah L, Jain M et al (2024) Morphology of softening and translucency in the stems of tomato and eggplant seedlings due to infection by Ralstonia pseudosolanacearum F1C1. bioRxiv 2024–09. https://doi.org/10.1101/2024.09.10.612199
18. Ray SK, Kumar R, Peeters N et al (2015) rpoN1, but not rpoN2, is required for twitching motility, natural competence, growth on nitrate, and virulence of Ralstonia solanacearum. Front Microbiol 6. https://doi.org/10.3389/fmicb.2015.00229
19. Jain M, Begum S, Bhuyan S et al (2024) The impact of fold-dilution and diluent volume on the estimation of colony forming units in a bacterial culture. bioRxiv 2024–11. https://doi.org/10.1101/2024.11.28.625891

Chapter 9

A Blotting Paper Technique for the Screening of Chickpea Genotypes Against Dry Root Rot Disease

Athimoolam Durgadevi, Prachi Pandey, and Muthappa Senthil-Kumar

Abstract

Dry root rot (DRR) disease is a major threat to chickpea production across the world. This disease is caused by a soil-borne necrotrophic fungal pathogen, *Macrophomina phaseolina*. The use of disease-resistant varieties paves the way to conquer the disease spread. Though chickpea germplasm with rich genetic diversity is available around the world, its response to DRR is still unexplored. In turn, this demands screening and identification of resistant genotypes for crop protection against the disease. Here we describe an improved blotting paper technique for the large-scale screening of chickpea genotypes for DRR resistance. The method is quick, cost-effective, less labour-intensive, and thus optimized for high-throughput screening and can be efficiently used to screen a large number of chickpea genotypes for resistance against DRR.

Key words Chickpea, Dry root rot, *Macrophomina phaseolina*, Disease resistance, Screening, Blotting paper technique

1 Introduction

Chickpea (*Cicer arietinum* L.) is an economically important winter legume, usually cultivated in arid and semi-arid tropical regions under rain-fed conditions. It is a robust, dryland, self-pollinated crop, and has a deep root system with numerous and finer lateral roots along with a central taproot system. In agricultural fields, dry root rot (DRR) is a major threat to chickpea production [1]. A soil-borne necrotrophic fungus, *Macrophomina phaseolina* (formerly known as *Rhizoctonia bataticola*), causes DRR in chickpea. The typical DRR symptoms are chlorosis and drooping of leaves and petioles, leading to dead and dry plant. The infected plant leaves become straw yellow, and the taproot becomes dark, brittle, and

Muthappa Senthil-Kumar and Hema Ramanna (eds.), *Plant Biotic Stress: Methods and Protocols*, Methods in Molecular Biology, vol. 2966, https://doi.org/10.1007/978-1-0716-4746-2_9,

devoid of lateral roots [2, 3]. Microsclerotia can be seen in the roots and pith when vertically cut [4]. The DRR disease incidence is increased when the plants experience drought stress. The disease is more prevalent in rain-fed areas and is predominantly seen in the reproductive stage, that is, flowering to podding stages [2, 5].

Utilizing diverse genotypes or multilines with varying levels of resistance and susceptibility can significantly reduce the intensity and severity of disease outbreaks, offering better protection than single resistant varieties, which are more vulnerable to epidemics [6]. The use of cultivars that are resistant to DRR is the key to overcoming the disease. This is broadly the safest and an effective method for protecting the crops from the disease [7]. The large and diverse germplasm availability and knowledge of both plant and pathogen biology are important for successful screening. In molecular breeding programs, screening and identification of resistant and susceptible cultivars are essential for crop improvement. Easy, cost-effective, and high-throughput disease assays are crucial for the large-scale screening of the genotypes. Resistance screening against DRR is significant to protect the crop from the disease [8]. Till date, several methods like blotting paper technique, sick pot method, and sick plot screening have been developed and standardized to screen germplasm and develop disease-resistant cultivars [5, 9].

Among the different methods, blotting paper technique is simple and can be used to screen a large number of genotypes in a short time, unlike pot and field screening methods. This method can be executed simply with a liquid inoculum of the fungus and chickpea seedlings [9]. This chapter describes a high-throughput technique to screen a large number of chickpea genotypes for DRR resistance. This technique is quick, less tedious, and thus optimized for high-throughput screening. The seedlings can be inoculated with the fungus using dip inoculation or placing the agar plugs with the fungal sclerotia on the surface of the roots. Dip inoculation method is more reliable compared to agar plug method as it ensures the uniform attachment of the fungal mycelia to the roots.

2 Materials

2.1 Seed Material

Chickpea genotype seeds (*see* **Note 1**).

2.2 Fungal Pathogen

Macrophomina phaseolina (ITCC 8635 with GenBank id, MH509971.1) (*see* **Note 2**).

2.3 Media and Chemical Reagents

1. Potato dextrose agar—PDA.
2. Potato dextrose broth—PDB.
3. Agar.

4. 2% Sodium hypochlorite for the surface sterilization of the chickpea seeds.
5. Soilrite, a combination of an equal ratio, that is, 1:1:1 of horticulture-grade expanded perlite, Irish peat moss, and exfoliated vermiculite, is used in the nursery to grow the seedlings.
6. Spirit.
7. CTAB (cetyl trimethyl ammonium bromide) extraction buffer: 3% CTAB,100 mM Tris-HCl, pH 8.0, 2 M NaCl, 25 mM EDTA (ethylenediamine tetra acetic acid), 4% ß-mercaptoethanol, 5% PVP (polyvinylpyrrolidone).
8. Chloroform.
9. Iso-amyl alcohol.
10. Isopropanol.
11. 70% ethanol.
12. TE buffer: 10 mM Tris-HCl, pH 8.0, 1 mM EDTA pH 8.0.
13. Agarose.
14. SYBR green.
15. 10% KOH solution
16. 1 M Tris-HCl, pH 7.0
17. 50 mM Tris-HCl, pH 7.0
18. WGA-FITC (wheat germ agglutinin-fluorescein isothiocyanate).

2.4 General Materials and Equipment

1. Petri dishes for the culturing of fungus.
2. Laminar airflow.
3. Falcon tubes (50 mL) for surface sterilization.
4. Vortex mixer.
5. Polythene bags with a dimension of 30 × 25 cm for raising the nursery.
6. 1000 mL conical flask for the preparation of broth culture
7. Autoclave for the sterilization of medium.
8. Sterilized glass or plastic funnel for filtering out fungal mycelia and microsclerotia.
9. Nylon mesh with a diameter of 1 mm for the filtration of fungal mycelia and microsclerotia.
10. Petri plates with a diameter of 20 cm.
11. Autoclaved RO (reverse osmosis) water.
12. Rectangular plastic trays with a dimension of 42 × 30 cm.

13. Autoclaved and oven-dried blotting papers.
14. Spray bottle.
15. Paintbrush.
16. Mortar and pestle.
17. Microcentrifuge tubes (2 mL and 1.5 mL).
18. Water bath/heating block.
19. Table-top centrifuge.
20. NanoDrop spectrophotometer.
21. Thermal cycler.
22. Autoclaved MQ water.
23. Nuclease-free water.
24. 96-well PCR plate
25. Fast RT-qPCR system.
26. Research stereomicroscope.
27. Fluorescence microscope.
28. Glass slides and coverslip.
29. Forceps.

3 Methods

3.1 Preparation of Fungal Plate

1. Prepare 100 mL of PDA medium in a conical flask and autoclave it at 120 lbs. for 15 min.
2. Let the medium cool down, pour it into the plates and let the medium solidify.
3. Cut the tip of hyphal growth and inoculate it in the fresh PDA plates.
4. Incubate the plates at 28 °C for 5 days.

3.2 Raising of Seedlings

Select chickpea genotype seeds with uniform shape and size (Fig. 2a).

3.2.1 Surface Sterilization of Seeds

1. Take the seeds in Falcon tubes and wash them with autoclaved RO water for a minute, three time, to remove the dust particles under laminar airflow.
2. Treat the washed seeds with 2% sodium hypochlorite (NaOCl) under intermittent shaking using a vortex mixer for about 2 min (Fig. 1a).
3. Then rinse the seeds with autoclaved RO water to remove the traces of NaOCl.

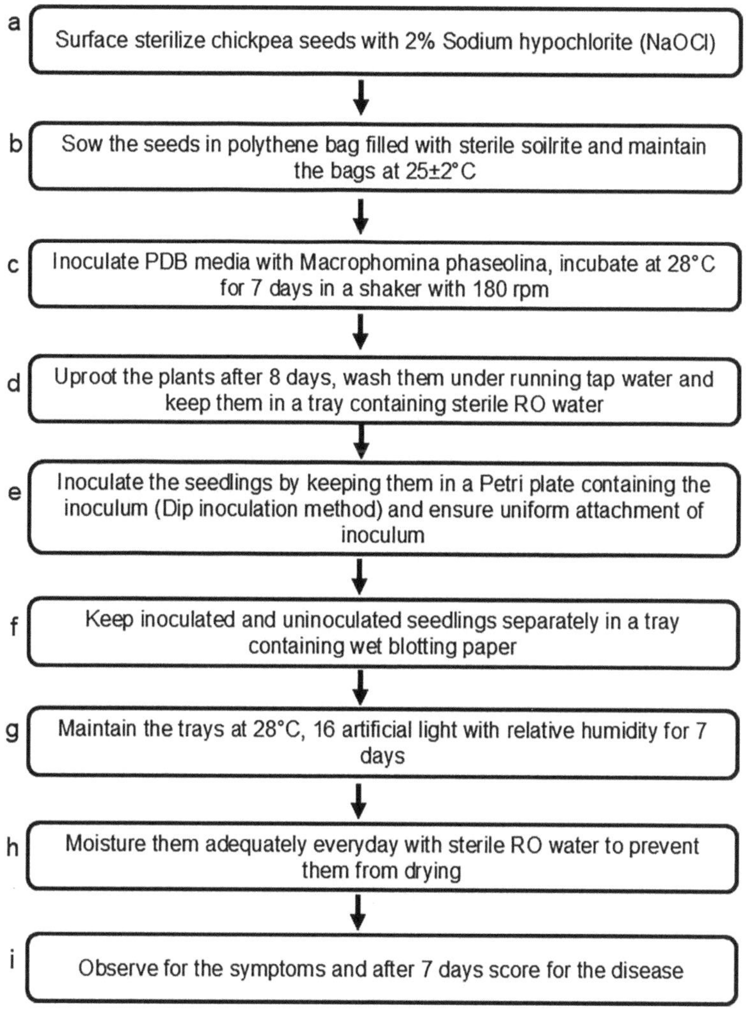

Fig. 1 Flowchart depicting the steps involved in high-throughput screening of chickpea genotypes for DRR resistance using blotting paper technique

3.2.2 Sowing

1. Fill polythene bags with autoclaved soilrite (Fig. 2b).
2. Sow the surface sterilized seeds in the bags with soilrite and keep them at 25 ± 2 °C, 16 h photoperiod with 70% humidity (Fig. 1b).
3. Uproot the plants 7 days after sowing and gently wash the roots (Fig. 1d) under running tap water to remove the soil particles completely without damaging the roots (Fig. 2c). Then keep the plants in a tray containing autoclaved RO water to allow infection (*see* **Note 3**).

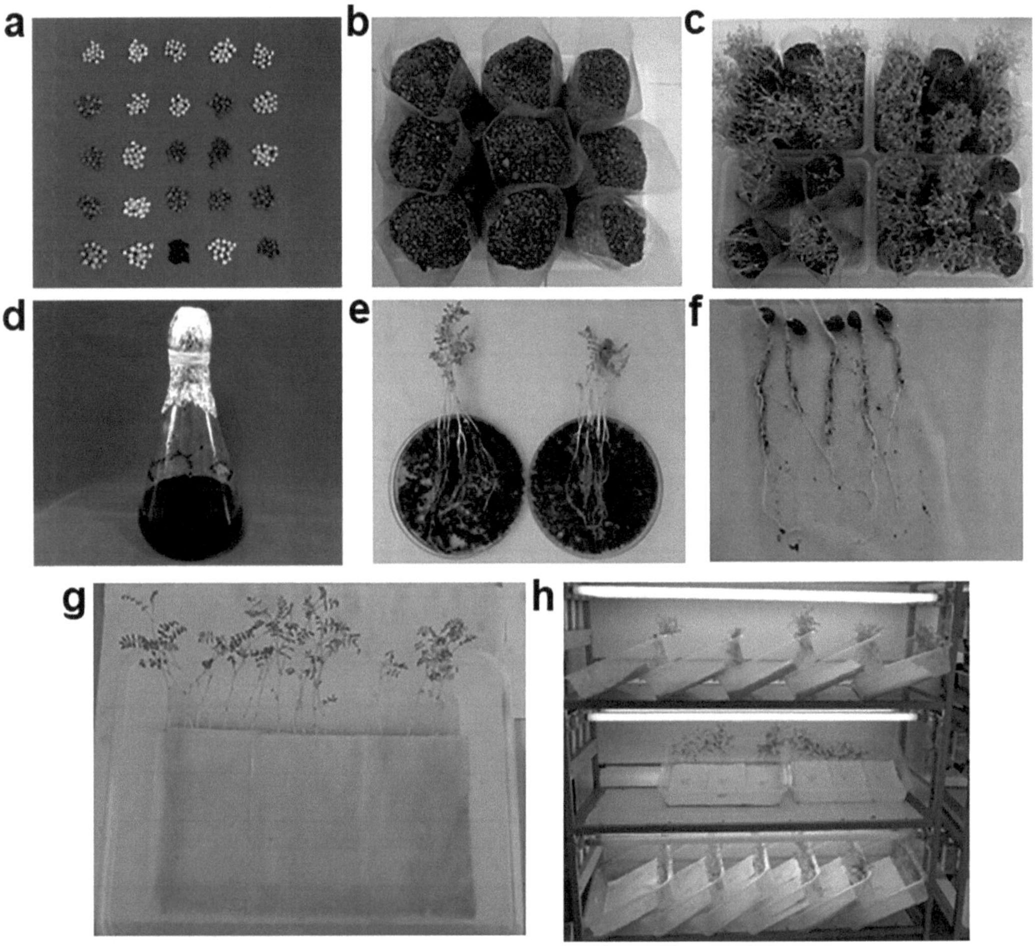

Fig. 2 Detailed methodology of blotting paper technique for DRR disease screening in chickpea. The seeds were collected (**a**) and surface-sterilized using sodium hypochlorite and rinsed with sterile RO water. The surface-sterilized seeds were sown in polythene bags filled with soilrite (**b**) and maintained at 28 °C, 16 h photoperiod with an RH of 60–70%. After 8 days (**c**), the plants were uprooted and washed under tap water gently to remove the soil particles. PDB media was prepared and inoculated with actively growing *Macrophomina phaseolina* and incubated at 28 °C at 180 rpm in dark. After 7 days (**d**), the fungal mycelia and microsclerotia were filtered out and dissolved in autoclaved RO water. Then the plant roots were dipped (**e**) in the inoculum for a min with intermittent shaking. The inoculated plants were then kept in a tray containing wet blotting sheet (**f**). The blotting sheet was then folded into half to cover the roots (**g**). Inoculated and uninoculated seedlings were kept in separate trays. Then all the trays were maintained at 28 °C, 16 h photoperiod with a relative humidity of 60–70% for 7 days (**h**). The plants were adequately moisturized throughout the 7 days and observed for disease symptoms. After 7 days, the plants were scored for the disease

3.3 Preparation of Liquid Inoculum

1. Prepare 300 mL of PDB medium in four 1000 mL conical flasks, and autoclave them at 120 lbs. for 15 min.
2. Let the medium cool down completely, then inoculate the media with 9 mm discs of agar plugs from a 5-day-old fungal agar plate with actively growing *M. phaseolina* culture (Fig. 1c).

3. Incubate the conical flasks at 28 °C in a shaker at 180 rpm in the dark for 7 days.
4. After 7 days (Fig. 2d), using an autoclaved glass or plastic funnel with multiple layers of nylon mesh with a 1 mm diameter pore size to filter out the fungal mycelia and microsclerotia.
5. Wash the mycelia thoroughly with autoclaved RO water to remove the medium completely and then use it as an inoculum.

3.4 Preparation of Blotting Sheets

1. Cut the blotting papers into pieces with a dimension of 30 × 24 cm in sufficient numbers to meet all the replications in large-scale screening.
2. Pack the sheets in a bunch in autoclavable polythene bags, and autoclave at 121 lbs. for 15 min.
3. Keep the blotting sheets in a hot air oven for drying till the moisture gets completely removed (*see* **Note 4**).

3.5 Inoculation

3.5.1 Dip Inoculation

1. Dissolve the washed mycelia (**step 5**, Subheading 3.3) in autoclaved RO water (400 mg/mL) in multiple petri plates with a diameter of 20 cm (*see* **Note 5**).
2. Dip the uprooted plants (**step 3**, Subheading 3.2.2) in the inoculum (Fig. 1e) for a minute in continuous movement to ensure uniform attachment of fungal mycelia on the roots (Fig. 2e).

3.5.2 Agar Plug

1. Take a 5-day-old *M. phaseolina* plate.
2. Cut 2 cm blocks using pipette tips and keep them on the surface of the roots from the point of seed attachment after transferring the seedlings to blotting paper (Subheading 3.5).

3.6 Transfer of Seedlings to Blotting Paper

1. Take a clean tray, wipe it thoroughly with spirit, and pour sterile (autoclaved) RO water, then keep the autoclaved blotting sheets (**step 3**, Subheading 3.4) in the water (*see* **Note 6**).
2. Arrange the wet blotting sheets in a clean, sterilized tray in two layers.
3. Place the inoculated seedlings side by side (five seedlings/replicate) on a wet blotting paper (Fig. 2f), and fold the sheet into half so that only the roots will be covered and the green foliage will be left outside (Fig. 2g).
4. Place the folded blotting papers with the seedlings of the same genotype (three replicates/genotype) one over the other.
5. Keep the plants of three different genotypes with all the replicates in a single tray.

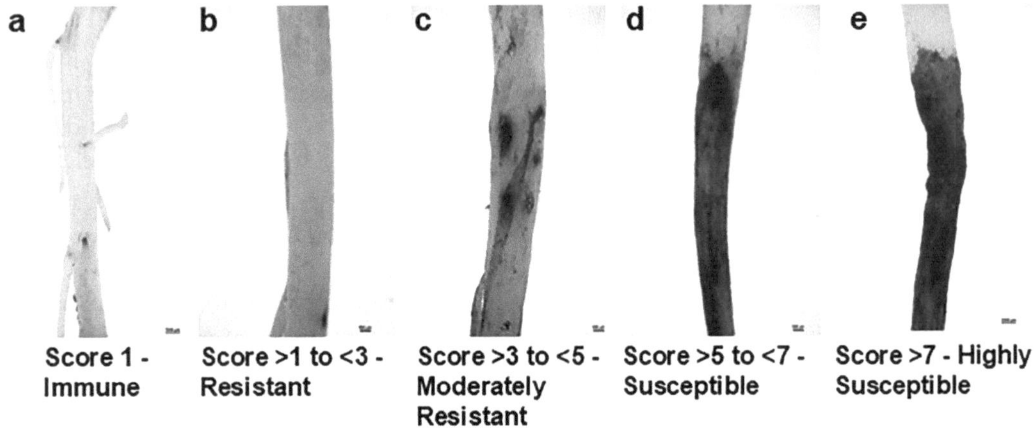

Fig. 3 Representative root images for disease scoring based on the severity of dry root rot. Root images (2 cm from the point of seed attachment) taken 7 days after infection using stereomicroscope SMZ25. Image shows (**a**) no infection on roots (score 1/immune), (**b**) very few lesions on roots (score > 1 to <3/resistant), (**c**) roots on the lesions are clear but small (score > 3 to <5/moderately resistant), (**d**) lesions on roots are clear and visible (score > 5 to <7/susceptible), (**e**) roots are infected and completely discolored (score > 7/highly susceptible). Scale bar—1000 μm

6. Place control and inoculated seedlings in separate trays (Fig. 1f), and maintain them at 28 ± 2 °C, 16 h photoperiod (Fig. 1g) with a relative humidity of 70% for 7 days (Fig. 2h).
7. Water all the trays adequately to prevent the roots from drying (Fig. 1h) (*see* **Note 7**).
8. Observe the plants after 7 days for DRR disease symptoms and score for the disease.
9. Keep a DRR susceptible and a resistant check with every batch (*see* **Note 8**).

3.7 Scoring for the Disease

1. Observe the plants for leaf yellowing and root necrotic symptoms (Fig. 1i) and, after 7 days, score for the disease (Fig. 3).
2. Represent the disease severity on a scale of 1–9, and give a qualitative classification based on the rating scale (Table 1).
3. At the end of the infection period, gently remove the fungal mycelia using autoclaved RO water and a paintbrush from each root surface to see the infection visibly. Assess the degree of disease severity using the disease rating scale given (Fig. 4).

3.8 In Planta Fungal DNA Quantification

3.8.1 DNA Isolation Using CTAB Method

1. Take 250 mg of frozen root sample and add 2 mL of CTAB extraction buffer.
2. Incubate at 70 °C for 1 h, and cool at room temperature.
3. Prepare chloroform: Isoamyl alcohol in a ratio of 24:1, and add equal volume of this to all the tubes.

Table 1
Disease severity on 1–9 rating scale for dry root rot of chickpea

Rating/score	Observation/symptoms	Disease reaction
1	No infection on roots	Immune
>1 and < 3	Very few small lesions on roots	Resistant
>3 and < 5	Lesions on roots clear but small; new roots free from infection	Moderately resistant
>5 and < 7	Lesions on roots; many new roots generally free from lesions	Susceptible
>7	Roots infected and completely discolored	Highly susceptible

4. Mix the tubes well and centrifuge at 10,000 g for 10 min at room temperature.
5. Transfer the aqueous phase to a fresh tube without disturbing, add 0.6 volume of isopropanol.
6. Centrifuge the tubes at 10,000 g for 10 min at 4 °C.
7. Discard the supernatant, wash the pellet with 70% ethanol, and centrifuge at 10,000 g for 10 min at 4 °C.
8. Air-dry the pellet and dissolve it in 30 μL TE buffer.
9. Measure the quality and quantity of genomic DNA using 0.8% agarose gel and NanoDrop spectrophotometer.
10. Confirm the presence of *M. phaseolina* using PCR with specific primers (*see* **Note 9**).

3.9 Quantification of *M. phaseolina* Using qPCR

3.9.1 DNA Concentration Normalization

1. Normalize the DNA concentration to 50 ng/μL.

3.9.2 Generation of Standard Calibration Curve

1. Collect the fungal mycelia from a 5-day-old fungal plate and isolate fungal genomic DNA using the protocol mentioned above.
2. Dilute the pure fungal genomic DNA to tenfold serially (from 100 to 1 pg/μL) with sterile autoclaved MQ.
3. Prepare the master mix of 9 μL containing 5 μL KAPA SYBR FAST Master Mix (2X) ABI PrismTM and 1 μL of each of the primers (Mph-F and Mph-R) (450 nm) and 3 μL nuclease-free water.
4. Aliquot 9 μL master mix to all the wells of the 96-well PCR plate.

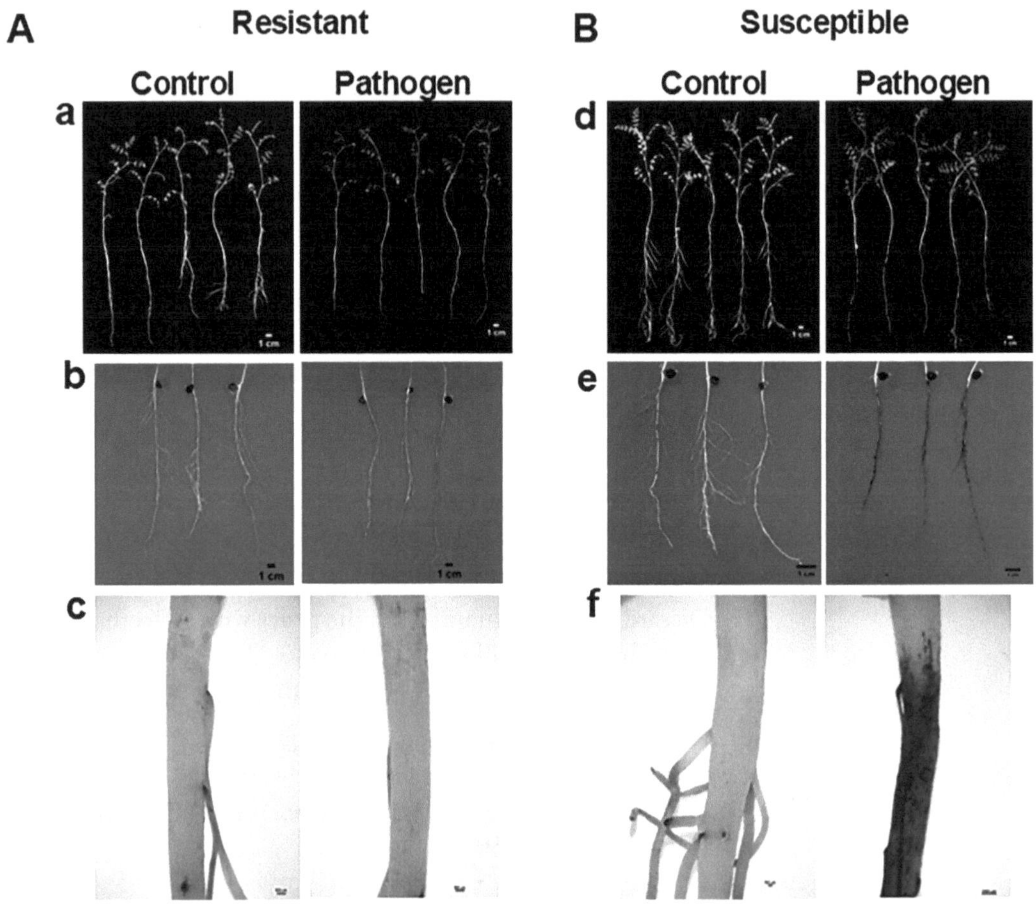

Fig. 4 DRR disease symptoms in resistant and susceptible chickpea genotypes under blotting paper technique. The plants were scored after 7 days of infection. Images shows (**a**) phenotype of resistant genotype, (a) whole plant image of the both control and pathogen, (b) root images of control and pathogen-infected plants, (c) single root image taken 2 cm from the point of seed attachment under research stereomicroscope SMZ25 of both control and pathogen-infected plants of resistant genotype. (**b**) phenotype of susceptible genotype. (d) whole plant image of both control and pathogen, (e) root images of control and pathogen-infected plants, (f) single root image taken 2 cm from the point of seed attachment under research stereomicroscope SMZ25 of both control and pathogen-infected plants of susceptible genotype. The pictures were captured using Nikon D5100. Scale bar = 1 cm, 1000 μm

5. Add 1 μL of genomic DNA of *M. phaseolina* to the plate, from 1 pg dilution to 100 ng dilution for standard curve preparation. Include three replicates per dilution.
6. Add 1 μL of DNA isolated from the root sample (50 ng/μL) to each well separately.
7. Add 1 μL water to three wells for the no template control (NTC).
8. Load the 96-well PCR plate containing samples into the 7900HT Fast RT-qPCR system.

9. Adjust the program to include an initial denaturation step for 5 min at 95 °C, followed by 40 cycles of denaturation at 95 °C for 10 s and annealing at 65 °C for 35 s, followed by heating to 95 °C, cooling to 65 °C for 1 min, and then heating to 97 °C at a rate of 1 °C/5 s [10].
10. Report the results as the mean C q value. Additionally, the target can be reported as fungal concentration per defined unit of the template.
11. Determine the concentration of fungal DNA present from the linear equation of the amplification using the target fungal gene C q values.
12. Generate the linear equation from the standard curve of known starting DNA concentrations containing the target gene (*see* **Note 10**).

3.10 Microscopy

1. Visualize the roots under a research stereomicroscope to observe the necrosis.

3.10.1 Staining

1. Cut the roots into 2 cm from the point of seed attachment.
2. Place them in a microcentrifuge tube (2 mL) containing 10% KOH for root clearing and keep them on a shaker for 48 h at room temperature (see **Note 11**).
3. Remove KOH and neutralize the root samples with 1.5 mL of 1 M Tris-HCl (pH 7.0).
4. Replace the buffer with 50 mM Tris-HCl (pH 7.0).
5. Place the root sample in a clean glass slide and add 100 μL of WGA-FITC (50 μg/mL) (*see* **Note 12**) over the root.
6. Place a coverslip on the root and press gently with forceps.
7. Remove excess stain using tissue paper and observe the roots under a fluorescence/confocal microscope with a bandpass of 485/535 nm [11].

4 Notes

1. The seeds can be procured from ICRISAT (International Crops Research Institute for the Semi-Arid Tropics), Patancheru, Hyderabad, Telangana.
2. The soil-borne necrotrophic fungus was isolated from the roots of an infected plant collected from a field in NIPGR (28.5307 ° N, 77.1665 °E), New Delhi. The pure culture was obtained using the hyphal tip method [7]. The same can be procured from ITCC.

3. After uprooting, do not keep the plants in water for more than 30 min. Once uprooted, keep them in the blotting sheet within 3 h.
4. Blotting sheets should be completely dried; otherwise, it may lead to secondary infection.
5. Mix 20 g of filtered fresh mycelia with 50 mL of autoclaved RO water to prepare 400 mg/mL (40%) inoculum.
6. To avoid secondary infection, use a freshly opened autoclaved polythene bag with blotting sheets for every batch.
7. Watering more can cause the rotting of the roots, and less watering may lead to the drying of roots. Spray water frequently (three times/day) to avoid these problems.
8. CSJ 515 (DRR-resistant) and BG 212/L550 (DRR-susceptible) can be used as check varieties.
   Mph-F—AAACACACTTTTCGCACTCCTGC.
   Mph-R—TATGCTCGCAGAGAAGAACACGA.
   are specific primers for the confirmation of *M. phaseolina*.
9. Ensure no amplification with target primers in the control plant samples or the NTC that indicate the absence of *M. phaseolina* DNA in the uninfected roots and no cross-contamination or primer dimers.
10. Root clearing helps in removing the cytoplasmic contents of the cell, which in turn ensures better visualization of the fungal mycelia.
11. Prepare 50 μg/mL working solution from a 1 mg/mL stock solution by diluting it.

Acknowledgments

Projects at the M.S.K lab are supported by the National Institute of Plant Genome Research core funding and DBT multi-institutional project (BT/Ag/Network/Chickpea/2019-20) "**Characterization of chickpea germplasm resource to accelerate genomics-assisted crop improvement**," under mission program of "Characterization of genetic resources." We thank Mr. Rahim H Tarafdar and Mr. Sunder Solanki for their help in the laboratory. We acknowledge DBT-eLibrary Consortium (DeLCON) and NIPGR Library for providing access to e-resources and NIPGR Plant Growth Facility for plant growth support/space.

References

1. Sharma M, Ghosh R, Pande S (2015) Dry root rot (Rhizoctonia bataticola (Taub.) Butler): an emerging disease of chickpea – where do we stand? Arch Phytopathol Plant Protect 48:13–16
2. Sharma M, Ghosh R, Sharma TR, Pande S (2012) Intra population diversity in Rhizoctonia bataticola causing dry root rot of chickpea (Cicer arietinum L.) in India. Afr J Microbiol Res 6. https://doi.org/10.5897/AJMR12.954
3. Rai A, Irulappan V, Senthil-Kumar M (2022) Dry root rot of chickpea: a disease favored by drought. Plant Dis 106:346–356. https://doi.org/10.1094/PDIS-07 21 1410 FE
4. Sinha R, Irulappan V, Mohan-Raju B et al (2019) Impact of drought stress on simultaneously occurring pathogen infection in field-grown chickpea. Sci Rep 9:5577
5. Nene YL, Haware MP, Reddy MV (1981) Chickpea diseases: resistance-screening techniques. Information Bulletin No. 10. Patancheru, A.P., India: International Crops Research Institute for the Semi-Arid Tropics
6. Zhu Y, Chen H, Fan J et al (2000) Genetic diversity and disease control in rice. Nature 406:718–722
7. Johnson R (1992) Past, present, and future opportunities in breeding for disease resistance, with examples from wheat. In: Johnson R, Jellis GJ (eds) Breeding for disease resistance. Developments in plant pathology, vol 1. Springer, Dordrecht
8. Mirchandani R, Irulappan V, Chilakala AR, Senthil-Kumar M (2023) Dry root rot disease: current status and future implications for chickpea production. Proceedings of the National Academy of Sciences, India Section B
9. Irulappan V, Senthil-Kumar M (2021) Dry root rot disease assays in chickpea: a detailed methodology. J Vis Exp 167:e61702
10. Irulappan V, Mali KV, Patil BS, Manjunatha H, Muhammad S, Senthil-Kumar M (2021) A sick plot-based protocol for dry root rot disease assessment in field-grown chickpea plants. Appl Plant Sci 9:e11445
11. Irulappan V et al (2022) Drought stress exacerbates fungal colonization and endodermal invasion and dampens defense responses to increase dry root rot in chickpea. Mol Plant-Microbe Interact 35(7):583–591

Chapter 10

CRISPR-Mediated Gene Editing for Inducing Thermosensitive Genic Male Sterility and Sheath Blight Resistance in Rice

Shanthinie Ashokkumar, Vignesh Ponnurangan, Kumar K. Krish, Arul Loganathan, Kokiladevi Eswaran, Paranidharan Vaikuntavasan, Sudhakar Duraialagaraja, and Varanavasiappan Shanmugam

Abstract

Recent advances in genome editing enable the researchers to focus more and more on the ability to manipulate genomes at specific sites. Efficient methods for genome editing further promote gene discovery and functional gene analyses in model plants as well as the introduction of novel desired agricultural traits in important species. CRISPR/Cas9 technology enables precise genetic modification through the creation of double-strand breaks in a target region and the generation of desired alterations during the repair process. In this chapter, we describe the cloning strategy, transformation protocols, triparental mating procedure, and characterization of genome-edited genetic male sterile mutants and sheath blight disease-resistant mutant plants.

Key words CRISPR/*Cas9*, Cloning, Rice, Transformation, TGMS, Sheath blight resistance

1 Introduction

The CRISPR/Cas system was adapted from the bacterial/archaeal adaptive immune system [1]. It is a two-component system containing small guide RNA and Cas9 nuclease. A small guide RNA (sgRNA) is the duplex of the crRNA (CRISPR-RNA) and tracrRNA (trans-activating RNA), which gets paired with Cas9 nuclease and targets the appropriate region in the host genome [2]. The protospacer adjacent motif (PAM), located downstream of the target DNA sequence, functions as a binding signal for Cas9 [3]. Cas9 chiefly makes double-strand breaks (DSBs) between the 17th and 18th nucleotide in the target site sequence. The DSBs are repaired either by the cell's natural repair pathways, such as nonhomologous end joining (NHEJ) or homology-directed repair

Muthappa Senthil-Kumar and Hema Ramanna (eds.), *Plant Biotic Stress: Methods and Protocols*, Methods in Molecular Biology, vol. 2966, https://doi.org/10.1007/978-1-0716-4746-2_10,

(HDR). NHEJ is the most popular and efficient cellular repair mechanism. But it is an error-prone mechanism that may introduce random indels at the DSB site, leading to frameshift mutation or premature stop codon (knockout). If the donor DNA template is present, it follows the homology-directed repair (HDR) pathway, resulting in gain-of-function mutations (knock-in) [4]. With the advantage of this distinct feature of the bacterial defense mechanism, CRISPR/Cas9 is being successfully employed in various organisms like plants, animals, etc. for precise editing of genes of interest. This chapter highlights the detailed step-by-step procedure, starting from choosing the target to characterization of the CRISPR mutants.

2 Materials

2.1 Enzymes and Cultures

1. Synthetic DNA (sgRNA) oligos.
2. pRGEB32 vector in *E. coli* (Addgene Plasmid #63142),
3. *Bsa*I-HF®v2 restriction enzyme.
4. *Xba*I restriction enzyme.
5. 10× CutSmart Buffer
6. T4 DNA ligase enzyme.
7. 10× T4 DNA ligase buffer
8. Nuclease-free water.
9. *E. coli* competent cells (*DH5α* strain).
10. *E. coli* helper plasmid (pRK2013 strain).
11. *Agrobacterium tumefaciens* (LBA4404 strain).
12. Pathogen inoculum—*Rhizoctonia solani* AG-1 IA strain.

2.2 Antibiotics Preparation

1. Kanamycin (100 mg/mL): Dissolve 100 mg of kanamycin sulfate in 1 mL of sterile distilled water and store at −20 °C.
2. Rifampicin (20 mg/mL): Dissolve 20 mg of rifampicin in 1 mL of DMSO or methanol. Stock will be dark red in color. Make aliquots and store at −20 °C.

2.3 Plant Materials

Immature embryos of CO51 and ASD16 rice cultivars.

2.4 Primers

1. M13-Reverse: 5′ CAGGAAACAGCTATGACC 3′.
2. *hpt*-Forward: 5′ GCTGTTATGCGGCCATTGGTC 3′,
3. *hpt*-Reverse: 5′ GCCTCCAGAAGAAGATGTTG 3′,
4. *Cas9*-Forward: 5′ CTTCTGGCGGTTCTCTTCAG 3′.
5. *Cas9*-Reverse: 5′ TGCTGTTTGATCCGTTGTTG 3′.

6. *Vir*-Forward: 5′ATCGACGCCTCCAAAGTGAC 3′.
7. *Vir*-Reverse: 5′AAACGCGAGGAGGAGAAGATTGA 3′.

2.5 Solutions and Medium

2.5.1 Sodium Chloride Solution (0.9%)

Dissolve 0.9 g of NaCl in 100 mL of sterile distilled water and store at room temperature.

2.5.2 Potassium Iodide Solution

Dissolve 1 g of potassium iodide and 0.5 g of iodine in 100 mL of distilled water, and store the solution in an amber bottle at 4 °C.

2.5.3 Potato Dextrose Agar (PDA) Medium

1. Mix 250 g of potato infusion and 20 g of dextrose in distilled water.
2. Adjust the pH to 5.6 ± 0.2.
3. Add 20 g of agar to the above mixture and autoclave.

3 Methods

3.1 Designing of sgRNA

1. Retrieve the target gene sequence from the genome database (EnsemblPlants) using the locus ID (BGIOSGA008868 and BGIOSGA004522).
2. Design the guide RNAs using the online tools. Select the desired guide RNAs with the highest on-target and lowest off-target effects (*see* **Note 1**).
3. During the sgRNA synthesis, add the appropriate adaptor sequences (5′ GGCA 3′ and 3′ CAAA 5′) to the top and bottom strands of the sgRNA sequence (20 bp), respectively, to enable the ligation of an oligo duplex at the *Bsa*I restriction site of the pRGEB32 vector (*see* **Note 2**).

3.2 Cloning Strategy

3.2.1 Synthesis of sgRNA Oligo Duplex (Insert)

1. Resuspend the DNA oligos in the appropriate volume of nuclease-free water to get a 100 pM/μL concentration (stock). Store the oligos at −20 °C for long-term use.
2. Dilute the primers (top and bottom strand oligos) to the working stock concentration (10 pM/μL) by adding 10 μL of the respective stock primer to the 90 μL of nuclease-free water.
3. Anneal the oligos by setting a 20 μL reaction mixture comprising 9 μL of top strand oligo (10 pM/μL), 9 μL of bottom strand oligo (10 pM/μL), and 2 μL of 10× T4 DNA ligase buffer in a thermocycler machine (Eppendorf) with the ramp down PCR (*see* **Note 3**) conditions, as indicated in Fig. 1.

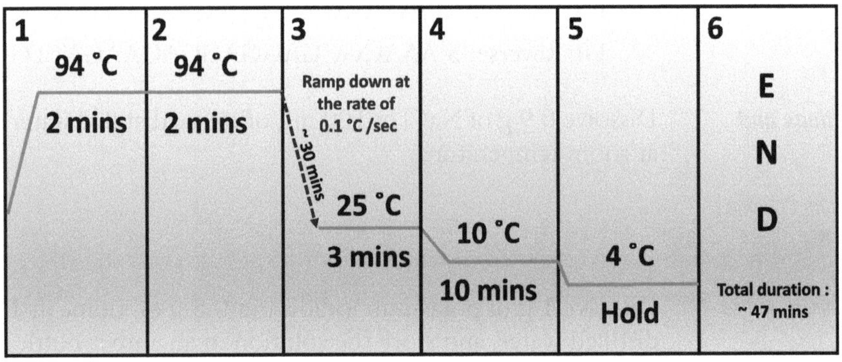

Fig. 1 Schematic representation of ramping down PCR conditions for sgRNA oligo-duplex formation

3.2.2 Preparation of a Linearized Plasmid (Vector)

1. Digest the pRGEB32 plasmid with the *Bsa*I restriction enzyme in a 20 μL reaction mixture (*see* **Note 4**).
2. Purify the digested and linearized plasmids using the standard PCR product cleanup kit, such as the NucleoSpin Clean-up Kit (Machery Nagel), by following the manufacturer's protocol to nullify the traces of excess buffers and enzymes that may also restrict the efficiency of the ligation reaction.

3.2.3 Ligation Protocol

1. Ligate the sgRNA oligo duplex into the linearized pRGEB32 plasmid to form a recombinant pRGEB32 plasmid (Fig. 2).
2. Set up the ligation reaction in an ice-cold condition by adding 4 μL of linearized pRGEB32 plasmid (50 ng/μL), 12 μL of sgRNA oligo duplex, 2 μL of 10× T4 DNA ligase buffer, 1 μL of T4 DNA ligase enzyme (5 units μL/L), and sterile nuclease-free water to the 20 μL reaction volume in a 0.2 mL PCR tube.
3. Incubate the ligation mixture in a thermocycler programmed as follows: 6 °C for 5 min, 8 °C for 3 min, 10 °C for 3 min, 14 °C for 3 min, 18 °C for 3 min, 22 °C for 15 min, and hold at 4 °C.
4. Store the ligated product at −20 °C until transformation.

3.3 Transformation of Recombinant Plasmid

1. Thaw the competent cells (100 μL approx.) contained in a 2 mL microcentrifuge tube on ice for 20 min.
2. Mix 5 μL of the ligated product in 100 μL of *E. coli* (DH5α strain) competent cells inside a laminar airflow chamber.
3. Incubate the above mixture on ice for 30 min.
4. Followed by heat shock treatment at 42 °C for 45 s.
5. Place the tubes immediately on ice for 2 min.
6. Add 1 mL of LB broth to the above tube.
7. Incubate the tube at 37 °C for 1 h with shaking at 180 rpm. Ensure that the tube is kept in a horizontal position.

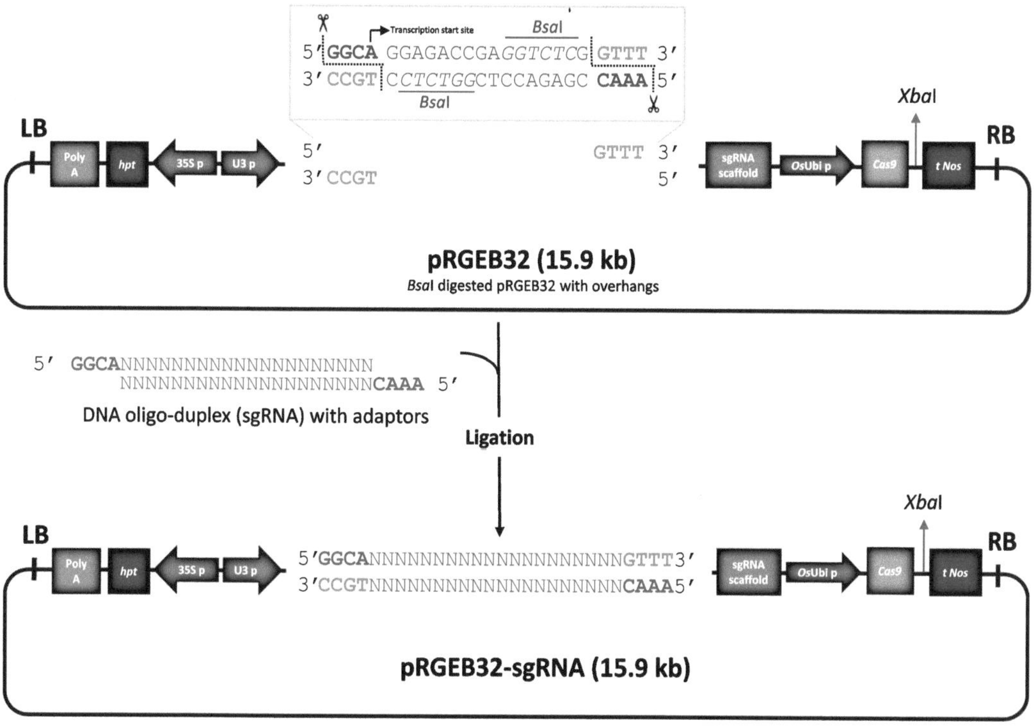

Fig. 2 Cloning strategy in the pRGEB32 vector. *Bsa*I enzyme cleaves at two closely located *Bsa*I restriction sites and generates overhangs (in green). The sgRNA oligo-duplex with adaptors (in blue) then ligates with *BsaI*-digested pRGEB32 with overhangs

8. Plate 100 μL of the cell suspension on a LB + kanamycin (50 mg/L) plate using flame-sterilized L rods.
9. Incubate the above plate at 37 °C for 12 to 15 h.
10. Randomly select five putative recombinant colonies and resuspend them in the LB broth containing the kanamycin (50 mg/L), and incubate them overnight at 37 °C (*see* **Note 5**).
11. Isolate the plasmid DNA from the overnight-grown liquid culture and check the concentration of the plasmid by agarose gel electrophoresis.

3.4 Methods for Screening of Recombinant Clones

3.4.1 Restriction Digestion

1. Perform the restriction digestion (*see* **Note 6**) of both the control plasmid (pRGEB32) and the recombinant (pRGEB32-sgRNA) plasmid.
2. Incubate the digestion mixtures (*see* **Note 7**) at 37 °C for 1 h.
3. Run the digested products on a 0.8% agarose gel to resolve the DNA fragments as expected (Fig. 3, *see* **Note 8**).

3.4.2 Sanger Sequencing

1. Subject the isolated recombinant plasmid along with the control plasmid to Sanger sequencing using M13 reverse primer (*see* **Note 9**).

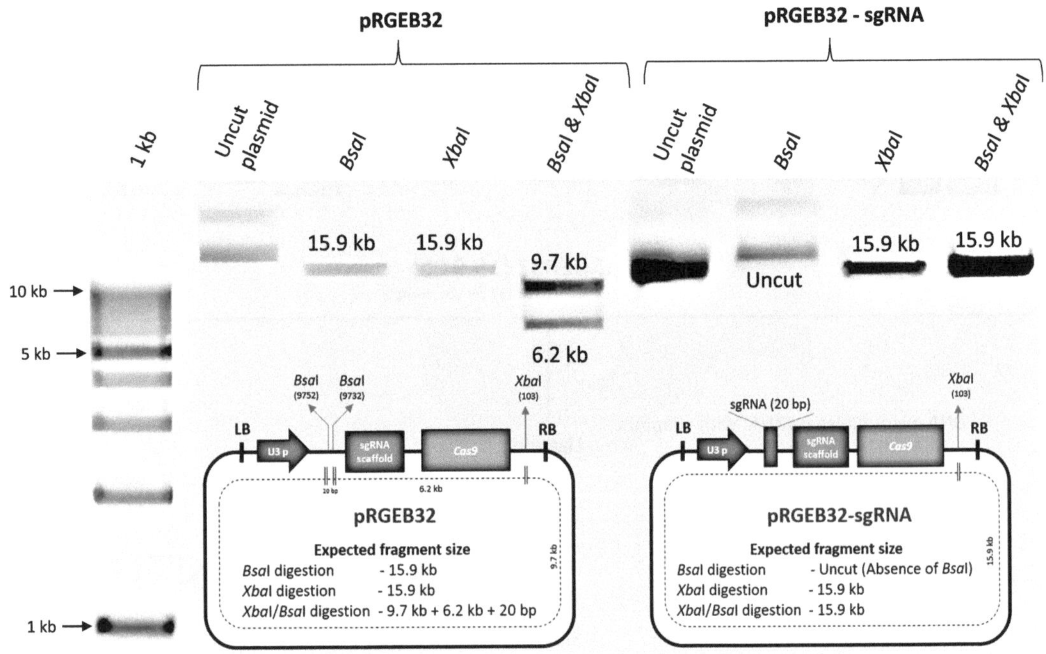

Fig. 3 Restriction digestion analysis. The pRGEB32 plasmid has two *Bsa*I sites (at positions 9732 and 9752) and one *Xba*I site (at position 103). During double digestion (*Bsa*I/*Xba*I), pRGEB32 releases three fragments (9.7 kb + 6.2 kb + 20 bp). In the recombinant clone (pRGEB32-sgRNA), *Bsa*I sites do not cleave; only *Xba*I will cleave and linearize the plasmid. The expected fragment sizes of the plasmids (pRGEB32 and pRGEB32-sgRNA) are outlined in the physical maps

2. Perform pairwise alignment of the control plasmid and the recombinant plasmid to confirm the correct sequence of the inserted sgRNA segment in the pRGEB32 plasmid.

3.5 Mobilization of Recombinant Plasmid into Agrobacterium via Triparental Mating

1. Place a 2-day-old culture of *Agrobacterium* and a 1-day-old *E. coli* culture possessing a recombinant clone and pRK2013 on the corners of a demarcated triangle on a plain LB agar plate (*see* **Note 10**).
2. Mix the three cultures with the help of sterile inoculation loops as indicated in the Fig. 4.
3. Incubate the plate at 28 °C in the dark for 16 h.
4. Serially dilute the fully grown culture in a sterile 0.9% NaCl solution up to 10^{-6} dilutions.
5. Plate 100 μL of the cell suspension (10^{-5} and 10^{-6}) on LB agar plates supplemented with antibiotics such as rifampicin (20 mg/L) and kanamycin (100 mg/L).
6. Incubate the plates at 28 °C for 2–3 days.

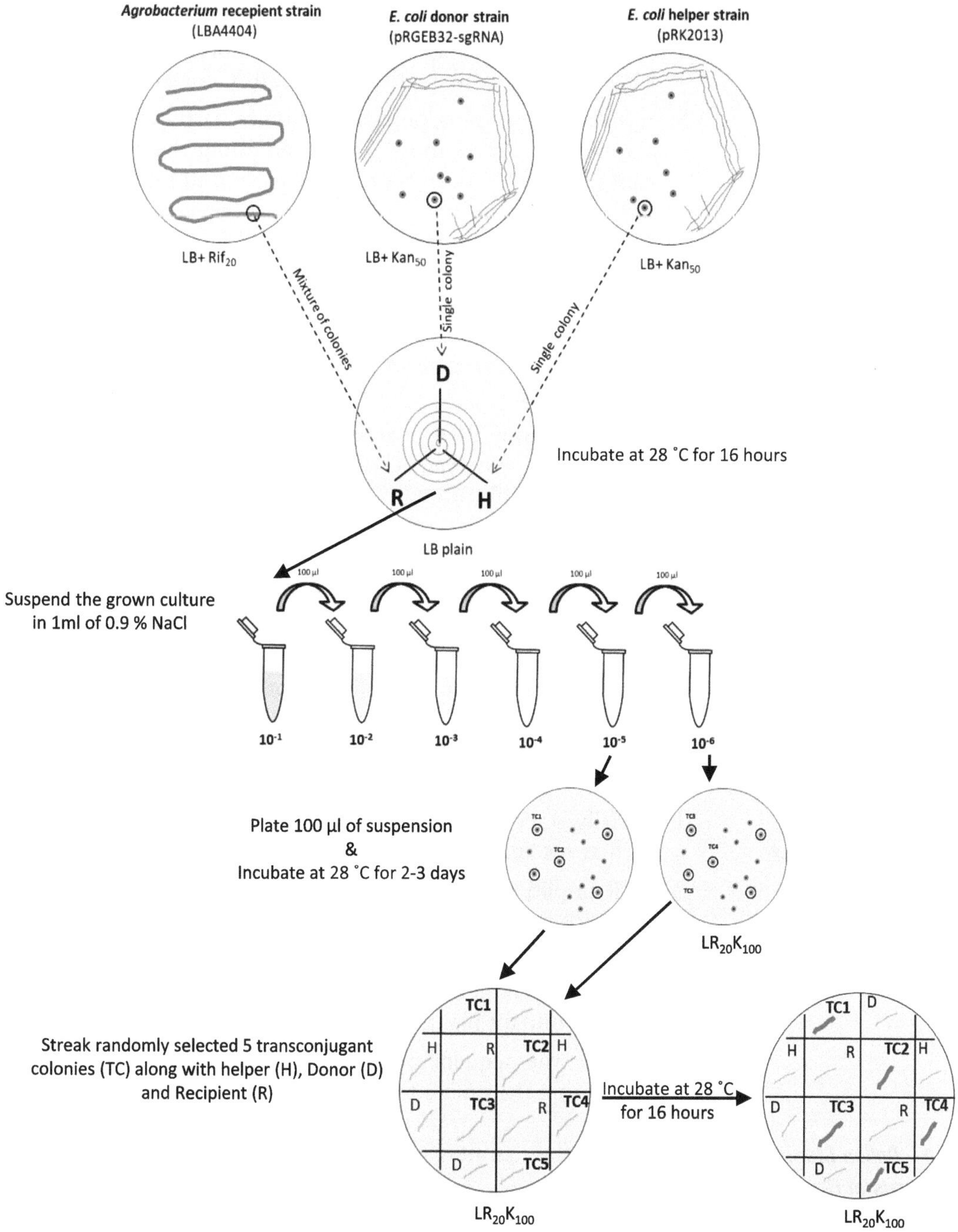

Fig. 4 Conceptual sketch of the triparental mating procedure. Streak the grown cultures (two *E. coli* and one *Agrobacterium*) on the demarcated triangle on a plain LB agar plate. Mix the three cultures and then incubate the plate at 28 °C. Serially dilute the fully grown culture up to 10^{-6} dilutions. Plate the cell suspension on $LR_{20}K_{100}$ plate and incubate it at 28 °C. Streak the selected colonies (including D, H, and R) on $LR_{20}K_{100}$ plate and incubate at 28 °C. After 16 h, *Agrobacterium* transconjugants can be distinguished from nongrowing donor, recipient, and helper strains

7. Short-streak the randomly selected five transconjugant colonies along with donor (D), helper (H), and recipient (R) on $LR_{20}K_{100}$ and incubate the plate at 28 °C for 16 h.
8. *Agrobacterium* transconjugant colonies harboring the donor plasmid alone grow in $LR_{20}K_{100}$, indicating that the recombinant plasmid has been successfully mobilized into *Agrobacterium*, while the *E. coli* donor, recipient, and helper *Agrobacterium* strains will not grow in $LR_{20}K_{100}$.

3.6 Screening Procedure for Transconjugant Clones

1. Prepare the bacterial suspension by picking a single colony using sterile pipette tips and twirling it in 20 μL of nuclease-free water. This suspension will be used as a template for colony PCR analysis.
2. Perform colony PCR analysis (*see* **Note 11**) for checking the presence of *vir*, *hpt*, *Cas9*, and sgRNA sequences using sequence specific primer pairs (*see* **Note 12**).
3. The transconjugant colony, which is positive for all the aforementioned genes, will serve as the inoculum for *Agrobacterium*-mediated rice genetic transformation.

3.7 Generation of Genome-Edited Mutant Rice Lines

1. Generate transgenic rice events through the *Agrobacterium*-mediated rice genetic transformation by exploiting immature rice embryos as explants (*see* **Note 13**).
2. Isolate the genomic DNA from the T_0 transgenic plants.

3.8 Molecular Characterization of Putative Genome-Edited Mutants

1. Initially, screen all the T_0 mutants using vector-specific primers such as *hpt* and *Cas9* (*see* **Note 14**).
2. PCR amplify the target sgRNA region in mutant plants (which were positive for *hpt* and *Cas9* genes) by using gene-specific primers encompassing the sgRNA region of a gene.
3. Subject the target-specific PCR amplicons of wild-type and putative mutant plants to Sanger sequencing.
4. Analyze the sequencing results and interpret the mutation types using the online tools (Table 1, *see* **Note 15**).

3.9 Pollen Sterility Analysis of Genome-Edited TGMS Mutants

1. Collect the mature pollen grains on the day of flowering in a petri dish.
2. Place the anthers on a clean glass slide.
3. Add 1–2 drops of potassium iodide solution and crush the anthers using a glass rod.
4. Observe the stained pollens under the microscope to detect the presence of sterile or fertile pollens (Fig. 5, *see* **Note 16**).

Table 1
Mutation types of genome-edited mutants

S. No .	Event name	DNA traces	Mutation type
1	TGMS - CO51-Event 1	Allele1:TCAGGAGCACGAG----GGA**GGG** Allele2:TCAGGAGCACGAG----GGA**GGG** Control:TCAGGAGCACGAGCTCAGGA**GGG**	4d/4d; Bi-allelic Homozygous
2	TGMS - CO51-Event 2	Allele1:TCAGGAGCACGAG----GGA**GGG** Allele2:TCAGGAGCACG------GGA**GGG** Control:TCAGGAGCACGAGCTCAGGA**GGG**	4d/6d; Bi-allelic Heterozygous
3	TGMS - CO51-Event 3	Allele1:TCAGGAGCACGAGCTC-AGGA**GGG** Allele2:TCAGGAGCACGAGCTCTAGGA**GGG** Control:TCAGGAGCACGAGCTC-AGGA**GGG**	WT/1i; Mono-allelic
4	ShB - ASD16-Event 1	Allele1:ATTTCTGGCTAGTTTCT---**TGG** Allele2:ATTTCTGGCTAGTTTCT---**TGG** Control:ATTTCTGGCTAGTTTCTAGC**TGG**	3d/3d; Bi-allelic Homozygous
5	ShB - ASD16-Event 2	Allele1:ATTTCTGGCTAGTTG---GC**TGG** Allele2:ATTTCTGGCTAGTTG---GC**TGG** Control:ATTTCTGGCTAGTTTCTAGC**TGG**	3d/3d,2s; Bi-allelic Heterozygous

"d" denotes deletion, "i" denotes insertion, "s" denotes substitution, "WT" denotes wild type and PAM sequence in bold.

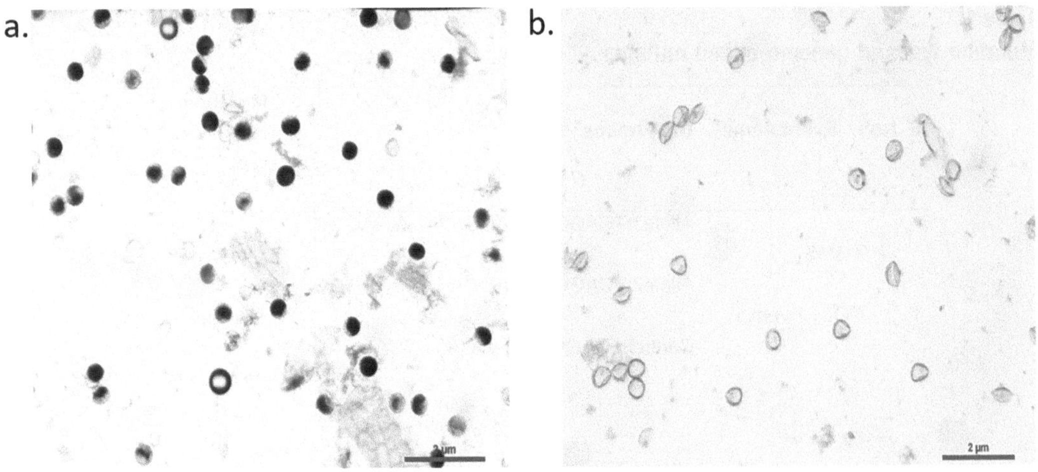

Fig. 5 Pollen sterility analysis. (**a**) The darkly stained wild-type pollen indicates pollen fertility. (**b**) The unstained TGMS mutant pollens indicate sterility

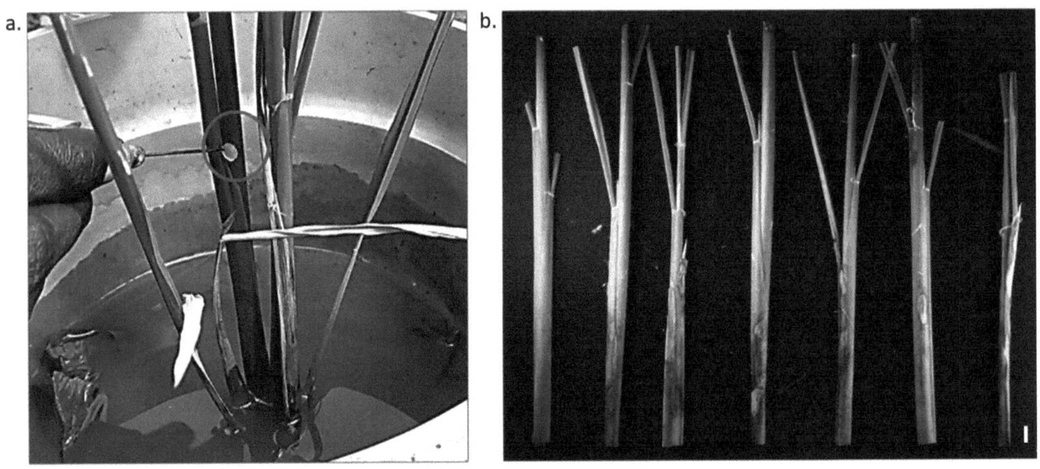

Fig. 6 Bioassay of the sheath blight pathogen. (**a**) Mycelial plug was inoculated on 40-day-old rice seedlings. (**b**) Disease symptoms after 4 dpi

3.10 Bioassay of Genome-Edited Putative Sheath Blight Resistant Mutants

1. Sterilize the soil and fill in earthen pots of the required size.
2. At the tillering stage, inoculate 3–4-day-old culture of the pathogen into the healthy tillers (both transgenic and non-transgenic plants) by making a patch with the help of sterilized forceps.
3. Cover the inoculated patch with aluminium foil for the development of disease symptoms.
4. Calculate the disease severity using the IRRI scale (Fig. 6, *see* **Note 17**).

4 Notes

1. The selection of the right guide RNA sequence is the primary and most important step in designing the CRISPR experiment. For gene knockout, it is critical to choose the sgRNA target in the exon, which is essential for protein function. For designing guide RNAs, various online tools such as CRISPOR [5], CRISPR-PLANT v2 [6], CRISPR-P 2.0 [7], and CRISPR-GE [8] can be used.
2. *Bsa*I is a Type IIS restriction enzyme that recognizes the following sequence information:

   5′...GGTCTC$(N)_1$...3′

   3′...CCAGAG$(N)_5$...5′

   In the pRGEB32 vector, two *Bsa*I sites are located in very close proximity at 9752 bp and 9732 bp. Digestion with the *Bsa*I restriction enzyme linearizes the plasmid, producing 5′ GTTT 3′ and 3′ CCGT 5′ overhangs (Fig. 2).
3. The PCR ramp rate refers to the duration required for a thermal cycler to transition from one temperature step to another. Ramp rate is typically expressed in °C/second. The term "ramp down" pertains to the gradual reduction of temperature in the thermal blocks of a PCR machine, starting from a higher temperature. This slow cooling facilitates the oligo duplex formation by aiding in the formation of new hydrogen bonds between the complementary nucleotide bases [9].
4. Linearize the vector using the restriction digestion mixture comprising 2 μL of pRGEB32 (500 ng/μL), 2 μL of 10× CutSmart Buffer, 2 μL of *Bsa*I-HF®v2 enzyme and sterile nuclease-free water to the 20 μL reaction volume. The use of higher concentration of *Bsa*I enzyme ensures the complete digestion of the plasmid, so that no undigested plasmid traces are present.
5. Mark at least 10 well-separated individual colonies by circling and numbering them using marker pen. Pick the marked colonies using sterile pipette tips, resuspend them in the LB broth containing the kanamycin (50 mg/L), and incubate them overnight at 37 °C. Also, incubate the control tube without inoculating the culture to ensure that the broth is not contaminated.
6. A precise and simple method for screening transformed colonies is to use restriction enzymes to confirm the presence and direction of our desired insert fragment. Initially, restriction mapping should be performed to determine the appropriate restriction enzymes that can detect the presence of the insert within the plasmid.

7. Subject the isolated plasmid DNA to a single digestion reaction with *Bsa*I-HF®v2 and *Xba*I enzymes separately and to double digestion using both enzymes. For single digestion with *Bsa*I-HF®v2 and *Xba*I enzymes, set the total reaction volume at 20 μL comprising 2 μL of pRGEB32 plasmid (300 ng/μL), 1 μL of the respective enzyme (20 units/μL), 2 μL of 10X CutSmart buffer, and 15 μL of nuclease-free water. For double digestion, set the total reaction volume at 20 μL comprising 2 μL of pRGEB32 plasmid (300 ng/μL), 1 μL of *Bsa*I-HF®v2 enzyme, 1 μL of *Xba*I enzyme (20 units/μL), 2 μL of 10× CutSmart buffer, and 14 μL of nuclease-free water.
8. Digesting the pRGEB32 plasmid with *Bsa*I and *Xba*I enzyme will release three DNA fragments (9.7 kb, 6.2 kb, and 20 bp). In the case of the recombinant plasmid (pRGEB32-sgRNA), both *Bsa*I sites get abolished due to the ligation of the sgRNA oligo duplex. Hence, during double digestion *Bsa*I restriction enzyme will not cut, only *Xba*I will make a cut. This confirms the successful integration of the sgRNA oligo duplex between the two *Bsa*I sites of the pRGEB32 plasmid (Fig. 3).
9. Sanger sequencing is the most reliable method for confirming the accuracy of recombinant plasmids. This is used to validate the exact insert sequence and its orientation in the vector. M13 reverse is a universal primer that is frequently used in DNA sequencing reactions.
10. Triparental mating is a simple and efficient method for the mobilization of recombinant plasmids from *E. coli* to *Agrobacterium* through a bacterial conjugation mechanism. Three bacterial strains and their growing conditions are briefed below:
    (a) *Donor strain* (*E. coli* harboring the recombinant plasmid to be mobilized) grows at 37 °C on LB agar plate supplemented with kanamycin (50 mg/L). This *E. coli* donor strain should carry a binary vector with an origin of replication to replicate in both *E. coli* and *Agrobacterium* strains.
    (b) *Conjugal helper strain* (*E. coli* harboring pRK2013 plasmid) grows at 37 °C on LB medium with kanamycin (50 mg/L). This plasmid facilitates the mobilization of binary plasmids into *Agrobacterium.*
    (c) *Recipient strain* (*Agrobacterium tumefaciens* LBA4404) grows at 28 °C on LB agar medium supplemented with rifampicin (20 mg/L).
11. Colony PCR is the quickest and most cost-effective method for screening the plasmid or culture for desired insert fragments. This method eliminates the need for plasmid isolation and purification procedures.

12. PCR amplification of the plasmid with the vector-specific primers (*hpt* and *Cas9*) and *vir* gene primer confirms the successful mobilization of the recombinant plasmid into *Agrobacterium*. To ensure that the insert is ligated in the correct orientation, colony PCR using an insert (sgRNA)-specific primer paired with a vector (pRGEB32)-specific primer can be carried out to obtain amplicon of appropriate size.
13. Collection of panicles (CO51 and ASD16) at dough stage, that is, 15 days after flowering. The immature rice embryos serve as the explant. The rice genetic transformation procedure suggested by [10] was followed.
14. To confirm the transgenicity of regenerated putative mutants, PCR amplification of the *hpt* and *Cas9* genes is the initial step in molecular characterization. The putative mutants that did not give amplification for the above genes should not be further studied.
15. Online tools such as DSDecodem [11], CRISP-ID [12], and Synthego ICE analysis [13] are being used to detect the mutations. The expected mutation patterns are monoallelic and biallelic. Monoallelic mutation refers to the condition when any one of the alleles gets mutated while the other allele remains unmodified. Biallelic mutation refers to the condition when both of the alleles are mutated. If the mutation in both alleles is similar, it is termed a homozygous mutation. And if the mutation is not similar, it is referred to as a heterozygous mutation.
16. The pollen fertility or sterility can be assessed by the potassium iodide staining of mature pollen grains. Pollens that are darkly stained are counted as fertile, whereas unstained pollens are counted as sterile.
17. A bioassay of the sheath blight pathogen will be conducted using the following methods described by Koch postulates. The virulent *R. solani* AG-1 IA strain is grown on potato dextrose agar (PDA) at 28 °C in the dark for a few days. The mycelia are collected and used for inoculation on rice plants. After 7–8 weeks after transplanting, seedlings attain tillering stage. Sterilized pins will be used to make holes in the sheaths and 5–7 days old culture of the pathogen will be inoculated into the healthy tillers. For the control treatment, plants are maintained by spraying with sterile distilled water. Disease severity calculations can be done according to the IRRI scale [14].

Acknowledgments

The authors are profoundly thankful to the Department of Plant Biotechnology, Centre for Plant Molecular Biology and Biotechnology, Tamil Nadu Agricultural University, Coimbatore, for providing facilities. This work was substantially funded by DBT, GOI programs (BT/PR2656/GET/119/2017; BT/PR40456/AGIII/103/1248/2020). AS and VP acknowledge DBT for the fellowship.

References

1. Doudna JA, Charpentier E (2014) Genome editing. The new frontier of genome engineering with CRISPR-*Cas9*. Science 346(6213):1258096
2. Lowder LG, Zhang D, Baltes NJ, Paul JW, Tang X, Zheng X, Voytas DF, Hsieh TF, Zhang Y, Qi Y (2015) A CRISPR/*Cas9* toolbox for multiplexed plant genome editing and transcriptional regulation. Plant Physiol 169(2):971–985
3. Gleditzsch D, Pausch P, Müller-Esparza H, Özcan A, Guo X, Bange G, Randau L (2019) PAM identification by CRISPR-Cas effector complexes: diversified mechanisms and structures. RNA Biol 16(4):504–517
4. Yang H, Ren S, Yu S, Pan H, Li T, Ge S, Zhang J, Xia N (2020) Methods favoring homology-directed repair choice in response to CRISPR/*Cas9* induced-double strand breaks. Int J Mol Sci 18:6461
5. Concordet JP, Haeussler M (2018) CRISPOR: intuitive guide selection for CRISPR/*Cas9* genome editing experiments and screens. Nucleic Acids Res 46(W1):W242–W245
6. Minkenberg B, Zhang J, Xie K, Yang Y (2019) CRISPR-PLANT v2: an online resource for highly specific guide RNA spacers based on improved off-target analysis. Plant Biotechnol J 17(1):5
7. Lei Y, Lu L, Liu HY, Li S, Xing F, Chen LL (2014) CRISPR-P: a web tool for synthetic single-guide RNA design of CRISPR-system in plants. Mol Plant 7(9):1494–1496
8. Xie X, Ma X, Zhu Q, Zeng D, Li G, Liu YG (2017) CRISPR-GE: a convenient software toolkit for CRISPR-based genome editing. Mol Plant 10(9):1246–1249
9. Fu G (2016) U.S. Patent No. 9,528,157. Washington, DC: U.S. Patent and Trademark Office
10. Hiei Y, Komari T (2008) *Agrobacterium*-mediated transformation of rice using immature embryos or calli induced from mature seed. Nat Protoc 3(5):824–834
11. Liu W, Xie X, Ma X, Li J, Chen J, Liu YG (2015) DSDecode: a web-based tool for decoding of sequencing chromatograms for genotyping of targeted mutations. Mol Plant 8(9):1431–1433
12. Dehairs J, Talebi A, Cherifi Y, Swinnen JV (2016) CRISP-ID: decoding CRISPR mediated indels by sanger sequencing. Sci Rep 6(1):28973
13. Roginsky J (2018) Analyzing CRISPR editing results: Synthego developed a tool called ICE to be more efficient than other methods. Genetic Eng Biotechnol News 38(11):S24–S26
14. Aye SS, Myint YY, Lwin T, Matsumoto M (2009) Stem rot of rice caused by Sclerotium hydrophilum isolated in Myanmar. Plant Pathol 58(4):799

Chapter 11

A Seedling Root Dip-Based Technique to Screen Chickpea for Resistance to *Fusarium* Wilt Disease

B. S. Patil, Vijaykumar K. N, K. S. Iliger, C. Bharadwaj, and Shailesh Tripathi

Abstract

Fusarium wilt poses a significant threat to chickpea cultivation, causing substantial yield losses. Developing resistant chickpea varieties is a crucial strategy for managing this devastating disease. Screening a large number of germplasm and breeding lines against the pathogen is necessary to achieve this goal. In this context, the seedling root dip method has emerged as an effective technique to differentiate between resistant and susceptible chickpea genotypes. This method offers the advantages of screening a large number of lines within a short time frame and limited space. Another critical aspect of breeding for disease resistance is the rapid and accurate identification of the pathogen. Traditional pathogen detection methods are labor-intensive and time-consuming. This chapter presents a detailed protocol for the seedling root dip method, enabling the screening of chickpea genotypes against *Fusarium oxysporum*. Additionally, a rapid approach utilizing ITS primers for identifying the pathogen is discussed, providing a precise and expedient tool for disease resistance breeding efforts.

Key words *Cicer arietinum*, *Fusarium oxysporum*, Pathogen detection, Molecular diagnosis, Seedling root dip, Soil-borne pathogen

1 Introduction

Chickpea (*Cicer arietinum* L.) is a cool season legume crop, considered to be one of the founder crops of modern agriculture [1]. It is grown throughout tropical, subtropical, and temperate regions in South and West Asia, East and North Africa, southern Europe, North and South America, and Australia [2]. Chickpea seeds are important source of protein, vitamins, and essential minerals for human diet, and the crop enhances soil fertility by fixing atmospheric nitrogen through symbiotic association [3, 4].

The realized productivity of this pulse crop is not appreciable due to various production constraints. Among the several biotic and abiotic constraints responsible for low productivity, diseases are

Muthappa Senthil-Kumar and Hema Ramanna (eds.), *Plant Biotic Stress: Methods and Protocols*, Methods in Molecular Biology, vol. 2966, https://doi.org/10.1007/978-1-0716-4746-2_11,

the most serious constraints causing up to 100% yield loss. The crop is infected by 172 causal agents of biotic stresses, about 67 fungi, 3 bacteria, 22 viruses, and 80 nematodes reported from 55 countries of the world [5]. Among the different fungal pathogens, chickpea is seriously abridged by *Fusarium* wilt in all chickpea-growing areas of the world [6, 7]. *Fusarium* wilt caused by *Fusarium oxysporum* f. sp. *ciceris* (Padwick) (*Foc*) is the most important soil−/seed-borne disease of chickpea throughout the world and particularly in the Indian subcontinent, the Mediterranean basin, and California [8–10]. It is more prevalent in lower latitudes (0–30 °N), where growing season is relatively dryer and warmer than in the higher latitudes (30–40 °N). The disease can occur at all the stages of plant growth from seedling to maturity [11]. There are eight races of *Foc* (0, 1A, 1B/C, 2, 3, 4, 5, and 6) identified by reaction on a set of differential chickpea cultivars [12, 13] and consists of two pathotypes (yellowing and wilting) [14].

The pathogen is facultative saprophyte and can survive in soil up to 6 years in the absence of susceptible host [15, 16]. In recent years, incidence of wilt in the farmers' fields is considerably increasing every year and its severity is directly related to the increasing density of the pathogen inoculum in the soil [17–19]. The fungus enters the vascular tissues of plant via roots. Symptoms of the disease can develop at any stage of plant growth, and affected plants may be grouped in patches or appear spread across a field. Infected seedlings collapse and lie flat on the ground retaining their dull green color. The affected adult plants show typical wilt symptoms of drooping of petioles, rachis, and leaflets, followed by yellowing and necrosis of foliage. When uprooted they show uneven shrinking of stem below the collar region. The roots of the wilting plants do not show any external rotting, but dark brown discoloration of internal xylem is seen [20]. Pods from the wilted plants look normal, but seeds are generally smaller, wrinkled, and discolored.

The chemical control of the disease is not economically feasible and not a permanent solution. Because of the diverse, widespread, soilborne nature and capacity to survive in soil for many seasons without host made the pathogen very difficult to manage. The screening and evaluation of available germplasm is the first step in successful disease resistance breeding program to identify the sources of resistance, which could be utilized further for development of resistant varieties [21]. Dekkers and Hospital (2002) [22] emphasized that laboratory screening based on molecular markers without phenotypic screening in field or glasshouse could not be effective enough in defining the level of resistance. The success of disease resistance breeding depends on the precision of the screening and evaluation method. Screening of breeding material against soilborne diseases like wilt is laborios, expensive, and time-consuming. One of the widely used methods of screening of genotypes under sick plots is likely to encounter problems like uneven distribution of inoculum and presence of other soilborne fungi. Therefore, in order to overcome the disadvantages associated with field screening of the large number of chickpea genotypes, "seedling root dip method" [23] is being regularly used in our breeding program to screen germplasm and breeding lines against the *Fusarium* wilt. The method provides uniform application of known concentration of inoculum to all the test lines. Our study conducted to screen large number of genotypes and breeding lines of chickpea shows that the inoculation of young seedlings by root dipping is a quick, effective, and reliable method to screen large germplasm and breeding lines against *Fusarium* wilt resistance in chickpea. Further identification of the pathogen is another important aspect of disease resistance breeding, particularly for soilborne pathogens. Generally, cultural, morphological, and chemical markers are used to identify the pathogen. Molecular detection of the pathogen using molecular markers is found to be rapid and reliable tool in identification of the pathogen. The ribosomal DNA gene cluster consisting of three main conserved regions, a large subunit, the 28 S gene (LSU), a small subunit, the 18 S gene (SSU) and the 5.8 S gene. The three genes within the repeat are separated by variable internal transcribed spacer regions, ITS 1 and ITS 2. The presence of conserved regions in the SSU and LSU has allowed the generation of universal primers for use in PCR that give rise to amplicons spanning regions of variability such as ITS 1 and ITS 2. Sequencing of such amplicons is used to identify the pathogen [24]. The detailed protocol of screening chickpea germplasm against *Fusarium oxysporum* and molecular identification of the pathogen is provided in this chapter.

2 Materials

2.1 Isolation and Purification of the Pathogen

1. Transparent polythene covers: 10 × 10 inches.
2. Stainless steel scalpel: sterilized.
3. 1% sodium hypochlorite
4. Distilled water: sterilized.
5. Tissue paper.
6. Potato dextrose agar.
7. Water agar.
8. Petri dishes (90 mm): sterilized.
9. Cork borer: sterilized.
10. Test tubes: sterilized.

2.2 Pathogenicity Test

1. Polythene bags (20 cm length × 6 cm width).
2. River sand and perlite (1:1 ratio): sterilized.
3. *Fusarium oxysporum* f. sp. *ciceris* (*Foc*) fungal culture.
4. Conical flask: 250 mL volume.
5. Potato dextrose broth.
6. Sterilized water.
7. Hemocytometer.
8. Micropipette.
9. 1% sodium hypochlorite solution
10. Chickpea seeds: susceptible genotype JG 62.
11. Potato dextrose agar.
12. Petri dishes (90 mm).

2.3 Molecular Identification of Foc

1. Liquid nitrogen.
2. Tris buffer.
3. EDTA.
4. NaCl 5 M.
5. SDS 10%.
6. CTAB/NaCl solution.
7. Micropipette.
8. Chloroform: isoamyl alcohol (24:1).
9. Eppendorf tubes: 2 mL volume.
10. Isopropanol.
11. Sodium acetate.
12. 70% ethanol
13. TE buffer.
14. RNAse.
15. ITS primers.
    ITS 1: (5′-TCCGTAGGTGAACCTGCGG-3′).
    ITS 4: (5′-TCCTCCGCTTATTGATATGC-3′)
16. PCR tubes.
17. dNTP,
18. TBE buffer 10×.
19. Taq DNA polymerase.
20. Deionized water.
21. Agarose gel 1.8%.
22. Gel loading dye.
23. 100 bp ladder.

2.4 Raising the Chickpea Seedlings

1. Seeds of chickpea genotypes (*see* **Note 1**).
2. Sterilized distilled water.
3. Sodium hypochlorite.
4. Polythene bags (20 cm length × 6 cm width).
5. River sand and perlite (1:1 ratio): sterilized.

2.5 Inoculum Preparation

1. *Fusarium oxysporum* f. sp. *ciceris* culture.
2. Cork borer: sterilized.
3. Conical flask: 250 mL volume.
4. Potato dextrose broth.
5. Hemocytometer.
6. Micropipette.
7. Sterilized water.
8. Sterilized distilled water.

2.6 Inoculation and Transplanting

1. Twelve-day-old chickpea seedlings.
2. Beaker: sterilized.
3. Plastic pots (12 cm diameter).
4. Red soil, sand, and FYM (3:1:1) mixture: sterilized.
5. Water-soluble fertilizer N:P:K = 19:19:19%.

2.7 Equipment/ Facilities

1. Laminar airflow cabinet.
2. Bio-oxygen demand incubator.
3. Refrigerator.
4. Compound microscope.
5. Mixer-grinder.
6. Autoclave.
7. Incubator cum shaker.
8. pH meter,
9. Pestle and mortar.
10. Centrifuge.
11. Hot water bath.
12. UV Spectrophotometer.
13. Thermocycler.
14. Gel electrophoresis unit.
15. Gel documentation system.
16. Deep freezer.
17. Polyhouse.

2.8 Preparation of Solutions/Media/Broth

1. *1% sodium hypochlorite*: Add 20 mL of sodium hypochlorite containing 4–6% chlorine and 80 mL distilled water to make up 100 mL solution.
2. *Potato dextrose agar (PDA)*: Suspend 39 g PDA granules in 1000 mL distilled water. Boil the solution to dissolve the granules completely. Sterilize it by autoclaving at 15 lbs. pressure (121 °C) for 15 min.
3. *Potato dextrose broth (PDB)*: Suspend 24 g PDB granules in 1000 mL distilled water. Boil the solution to dissolve the granules completely. Sterilize by autoclaving at 15 lbs. pressure (121 °C) for 15 min.
4. *Water Agar 2% (WA)*: Suspend 20 g agar-agar powder in 1000 mL distilled water. Boil the solution to dissolve the powder completely. Sterilize by autoclaving at 15 lbs. pressure (121 °C) for 15 min.
5. *DNA extraction buffer*: Mix 50 mM Tris-HCl, pH 8.0, 100 mM EDTA, and 150 mM NaCl in double-distilled water, and make up the volume to 100 mL with distilled water.
6. *5 M NaCl*: Dissolve 29.22 g NaCl in 100 mL distilled water (do not add NaCl at once, as it will never go into solution). Autoclave and store at room temperature for no longer than 1 year.
7. *N-cetyl-N,N,N,-trimethyl ammonium bromide (CTAB) solution*: Dissolve 4.1 g NaCl in 80 mL distilled water. While stirring, add 10 g CTAB and heat the solution at 65 °C. Adjust the volume to 100 mL with distilled water.
8. *Sodium dodecyl sulfate 10% (SDS)*: Dissolve 10 g SDS in 100 mL distilled water.
9. *Chloroform: isoamyl alcohol (24: 1 v/v)*: Mix 24 volumes of chloroform with 1 volume of isoamyl alcohol. Store at room temperature for no longer than 1 year.
10. *Ethanol 70%:* Mix 70 mL of absolute alcohol with 30 mL distilled water and store at −20 °C.
11. *Ethylene diamine tetra acetic acid disodium salt (EDTA) 0.5 M*: Add 18.612 g of EDTA in double-distilled water, adjust pH to 8.0 by adding pellets of NaOH or HCl, and make up the volume to 100 mL.
12. *Tris-EDTA buffer (TE) 10×*: Mix 10 mM Tris-HCl (0.121 mL), pH 8.0, and 1 mM EDTA (0.03 mL) in double-distilled water, and make up the volume to 100 mL.
13. *Tris base 1 M (pH 8.0)*: Dissolve 121.1 g of Tris base in 800 mL of distilled water, and adjust pH to 8.0 by adding 42 mL of concentrated HCl. Make up the volume to 1 L with distilled water and sterilize using an autoclave.

14. *Tris boric acid EDTA buffer (TBE) 10×*: Add 108 g of Tris base, 55 g boric acid, and 40 mL of 0.5 M EDTA (pH 8.0), and make up to 1 L with sterile distilled water. To make a 0.5X working solution, do a 1:20 dilution of the 10 X TBE buffer.
15. *Agarose gel (1.8%)*: Add 1.8 g of agarose in 100 mL of 0.5× TBE buffer. Heat it in a microwave for 2 min to completely dissolve the agarose in buffer solution. Allow to cool, and add 2.5 μL of ethidium bromide; stir to mix.

3 Methods

3.1 Isolation and Purification of the Pathogen

3.1.1 Collection and Preparation of Diseased Plants for Laboratory Analysis

1. Collect chickpea plants displaying typical symptoms of *Fusarium* wilt in transparent polythene covers (*see* **Note 2**). Properly tag the plants and transport them to the laboratory for tissue isolation.
2. Rinse the infected root samples under running tap water to eliminate soil particles. Using a sterilized scalpel, cut the roots into small pieces measuring approximately 0.5–1 cm in size.

3.1.2 Surface Sterilization

1. Immerse the root bits in a 1% sodium hypochlorite solution for 1 min, ensuring complete coverage, and then rinse them twice with sterilized distilled water to remove any residual sodium hypochlorite (*see* **Note 3**).
2. Transfer the surface-sterilized root pieces onto sterilized tissue paper, and allow them to air-dry for 2 min.

3.1.3 Preparation of PDA Plates

1. In a laminar airflow cabinet, pour sterilized PDA media into sterilized and dried petri dishes.
2. Allow the PDA plates to solidify for approximately 10–15 min.

3.1.4 Transfer of Root Bits to Media and Incubation

1. Under the laminar airflow cabinet, transfer three to four root bits onto petri dishes containing PDA.
2. Incubate the petri dishes in a BOD incubator at 27 ± 1 °C for 3–4 days.
3. Transfer the early growing fungal mycelia to another set of PDA plates, and incubate for 7 days at 27 ± 1 °C.

3.1.5 Single Spore Isolation

1. Observe the cultural and morphological characteristics of the *Fusarium* growth (*see* **Note 4**). Select culture plates exhibiting typical *Fusarium* mycelial growth (Fig. 2a).
2. Prepare a spore suspension by mixing a fungal disc from a 7-day-old culture with 1 mL of sterilized distilled water (*see* **Note 5**).

3. Pour 0.1 mL of the suspension onto solidified water agar, and spread it evenly using a glass spreader.
4. Incubate the agar plates for 6–12 h, then observe and select a macroconidial agar disc. Transfer the disc onto PDA plates using a cork borer.
5. Incubate the inoculated PDA plates for 7–10 days. Store the pure culture at 4 °C in a refrigerator for future use.

3.2 Pathogenicity Test

1. Conduct pathogenicity test in polythene bags with sterilized sand-perlite mixture (1:1 ratio).
2. Prepare the inoculum of *Fusarium oxysporum* f. sp. *ciceris* by inoculating a disc of actively growing culture of *Foc* into a 100 mL of potato dextrose broth (PDB) in 250 mL flask (*see* **Note 6**).
3. After 7 days, homogenize the mycelial culture by adding 50–60 mL of sterilized water and blending it in a mixer-grinder for 1 min.
4. Using a pipette, extract 4 μL of the suspension from the homogenized mycelial culture onto a hemocytometer.
5. Under a compound microscope, observe and record the spore count. Adjust the inoculum concentration to 6.5×10^5 conidia/mL (*see* **Note 7**).
6. Add 20 mL of the prepared fungus inoculum to each polythene bag and thoroughly mix it to ensure uniform distribution.
7. Sow six to eight surface-sterilized seeds of the susceptible chickpea genotype into the pathogen-inoculated bags. Regularly irrigate the bags to maintain adequate moisture for the plants.
8. After 15 days of sowing, observe the plants for typical symptoms of *Fusarium* wilt. To confirm the presence of the pathogen, re-isolate it from the roots and/or stems of infected plants.

3.3 Molecular Identification of Foc

3.3.1 DNA Extraction

1. Total DNA is extracted by CTAB (N-cetyl-N,N,N,-trimethyl ammonium bromide) method.
2. Approximately 2 g of freshly harvested fungal mycelium is ground into a very fine powder in liquid nitrogen with a mortar and pestle.
3. Powdered mycelium is suspended in 10 mL of DNA extraction buffer (50 mM Tris buffer pH 8.0, 100 mM EDTA, 150 mM NaCl).
4. Add 1 mL of 10% SDS to the suspension and shake it to ensure proper mixing of SDS with the mycelium suspension. Incubate it for 1 h at 37 °C.

5. Add 1.5 mL of 5 M NaCl and 1.25 mL of CTAB solution (10% CTAB and 0.7 M NaCl) to it and incubate at 65 °C for 20 min in an incubator shaker at 60 rpm.
6. Extract DNA by adding an equal volume of chloroform: isoamyl alcohol (24: 1 V/V), and mix thoroughly but gently and then centrifuge at 12298 g for 12 min at 10 °C.
7. Remove aqueous viscous supernatant to a fresh tube. Add 0.6 volume of ice-cold isopropanol and 0.1 volume of sodium acetate to it. Keep it in the freezer at −20 °C for 8 to 10 h.
8. Centrifuge the mixture at 12298 g for 10 min at 10 °C and collect and wash the pellet with 70% ethanol.
9. Dry the tube completely and dissolve the pellet in minimum amount of TE buffer.
10. Treated with RNAse to purify the DNA. Quantification of DNA is done using UV spectrophotometer.
11. Variations in the steps for the DNA isolation from the fungus is also described in literature [24]

3.3.2 Primers, PCR Amplification, and Gel Electrophoresis

1. Prepare 50 μL of PCR reaction mixture using 2 μL dNTP (250 μM each dNTP), 1 μL each of forward (ITS 1) and reverse (ITS 4) primer (20 ng/μL each), 1 μL template DNA (30 ng/μ L), 2.5 μL TBE buffer (10×), 2.5 U Taq DNA polymerase, and volume is adjusted to 50 μL using deionized water.
2. Perform PCR reactions for ribosomal DNA amplification with 30 cycles (1 min denaturing at 94 °C, 30 s annealing at 54 °C, and 1 min polymerization at 72 °C).
3. After completion of PCR amplification, add 3 μL of gel loading dye to each sample, and resolve the 25 μL total volume on 1.8% agarose gel in 0.5× TBE buffer.
4. Size of the amplified DNA fragments is estimated with 100 bp ladder (*see* **Note 8**).

3.3.3 Sequencing and Bioinformatics Analysis

1. For sequencing amplified rDNA regions of *Foc*, prepare PCR products as per the instructions of outsourcing agency.
2. Sequence information obtained is used for performing Basic Local Alignment Search Tool for Nucleotide Sequences (BLASTN) analysis for homologous sequences at National Center for Biotechnology Information (NCBI).
3. Confirmation of the pathogen is based on percent sequence similarity with rDNA regions of *Foc* data bases in NCBI.

3.4 Seedling Root Dip Method

3.4.1 Raising of Seedlings

1. Immerse the chickpea seeds in 5% sodium hypochlorite for 1 min, ensuring complete coverage, and then rinse them twice with sterilized distilled water to remove any residual sodium hypochlorite.

2. Fill the polythene bags (20 cm length × 6 cm width) with a mixture of sterilized river sand and perlite (1:1 ratio).
3. Sow the chickpea seeds of the test entries, resistant and susceptible genotypes, in polythene bags and keep under controlled conditions in a polyhouse maintained at 25 ± 1 °C (Fig. 2b).

3.4.2 Inoculum Preparation

1. Inoculate 7 mm disc of actively growing pure culture in 100 mL of potato dextrose broth (PDB) in 250 mL flask to multiply the culture (*see* **Note 9**).
2. Incubate the inoculated flasks at 25 °C in an incubator shaker at 125 rpm for 10 days.
3. Observe inoculated flasks regularly for any bacterial contamination.
4. After 7 days, homogenize the mycelial culture by adding 50–60 mL of sterilized water and blending it using a mixer-grinder for 1 min.
5. Pipette out 4 μL of suspension from the homogenized mycelia culture on to a hemocytometer.
6. Under compound microscope, observe and record the spore count. Adjust the inoculum concentration of 6.5×10^5 conidia/mL.

3.4.3 Inoculation and Transplanting

1. Ten-to-twelve-days-old chickpea seedlings are chosen for root inoculation (*see* **Note 10**).
2. Uproot and clean the previously grown 10 to 12 days old chickpea seedlings under running tap water.
3. Dip the roots in inoculum for 10 min (*see* Fig. 2c).
4. Transplant inoculated seedlings in pre-irrigated and sterilized red soil, sand, and FYM (3:1:1) mixture filled in plastic pots (12 cm diameter) (Fig. 2d).
5. Maintain four test plants with resistant (WR 315) and susceptible (JG62) checks per pot/replication (Fig. 2d).
6. Incubate the inoculated pots in a greenhouse maintained at 25 ± 3 °C and irrigated as need (Fig. 2e).
7. Spray water-soluble fertilizer at 7 days interval.

3.4.4 Observations

1. Observe the inoculated seedlings for initiation of symptoms and mortality up to 60 days after inoculation (Fig. 1).
2. Estimate the incubation period, latent period, and disease mortality based on above observations (*see* **Note 11**).
3. Record the disease incidence using 1–9 rating scale given by ICRISAT where, 1 = asymptomatic (0% plants wilted), 3 = resistant (0–10% plants mortality), 5 = moderately

Fig. 1 Chickpea seedling showing typical wilting symptom

resistance (10.1–20% mortality), 7 = susceptible (20.1–40% mortality), and 9 = highly susceptible (more than 40% mortality) up to 60 days after inoculation.

4 Notes

1. The chickpea genotypes resistant and susceptible to *Fusarium* wilt are WR 315 and JG 62, respectively.
2. When isolating *Foc*, select chickpea plants displaying typical wilt disease symptoms, characterized by dark brown discoloration of the vascular system.
3. It is essential to wash the roots twice to ensure the complete removal of sodium hypochlorite residue, as it can hinder the mycelial growth of *Foc.*
4. Morphological identification of *Foc* spores includes the observation of sickle-shaped macroconidia with three septa, oval-shaped microconidia without septa, and abundant chlamydospores production.
5. The single spore isolation method is recommended for achieving pure *Foc* cultures.
6. Culture plates of *Foc* can be identified initially by the presence of white mycelial growth with pinkish pigmentation.

Fig. 2 Isolation of pathogen and steps involved in raising of chickpea seedlings, root dip inoculation and testing genotypes. (**a**). Agar plate containing 7-day-old pure culture of *Foc* showing white mycelial growth with pinkish pigmentation. (**b**). Raising of chickpea seedlings in polybags with a mixture of sterilized river sand and perlite. (**c**). Roots of 10-to-12 day old chickpea seedlings dipped in *Foc* inoculum (6.5×10^5 spores/mL) for 10 min. (**d**). Root-inoculated seedlings transplanted in a pot (a) susceptible check (JG 62), (b) resistant check (WR 315), (c–f) test lines. (**e**). General view of incubation of inoculated seedlings in polyhouse

7. To ensure better infection by *Foc*, maintain an inoculum concentration of 6.5×10^5 conidia/mL.
8. The size of the amplified DNA fragment using ITS primers for *Foc* is between 550 and 600 bp.
9. Sowing seeds in poly bags and inoculating them with actively growing *Foc* culture in PDB should be performed on the same day to ensure that the culture is ready for transplantation.
10. Normally, seedlings aged 10–12 days are selected for inoculation using the root dip method, but the vigor of the seedlings should be considered. Choose seedlings with at least 4–5 fully opened leaves.
11. Incubation period is the number of days taken from inoculation to initiation of symptoms; latent period is the number of days taken from initiation of the symptoms to death of the seedling, and disease mortality is the sum total of incubation period and latent period.

Acknowledgments

We acknowledge the Department of Biotechnology (DBT), Government of India for their financial support to carry out the research work.

References

1. Zohary D, Hopf M (2000) Domestication of plants in the old world: the origin and spread of cultivated plants in West Asia, Europe and the Nile Valley, 3rd edn. Oxford University Press, Oxford
2. FAOSTAT (Food and Agriculture Organization of the United Nations Statistical Database) (2014) FAOSTAT production statistics of crops. Available: http://faostat.fao.org/site/567/default.aspx#ancor
3. Ahmad MA, Iqbal SM, Ayub N et al (2010) Identification of resistance sources in chickpea against Fusarium wilt. Pak J Bot 42(1): 417–426
4. Jukanti AK, Gaur PM, Gowda CLL et al (2012) Nutritional quality and health benefits of chickpea (Cicer arietinum L.): a review. Br J Nutr 108:S11–S26
5. Nene YL, Sheila VK, Sharma SB (1996) A world list of chickpea and pigeonpea pathogens, 5th edition ICRISAT. Pantancheru 3(9):19–20
6. Cortes JA, Hav BA (2000) Yield loss in chickpea in relation to development of Fusarium wilt epidemics. Phytopathology 90:1269–1273
7. Chattopadhyay C, Mohapatra SD (2015) Perception of constraints in chickpea production in India. Indian J Agri Sci 85:287–289
8. Haware MP (1990) Fusarium wilt and other important diseases of chickpea in the Mediterranean area. Options Méditerr Sér Sémin 9:61–64
9. Jalali BL, Chand H (1992) Chickpea wilt. In: Singh US, Mukhopadhayay AN, Kumar J et al (eds) Plant diseases of international importance Vol 1. Diseases of cereals and pulses. USA Prentice Hall, Englewood Cliffs, pp 429–444
10. Nene YL, Reddy MV (1987) In: Saxena MC, Singh KB (eds) Chickpea diseases and their control. CAB International, Wallingford, pp 233–270
11. Jimenez-Diaz RM, Trapero-Casas A, Colina JCDL (1989) Races of Fusarium oxysporum f. sp. ciceris infecting chickpea in southern

Spain. In: Tjamos EC, Beckman CH (eds) Vascular wilt diseases of plants, vol H28. ASI Series Springer Verlag, Berlin, pp 515–520

12. Jimenez-Gasco MM, Jimenez-Diaz RM (2003) Development a specific polymerase chain reaction-based assay for the identification of Fusarium oxysporum f. sp. ciceris and its pathogenic races 0, 1A, 5, and 6. Phytopathology 93:200–209
13. Haware MP, Nene YL (1982) Races of Fusarium oxysporum f.sp. ciceris. Plant Dis 66:809–810
14. Jimenez-Gasco MM, Perez-Artes EA (2001) Identification of pathogenic races O, 1B/C, 5 and 6 of Fusarium oxysporum f.sp. ciceris with random amplified polymorphic DNA (RAPD). Eur J Plant Pathol 107:237–243
15. Haware MP, Nene YL, Rajeshwari R (1978) Eradication of Fusarium oxysporum f. sp. ciceris transmitted in chickpea seed. Phytopathology 68:1364–1367
16. Haware MP, YL Nene, Natrajan M (1986) Survival of Fusarium oxysporium f. sp. ciceris in soil in absence of chickpea. Paper presented in the national seminar on management of soil-borne diseases of crop plants, TNAU, Coimbatore, p 8–10
17. Bhatti MA, Kraft JM (1992) Effects of inoculum density and temperature on root rot and wilt of chickpea. Plant Dis 76:50–54
18. Sugha SK, Kapoor SK, Singh BM (1994) Factors influencing Fusarium wilt of chickpea (Cicer arietinum L.). Indian J Mycol Plant Pathol 24:97–102
19. Zote KK, Khalikar PV, Dandnaik BP (1986) Reactions of chickpea varieties to Fusarium wilt. Indian J Mycol Plant Pathol 16(1):80–81
20. Nene YL, Reddy MV, Haware MP et al (1991) Field diagnosis of chickpea diseases and their control. In: Information Bulletin no. 28. ed. by international crops research institute for the semi arid tropics, Patancheru, India
21. Sharma KD, Chen W, Muehlbauer FJ (2005) Genetics of chickpea resistance to five races of *Fusarium* wilt and a concise set of race differentials for Fusarium oxysporum f. sp. ciceris. Euphytica 89:385–390
22. Dekkers JCM, Hospital F (2002) The use of molecular genetics in improvement of agricultural populations. Nat Rev Genet 3:22–32
23. Pande S, Sharma M, Nagavardhini A et al (2012) High throughput phenotyping of chickpea diseases: stepwise identification of host plant resistance. Information Bulletin No. 92. International crops research institute for the semi arid tropics, Patancheru, Andhra Pradesh, India, p 56
24. White TJ, Bruns TD, Lee SB et al (1990) Amplification and direct sequencing of fungal ribosomal RNA genes for phylogenetics. In: Innis MA, Gelfand DH, Sninsky JJ et al (eds) PCR protocols: a guide to methods and applications. Academic Press, New York, pp 315–322

Chapter 12

Oligo-dT Anchored cDNA SCoT: An Innovative Approach for Deciphering *Fusarium* Wilt Resistance-Related Differential Gene Expression in Chickpea

S. L. Abhang, P. Sharma, S. N. Datarkar, M. P. Moharil, H. Kudapa, A. P. Ingle, and P. V. Jadhav

Abstract

The cDNA start codon targeted marker (cDNA-SCoT) methodology is a unique, simple, quick, and efficient method for investigating differential gene expression. The differentially expressed fragments (DEFs) in resistant and susceptible cultivars were identified using cDNA-SCoT profiling. Sanger sequencing and the nBLAST homology search program were used to characterize the functional properties of DEFs. As a result, this methodology is innovative and would be utilized in future studies to identify potential functional genes, which can then be used in marker creation, marker-assisted selection, and uncovering the regulatory role of the relevant genes.

Key words Gene expression, cDNA-SCoT, *Fusarium oxysporum*, Wilting disease

1 Introduction

A method for gene differential expression in plants based on the start codon targeted polymorphism (SCoT) DNA marker technique, called cDNA SCoT [1]. The cDNA-SCoT is an advantageous method compared to all other existing ones because it is relatively more efficient, cheap, faster, and simple to operate, and the results can be easily reproduced [2]. A cDNA-SCoT (cDNA starts codon-targeted) marker is used for determining the gene expression and assessment of genetic diversity in various plant species.

Start codon targeted (SCoT) polymorphism is a novel, simple, and reliable marker technique based on the translation start codon, which is a gene-targeted marker that can generate information correlated with biological traits compared with random DAN markers [3, 4]. The principle of SCoT primers is demonstrated in (*see*

Muthappa Senthil-Kumar and Hema Ramanna (eds.), *Plant Biotic Stress: Methods and Protocols*, Methods in Molecular Biology, vol. 2966, https://doi.org/10.1007/978-1-0716-4746-2_12,

Fig. 2). SCoT analysis is based on the short conserved region in plant genes, which are surrounded by the initiation or ATG translation start codon that has been well-characterized in differential gene expression and screening of stress-related genes [1]. One of the main purposes of this method is to search for the differential expression in plants and to provide valuable information to determine gene functions. In principle, this marker system is comparable to ISSR, RAPD, or single primer amplification reaction, as it also utilizes a single primer as forward and reverse primer. Amplification products can be easily pictured by ordinary gel electrophoresis using agarose gel and staining, making this practice more pertinent to the massive plant research with standard equipment [5].

2 Materials

2.1 Plant Material

1. *Fusarium* wilt resistance and susceptible chickpea seeds.

2.2 Fungal Culture

1. *Fusarium oxysporum f.* species ciceris.

2.3 10× MOPS Buffer

1. 0.2 M MOPS (pH 7.8)
2. 20 mM sodium acetate
3. 10 mM EDTA.

2.4 Potato Dextrose Agar (PDA) Medium (for 1000 mL Preparation)

1. 200 g of potato, 20 g of dextrose, 20 g of agar, 1 L of distilled water.

2.5 Potato Dextrose Broth (PDB) (for 1000 mL Preparation)

1. 200 g of potato, 20 g of dextrose, 1 L of distilled Water.

2.6 First-Strand cDNA Synthesis Reaction (Total Volume: 20 μL)

1. 2 μL of control GAPDH RNA (50 ng/μL), 1 μL of oligo (dT) primer, 4 μL of 5× reaction buffer, 1 μL of RiboLock RNase Inhibitor (20 U/μL), 2 μL of 10 mM dNTPs Mix, 1 μL of RevertAid RT (200 U/μL), and 9 μL of nuclease-free water.

2.7 PCR Reaction Mixture (Total Volume: 20 μL)

1. 1 μL of template (50 ng/μL), 2 μL of PCR buffer with 17.5 mM $MgCl_2$, 0.5 μL of 10 mM dNTPs Mix, 2 μL of 10 pmol primer, 0.3 μL of Taq polymerase (5 U/μL) and 14.2 μL of nuclease-free water.

2.8 Others

1. The ReverseAid First Strand cDNA Synthesis Kit.
2. Inoculation loop.
3. Scissor.
4. Scalpel blade.

3 Methods

3.1 Fungal Culture Preparation

1. Prepare the potato dextrose agar (PDA) media by weighing and dissolving the chemical constituents, as given in Subheading 2.4.
2. Sterilize the media by autoclaving at 121 °C/15 psi for 20 min.
3. After autoclaving, pour each 25 mL media into a sterile petri dish under aseptic conditions in a laminar airflow cabinet.
4. Subculture the pure culture of *Foc* on PDA media by using a sterile loop and incubating the plates at 28 °C for 5 to 6 days.
5. Purify the subcultured media by single spore suspension method.
6. Study the colony morphology of the *F. oxysporum* under a microscope.

3.2 Plant Material, Growth, and Inoculation

1. Collect the *Fusarium* wilt resistance and susceptible chickpea seeds.
2. Sterilize the surface of healthy seeds of chickpea with 1% sodium hypochlorite solution and soak them overnight in sterile distilled water.
3. Wrap the seeds in a wet sterile muslin cloth till sprouting, and transfer to surface-sterilized plastic cups containing autoclaved soilrite (mixture of 75% Irish peat moss and 25% horticulture grade).
4. Allow the plants to grow for 1 week in a growth chamber (14 h light/10 h dark, 22–25 °C, 50–60% relative humidity), and inoculate by root-clipping with freshly prepared spore suspension (1×10^6 spore/mL) of *Foc* races 4.
5. Cut the tips of the top and lateral roots and dip the entire root system in spore suspension for 5 min.
6. Plants dipped in sterile distilled water served as the control, and other plants served as the infected with *Foc* (*see* Fig. 1).

3.3 Collection of Infected Tissue

1. Take a premature tissue sample.
2. Clean the tissue with DEPC (diethyl pyrocarbonate)-treated water and wash it with 70% ethanol.
3. Again, clean the tissue with DEPC-treated water (*see* **Note 1**).

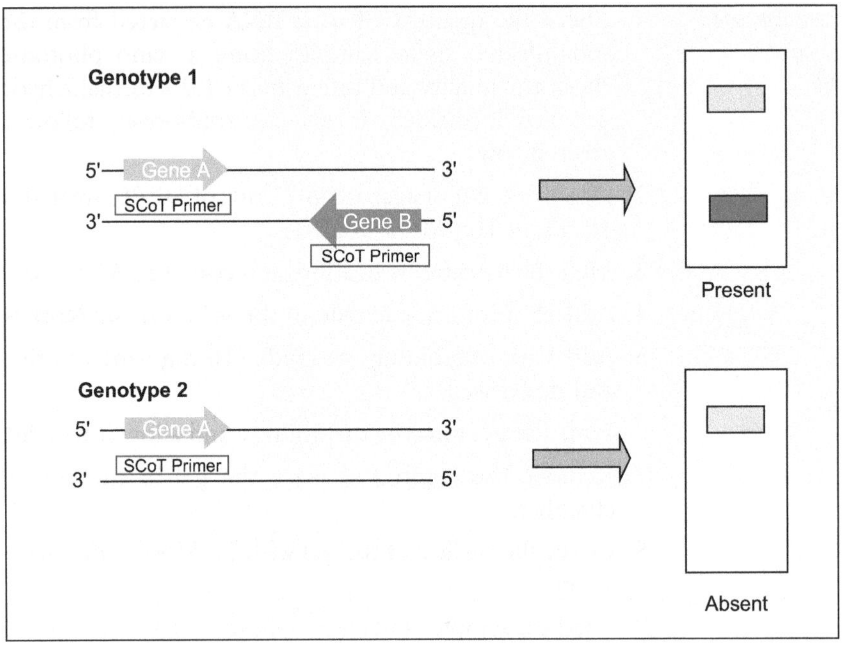

Fig. 2 Principle of SCoT based PCR amplification

2. The cDNA dilutes up to 50 ng for PCR amplification in RNase-free water.
3. Prepare the reaction mixture by adding the components (*see* Subheading 2.7).
4. Set the PCR cycling profile at 94 °C for 3 min, 35 cycles at 94 °C for 30 s, 48 °C for 30 s, and a 72 °C for 45 s and 72 °C for 5 min (*see* **Note 11**).
5. Prepare the 2% TBE agarose gel (*see* **Note 12** and **Note 13**)
6. Load the amplified product along with 6× loading dye (50% glycerol and bromophenol blue) on agarose gel and use the 100 bp DNA ladder as a standard (*see* **Note 14** to **Note 18**)
7. Compare the oligo-dT anchored cDNA-SCoT pattern for each infected and control (Fig. 3).

3.7 Isolation and Sequencing of Oligo-dT Anchored cDNA-SCoT Fragments

3.7.1 Elution of Bands from the Gel

1. After adequate separation of bands on the agarose gel, excise the DNA bands from the agarose gel with a sharp scalpel blade and place them in clean 2.0 mL centrifuge tubes.
2. Determine the weight of the gel slices and add three volumes of gel bind buffer (HG) accordingly as per gel slice volume. Incubate the mixture at 55–60 °C for 7 min or until the gel completely melts.
3. Mix the contents of the tube after every 2–3 min so that the agarose is completely dissolved (*see* **Note 19**).

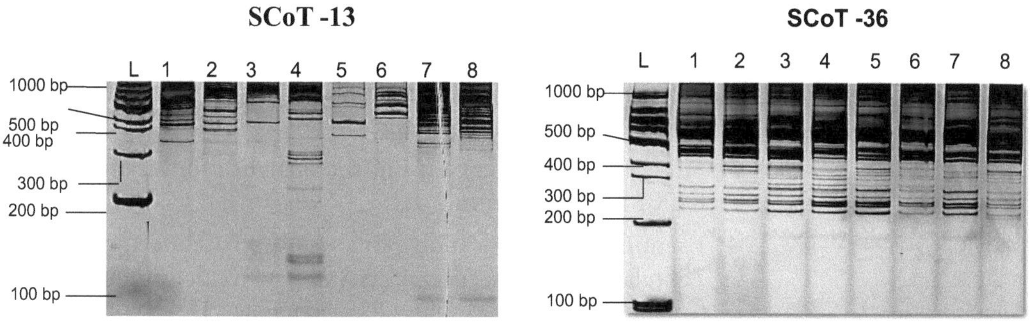

Fig. 3 cDNA SCoT Profiling L- Ladder, 1- Susceptible Control plant, 2- Susceptible (2DAI), 3- Susceptible (8DAI), 4- Susceptible(16DAI), 5- Resistant Control plant, 6- Resistant (2DAI), 7 - Resistant (8DAI), 8- Resistant (16DAI)

4. Apply the gel and gel binding buffer mixture (obtained from the above step) to a HiElute Miniprep Spin Column (Capped), and centrifuge at 12,000 g for 1 min at room temperature (15–25 °C) (*see* **Note 20**).
5. Discard the flow-through and place the column back into the same collection tube.
6. Add 300 μL of Gel Bind Buffer (HG) into the column and centrifuge for 1 min at 12,000 g at room temperature (15-25 ° C) to wash the membrane. Discard the flow-through and reuse the collection tube. For the washing step, place the column into the same collection tube and add 700 μL of diluted gel wash buffer (GWB). Centrifuge for 1 min at 12,000 g at room temperature (15–25 °C). Discard the flow-through and reuse the collection tube.
7. Centrifuge the empty column for 2 min at maximum speed, that is, 13,000 g to dry the column temperature (*see* **Note 21**).
8. Place the columns into a new 2.0 mL uncap collection tube, and add 30–50 μL of elution buffer (EB) (10 mM Tris-Cl, pH 8.5) (depending on the desired concentration of the final product) directly onto the column membrane. Incubate at room temperature (15–25 °C) for 1 min. Centrifuge for 1 min at a maximum speed of 13,000 g. Transfer elute to a fresh capped 2.0 mL collection tube (*see* **Note 22**).

3.7.2 Cloning and Sequencing of cDNA-SCoT Amplified P

1. Perform the cloning of the amplified product in the pGEM-t easy vector [6].
2. Sequence the product on the Applied Biosystem® platform by dideoxy termination method, and determine the sequences using an automated DNA sequence.
3. Use SQUAT software for the assessment of the quality of the sequence.

3.7.3 Differential Expression Using qRT-PCR

1. To track dsDNA synthesis, conduct PCR reaction in a real-time thermal cycler using iTaq Universal SYBR Green dye.
2. Prepare a total volume of reaction 10 μL (1.5 μL diluted cDNA template, 0.2 M each primer).
3. Study the fluorescence using the analytical program Bio-Rad CFX Manager version 2.1.
4. Examine all amplification plots using a baseline threshold of 75 relative fluorescence units (RFU), to determine the Cq (quantification cycle) values for each gene-cDNA combination [7] (*see* **Note 23**).

3.8 Expression Analysis

1. Perform the homology searches within Genbank[@] no redundant databases using the BLAST 2.2.8 (Basic Alignment Search Tool) algorithm at http://blast.ncbi. nlm.nih.gov/Blast.cgi of the National Center for Biotechnology Information (NCBI), with the program BLASTN.
2. Sequence analysis is helpful to find out some gene network analysis and protein-protein interaction.

4 Notes

1. For the preparation of 1 L of DEPC-treated water, add 1 mL of new DEPC. Give the DEPC a good shake to mix with the water. To inactivate the residual DEPC, incubate the sample for at least 12 h at 37 °C or autoclave it for 20 min at 15 psi on the liquid cycle.
2. Keep everything on ice. Make sure that every step of the RNA extraction process, and all of the tools used in the process, is kept cold. This is necessary to inhibit any RNase activity.
3. The TRIzol reagent is poisonous, when applied to the skin or when swallowed. Burns will result. When using the TRIzol reagent, be careful to use a lab coat, gloves, and safety glasses. Wash with plenty of soap and water if skin contact occurs.
4. This will result in the development of three layers including the lower red layer (phenol-chloroform), middle white layer, and upper aqueous layer.
5. Chilled isopropanol is used for the precipitation of RNA.
6. Add ethanol according to the pellet, about 30–50 μL.
7. The gel running buffer (MOPS) yellows with age if exposed to light or autoclaved. Discoloration does not affect its performance appreciably. Formaldehyde (F.W. = 30.03) is usually obtained as a 37% solution greater than 4.0. The concentrated solution should be handled and stored in a chemical hood.

8. Formaldehyde is toxic. The skin, throat, lungs, and eyes might get irritated when exposed to formaldehyde. Continual exposure to formaldehyde may cause cancer. Formaldehyde exposure may be harmful to workers. The dose, duration, and type of work determine the exposure level. To reduce formaldehyde vapor inhalation, formaldehyde should always be used with sufficient ventilation, ideally in a fume hood. When handling formaldehyde, wear chemical goggles or a face shield to reduce the risk of corneal damage from even a minor splash or vapor exposure.
9. Freeze/thaw cycles can play havoc with your reagents if you reuse an aliquot too many times, damaging enzymes and dNTPs via recrystallization. When preparing samples, ensure that you keep track of the number of times each aliquot has been used, ideally using a LIMS, which can manage your entire inventory, including the number of times samples and reagents have been frozen/thawed.
10. An RNase-free environment is necessary for successful cDNA synthesis. Use only DEPC-treated water, autoclaved or sterile-filtered solutions, and only RNase-free autoclaved tubes and aerosol-free pipette tips.
11. Another way errors can appear in your PCR results is by using the wrong annealing temperature for a given primer set. If the reaction does not work as intended, it is tempting to lower the annealing temperature to increase the chances of having a successful reaction. However, the lower the temperature, the more likely primer dimers, and false positives are to appear.
12. When measuring acrylamide, wear a mask. Cover the weigh boat carrying the acrylamide that has been weighed with another weigh boat (of a size equivalent to carrying the weighted acrylamide to the fume hood) from the initial weigh boat. Place a stirrer inside the fume hood and add the weighted acrylamide to the cylinder within. Stir because unpolymerized acrylamide is toxic to the nervous system, caution should be taken to prevent skin contact.
13. A glass or plastic tray has to be properly rinsed after being cleaned with detergent.
14. Prepare the silver solution by dissolving 0.3 g of $AgNO_3$ powder in 150 mL of 10% methanol solution with 750 μL glacial acetic acid.
15. You should never handle the gel bands directly, and always wear fresh, disposable gloves.
16. Prepare the developing solution by dissolving 9 g NaOH pellets in 300 mL distilled water with 900 μL of formaldehyde.

17. Stain the gel for as long as is necessary to find the bands of interest (often a few minutes).
18. When applying and discarding the stain, caution must be observed. Never empty a silver stain into the washbasin. It must be properly disposed of following the advised technique in a suitable trash container.
19. For 100 mg of agarose gel slice add 300 μL of Gel Bind Buffer (HG).
20. Load approximately 700 μL of sample mixture at a time in the HiElute Miniprep Spin Column (capped). If the sample volume is larger than 700 μL, then discard the flow-through; add the rest of the lysate to the column and repeat the spin.
21. Centrifuge the empty column for the removal of residual ethanol completely.
22. Avoid the repeated freezing and thawing of the sample because it may cause denaturation of DNA.
23. Validate the quality of your cDNA for the downstream application. If running qPCRs, run a normalizing set of reactions to establish the Cq values of a standard reference gene in all samples. You can compare this value to previous batches of cDNA and dilute your samples appropriately to ensure all samples contain comparable concentrations of cDNA if required. This approach may not be suitable to confirm successful reverse transcription of long transcripts, because qPCR targets are kept <0.2 kb. Thus, if you need to ensure the presence of long transcripts >0.5 kb, run second strand DNA synthesis (usually standard PCR with appropriate primers) and run part of your cDNA sample on an agarose gel. You should see a smear ranging from about 500 bp to 3–4 kb, verifying that full-length transcripts have been reverse transcribed.

Acknowledgments

We gratefully acknowledge the Biotechnology Centre, Dr. PDKV Akola, for supporting and providing their facilities in this work. In addition, we would like to thank Pulses Research Unit Dr. PDKV, Akola, and ICRISAT, Hyderabad for providing seeds.

References

1. Wu J-M et al (2013) cDNA-SCoT: a novel rapid method for analysis of gene differential expression in sugarcane and other plants. Aust J Crop Sci 7(5):659–664
2. Luo C, He X-H, Hu Y, Yu H, Ou S-J, Fang Z-B (2014) Oligo-dT anchored cDNA–SCoT: a novel differential display method for analyzing differential gene expression in response to several stress treatments in mango (Mangifera indica L.). Gene 548(2):182–189
3. Collard BCY, Mackill DJ (2009) Start codon targeted (SCoT) polymorphism: a simple, novel

DNA marker technique for generating gene-targeted markers in plants. Plant Mol Biol Report 27:86–93

4. Xiong FQ, Tang RH, Chen ZL, Pan LH, Zhuang WJ (2009) SCoT: a novel gene targeted marker technique based on the translation start codon. Mol Plant Breed 7(3):635–638
5. Bhawna MZ, Abdin LA, Verma M (2017) Use of SCoT markers to assess the gene flow and population structure among two different populations of bottle gourd. Plant Gene 9:80–86. https://doi.org/10.1016/j.plgene.2016.09.001
6. Nimbalkar SB, Harsulkar AM, Giri AP, Sainani MN, Franceschi V, Gupta VS (2006) Differentially expressed gene transcripts in roots of resistant and susceptible chickpea plant (Cicer arietinum L.) upon fusarium oxysporum infection. Physiol Mol Plant Pathol 68(4–6):176–188
7. Caballo C, Castro P, Gil J, Millan T, Rubio J, Die JV (2019) Candidate genes expression profiling during wilting in chickpea caused by fusarium oxysporum f. sp. ciceris race 5. PLoS One 14(10):e0224212

Chapter 13

Extracellular Acidification Assay to Evaluate the Effectiveness of Antifungal Agents on the Pathogenicity of *Macrophomina phaseolina*

Krishna Ramakrishnan Acharya, Aswin Reddy Chilakala, and Muthappa Senthil-Kumar

Abstract

Macrophomina phaseolina is a fungus that causes dry root rot disease and considerable yield loss worldwide. Fungi exhibit various ways of absorbing nutrients through their plasma membrane, such as free or facilitated diffusion, diffusion channels, or active transport. Glucose, as a preferred carbon source, activates the plasma membrane H^+-ATPase, resulting in the release of protons. Consequently, the protons, along with the organic acid metabolites released into the extracellular environment, acidify the cell surroundings. This decrease in pH cues the fungus to shift from saprotrophic to necrotrophic growth, facilitating host invasion. Sustainable dry root rot disease management often relies on the employment of antifungal agents from various biological sources. Despite the discovery of numerous antifungal agents, only a limited number have been evaluated for their efficacy against this phytopathogenic fungus. This scarcity of testing is primarily due to the limitations of existing methods, which often lack standardisation and reproducibility. This chapter introduces a rapid and sensitive method to assess the antifungal activity of various agents against *M. phaseolina*. By measuring extracellular pH changes after treatment in the presence of a nutrient source, we can determine the inhibitory concentrations of these agents and evaluate their potential for controlling fungal pathogenicity in plants.

Key words *Macrophomina phaseolina*, Extracellular acidification, Phytopathogenesis, Microsclerotia, Fungal physiology, Antifungal agent screening, pH regulation

1 Introduction

1.1 Plasma Membrane Physiology

In fungal physiology, membrane transport systems are crucial because they control effective cell reproduction, development, and metabolism. The selectively permeable membrane allows nutrient absorption and release of metabolites from the fungal cell. Fungi exhibit several active and passive absorption mechanisms at the plasma membrane namely, free diffusion, facilitated diffusion, diffusion channels, and active transport. In filamentous fungi,

Muthappa Senthil-Kumar and Hema Ramanna (eds.), *Plant Biotic Stress: Methods and Protocols*, Methods in Molecular Biology, vol. 2966, https://doi.org/10.1007/978-1-0716-4746-2_13,

active transport of nutrients including sugars, amino acids, nitrate, ammonium, sulfate, and phosphate demands spatial separation of the ion pumps, which are mostly located behind the apex while the symport proteins are active near the tip. Thus, when the hyphal tip continually digs into new resources, it absorbs nutrients, and the mitochondria located behind the apex provide ATP to fuel the ion pump and produce proton motive force [1].

1.2 Cellular pH Regulation and Glucose Uptake

In almost all cells, pH control is mainly achieved through the combined activity of pumps, exchangers, and buffers, supported by multiple layers of regulation [2]. Figure 1 shows proton pumps, such as Pma1 (P-type H^+-ATPase) at the fungal plasma membrane, which determine membrane potential by electrogenic transport of H^+ without counterions and highlights V-ATPases on fungal organelle membranes, which couple ATP hydrolysis with proton transport to establish pH gradients [3]. ATP-binding cassette (ABC) transporters protect *M. phaseolina* from host-derived secondary metabolites and supply essential nutrients [4].

Under conditions of abundant glucose, their preferred carbon source, *S. cerevisiae* cells will undergo rapid fermentative growth, producing ethanol, CO_2, and organic acids [5, 6]. Cells grown in glucose rapidly acidify their medium and require robust mechanisms to maintain cytosolic pH during growth, and cytosolic pH decreases as cells reach the stationary phase [3]. *Aspergillus niger* induces the expression of pH-dependent monosaccharide/H^+ symporter upon carbon starvation [7]. The growth of *Penicillium*

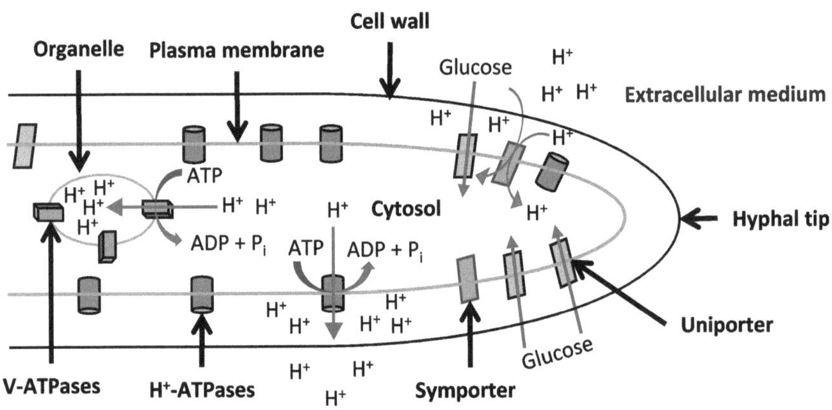

Fig. 1 Representation of H^+ and glucose transport in active hyphae. The figure shows a section of fungal hyphae with different cell layers and important membrane transporters involved in cellular homeostasis. Upon the recognition of extracellular nutrients (e.g., glucose), the metabolic activity in the growing end of the hyphae is upregulated and causes the secretion of catabolic enzymes. Eventually, the H^+ ions generated with other organic acid by-products create an acidic extracellular environment. This favors the uptake of the nutrients and invasion into host tissues

Orange cylinders: H^+-ATPases, yellow parallelograms: glucose uniporter, gray parallelograms: glucose/H^+ symporter, green cuboids: V-type ATPases

ochrochloron under glucose-limited conditions allowed for immediate glucose intake at higher rates from a glucose pulse compared to steady-state conditions [8]. In addition to the efflux of acidic metabolites, activation of plasma membrane H^+-ATPase by glucose contributes to about one-half of extracellular acidification [9]. Inhibition of plasma membrane Pma1 (H^+-ATPase) disrupts the electrochemical gradient, nutrient transporters, and extracellular acidification in many fungi [10].

1.3 Phytopathogenic Aspects of Extracellular Acidification

Pathogenic fungi such as *Sclerotinia* and *Botrytis* sp. produce an acidic environment that causes a shift from saprotrophic to necrotrophic growth in plants [11]. In *Fusarium sp*, fusaric acid acts as a virulence factor during the early stages of the disease [12]. Growth of *M. phaseolina* in the presence of host tissue triggers azelaic acid production, potentially aiding in disease establishment [13]. Endoglucanases catalyzing the hydrolysis of cellulose to glucose, and malate dehydrogenase causing organic acid production, were found in the secretome study of *M. phaseolina* [14]. Additionally, several investigations demonstrated the significance of plasma membrane H^+-ATPases during infection and pathogenesis [15, 16].

1.4 Dry Root Rot Disease in Chickpea

M. phaseolina propagates through dark-colored, branching hyphae with black microsclerotia [13]. Infected roots when split exhibit characteristic dark spots in the infected regions, which are microsclerotia and serve as the best parameter to identify the disease under the field conditions [17]. Several surveys and field investigations indicate that drought and high temperatures increase the prevalence and severity of DRR disease [17–21]. Moreover, they usually occur as mixed infections usually with wilt [22]. Dry root rot (DRR) of chickpea had its impact over the 13 million hectares of chickpea-growing regions worldwide leading to a significant reduction in yield of 30 to 50 per cent under moderate infection, and up to 80 per cent yield reductions in susceptible genotypes [17, 18, 21, 23].

High resistance levels are challenging to establish because of the pathogen's wide host range. None of the lines were fully resistant to DRR, according to a search for particular resistance in chickpea [24]. Therefore, the reliability of the effective sustainable metabolites and proteins increases in disease control. Although numerous fungi, bacteria, and plants have been found to contain antifungal metabolites/proteins that are effective against phytopathogenic fungi, their characterization is limited by diverse testing methods usually lacking standardized conditions and reproducibility.

5. MQ in wash bottle and tissue paper.
6. pH meter (*see* **Note 3**).

3 Methods

3.1 M. phasolina *Agar Culture*

1. Autoclave the desired number of test tubes with 10 mL of PDA medium (*see* **Note 4**).
2. Perform the subsequent steps inside a laminar airflow chamber with all equipment wiped with 70% ethanol.
3. Pour the molten PDA agar into the sterile petri plates, and let it solidify and dry properly.
4. Ensure that the surface of the pure culture on the medium is uniform (*see* **Note 5**).
5. Pick the microsclerotia each time from a definite radius of the dark-colored region to ensure uniform inoculum.
6. Stab the pure culture using a 1 mL tip with a micropipette in a direction perpendicular to the agar surface (*see* **Note 6**).
7. Depress the plunger by keeping the tip close to the center of the freshly prepared agar plate (*see* **Note 7**).
8. Seal the petri plate twice with parafilm to avoid contamination and incubate it at 28 °C for 3 days.
9. Perform the above inoculation three times to obtain biological replicates.

3.2 M. phaseolina *Broth Culture*

1. Perform the subsequent steps inside a laminar airflow chamber with each equipment wiped with 70% ethanol.
2. Follow **steps 4** and **5** from Subheading 3.1.
3. Stab the pure culture using a 200 μL tip in a direction perpendicular to the agar surface (*see* **Note 6**).
4. Depress the plunger to release the micro-agar plug into the wells of the 12-well culture plate (*see* **Note 7**).
5. After inoculation, pour 3 mL of sterile PDB into each well and wrap it with micropore tape.
6. Incubate the culture plate at 28 °C for 24 h at 180 rpm shaking incubator.

3.3 Treatment of M. phaseolina *with the AFA and Controls*

1. Perform the subsequent steps inside a laminar airflow chamber with each equipment wiped with 70% ethanol.
2. Select the germinated micro-agar plugs that have similar mycelial morphology.
3. Remove the broth from the culture plate using a 1 mL pipette (*see* **Note 8**).

ochrochloron under glucose-limited conditions allowed for immediate glucose intake at higher rates from a glucose pulse compared to steady-state conditions [8]. In addition to the efflux of acidic metabolites, activation of plasma membrane H^+-ATPase by glucose contributes to about one-half of extracellular acidification [9]. Inhibition of plasma membrane Pma1 (H^+-ATPase) disrupts the electrochemical gradient, nutrient transporters, and extracellular acidification in many fungi [10].

1.3 Phytopathogenic Aspects of Extracellular Acidification

Pathogenic fungi such as *Sclerotinia* and *Botrytis* sp. produce an acidic environment that causes a shift from saprotrophic to necrotrophic growth in plants [11]. In *Fusarium sp*, fusaric acid acts as a virulence factor during the early stages of the disease [12]. Growth of *M. phaseolina* in the presence of host tissue triggers azelaic acid production, potentially aiding in disease establishment [13]. Endoglucanases catalyzing the hydrolysis of cellulose to glucose, and malate dehydrogenase causing organic acid production, were found in the secretome study of *M. phaseolina* [14]. Additionally, several investigations demonstrated the significance of plasma membrane H^+-ATPases during infection and pathogenesis [15, 16].

1.4 Dry Root Rot Disease in Chickpea

M. phaseolina propagates through dark-colored, branching hyphae with black microsclerotia [13]. Infected roots when split exhibit characteristic dark spots in the infected regions, which are microsclerotia and serve as the best parameter to identify the disease under the field conditions [17]. Several surveys and field investigations indicate that drought and high temperatures increase the prevalence and severity of DRR disease [17–21]. Moreover, they usually occur as mixed infections usually with wilt [22]. Dry root rot (DRR) of chickpea had its impact over the 13 million hectares of chickpea-growing regions worldwide leading to a significant reduction in yield of 30 to 50 per cent under moderate infection, and up to 80 per cent yield reductions in susceptible genotypes [17, 18, 21, 23].

High resistance levels are challenging to establish because of the pathogen's wide host range. None of the lines were fully resistant to DRR, according to a search for particular resistance in chickpea [24]. Therefore, the reliability of the effective sustainable metabolites and proteins increases in disease control. Although numerous fungi, bacteria, and plants have been found to contain antifungal metabolites/proteins that are effective against phytopathogenic fungi, their characterization is limited by diverse testing methods usually lacking standardized conditions and reproducibility.

1.5 Common Methods Used for Screening Antifungal Compounds

The search for natural agents with antifungal activity has increased, but testing methods present many challenges. Techniques like agar diffusion, macrodilution, and microdilution are used to determine antifungal activity. Agar diffusion is semi-quantitative and involves applying a sample to an agar surface inoculated with fungal cells. Macrodilution and microdilution provide quantitative data about the Minimum Inhibitory Concentration of a compound. Colorimetric methods like MTT (3-(4,5-dimethylthiazol-2-yl)-2,5-diphenyltetrazolium bromide) and XTT (2,3-bis-(2-methoxy-4-nitro-5-sulphenyl)-(2H)-tetrazolium-5-carboxanilide) assays, are also used to estimate fungal biomass and metabolic activity.

Obtaining intact *M. phaseolina* inoculum across treatments is difficult due to its nature of propagation in synthetic media, making subsequent impact assessment challenging. Moreover, the existing methods are usually expensive, time-consuming, and cause stress to the fungus during sample preparation and processing. This assay tests the effect of an antifungal agent (AFA) on cultured mycelium using extracellular pH measurements after treatments in the presence of glucose (Fig. 2). The pH reductions can be monitored using a laboratory pH meter [25].

2 Materials

2.1 Fungal Sample Preparation

1. Potato dextrose agar (PDA).
2. 90 mm sterile disposable petri plates
3. 20 mL test tubes
4. Standard 1 mL micropipette and tips.
5. Parafilm.
6. 70% ethanol
7. Pure culture of *M. phaseolina* in PDA.
8. Laminar airflow chamber.
9. 28 °C incubator.

2.2 Growth in Culture Plate

1. Potato dextrose broth (PDB).
2. Flat bottom 12-well culture plate.
3. Standard 200 μL micropipette and tips.
4. L-rod.
5. Micropore tape.
6. 28 °C shaking incubator
7. Laminar airflow chamber.

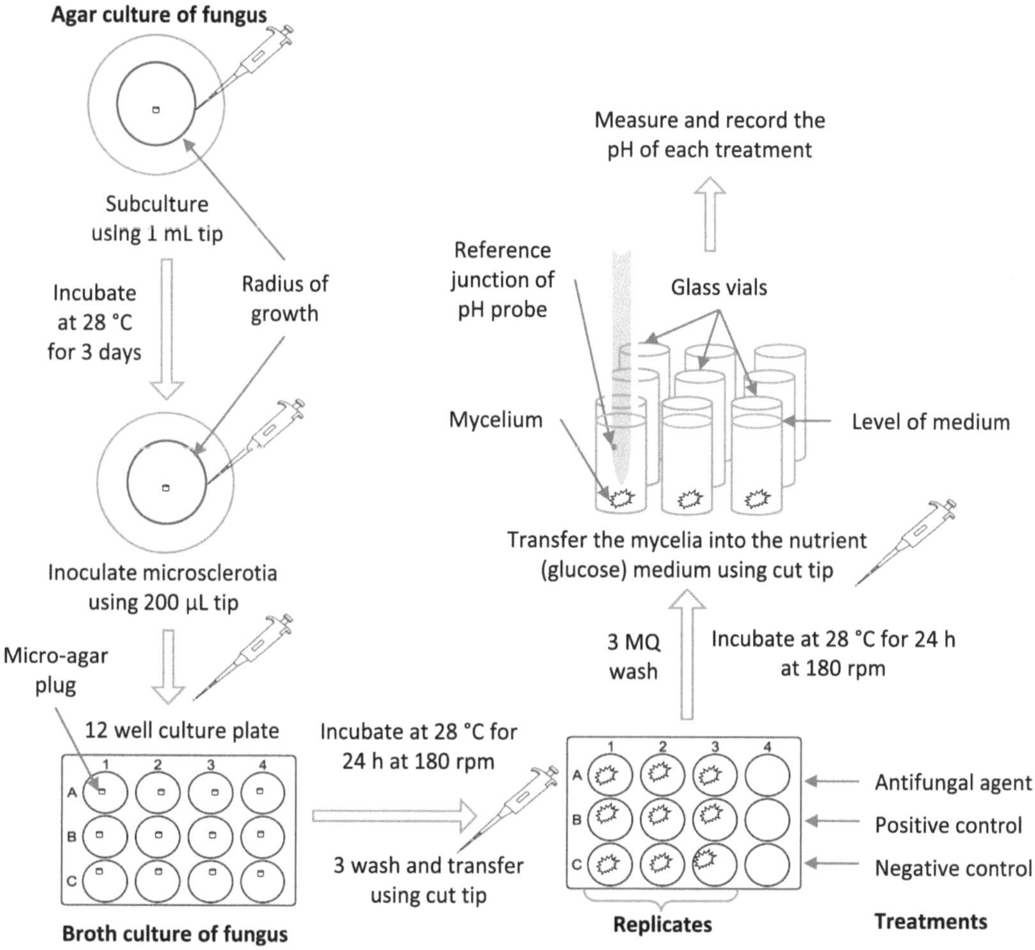

Fig. 2 Extracellular acidification assay protocol. The uniformity of agar culture ensures an equal size of micro-agar plug inoculum during subcultures. The microsclerotia collected from a fixed radius of the agar culture are allowed to germinate in the subsequent broth culture. The washing step before and after the treatments removes undesired residual matter, and the cut tips ensure the mycelial integrity between transfers. The mycelium is suspended in an appropriate medium for measuring the extent of acidification using a pH meter. MQ: Milli Q water, rpm: revolutions per minute

2.3 Treatments and AFA to Be Tested

1. Test AFA (*see* **Note 1**).
2. 0.2 μm filter compatible with the AFA used and 10 mL syringe
3. 1 mL micropipette and tip
4. Micropore tape.
5. Positive control and Negative control (*see* **Note 2**).

2.4 Nutrient Source and pH Measurement

1. 2% D—glucose (anhydrous) solution in Milli-Q® (MQ) water
2. Autoclaved MQ.
3. 0.2 μm filter and 10 mL syringe
4. Autoclaved 15 mL glass vials.

5. MQ in wash bottle and tissue paper.
6. pH meter (*see* **Note 3**).

3 Methods

3.1 M. phasolina *Agar Culture*

1. Autoclave the desired number of test tubes with 10 mL of PDA medium (*see* **Note 4**).
2. Perform the subsequent steps inside a laminar airflow chamber with all equipment wiped with 70% ethanol.
3. Pour the molten PDA agar into the sterile petri plates, and let it solidify and dry properly.
4. Ensure that the surface of the pure culture on the medium is uniform (*see* **Note 5**).
5. Pick the microsclerotia each time from a definite radius of the dark-colored region to ensure uniform inoculum.
6. Stab the pure culture using a 1 mL tip with a micropipette in a direction perpendicular to the agar surface (*see* **Note 6**).
7. Depress the plunger by keeping the tip close to the center of the freshly prepared agar plate (*see* **Note 7**).
8. Seal the petri plate twice with parafilm to avoid contamination and incubate it at 28 °C for 3 days.
9. Perform the above inoculation three times to obtain biological replicates.

3.2 M. phaseolina *Broth Culture*

1. Perform the subsequent steps inside a laminar airflow chamber with each equipment wiped with 70% ethanol.
2. Follow **steps 4** and **5** from Subheading 3.1.
3. Stab the pure culture using a 200 μL tip in a direction perpendicular to the agar surface (*see* **Note 6**).
4. Depress the plunger to release the micro-agar plug into the wells of the 12-well culture plate (*see* **Note 7**).
5. After inoculation, pour 3 mL of sterile PDB into each well and wrap it with micropore tape.
6. Incubate the culture plate at 28 °C for 24 h at 180 rpm shaking incubator.

3.3 Treatment of M. phaseolina *with the AFA and Controls*

1. Perform the subsequent steps inside a laminar airflow chamber with each equipment wiped with 70% ethanol.
2. Select the germinated micro-agar plugs that have similar mycelial morphology.
3. Remove the broth from the culture plate using a 1 mL pipette (*see* **Note 8**).

4. Wash the mycelia 3 times in the culture plate for 5 minutes at a 180 rpm shaker and suspend them in 3 mL of respective solutions (*see* **Note 9**).
5. Transfer the mycelia into a new 12-well culture plate using a 1 mL cut tip (*see* **Note 10**).
6. Add 3 mL each of filter-sterilized test AFA, positive control, and negative control solutions with three replicates (*see* **Note 11**).
7. Seal the culture plate with micropore tape and incubate the culture plate at 28 °C for 24 h at 180 rpm shaking incubator.

3.4 Measurement of the Glucose-Induced Acidification

1. Remove the solutions from the culture plate and wash the treated mycelia three times with 3 mL of MQ as in **steps 3** and **4** of Subheading 3.3, and suspend them in MQ until further process.
2. Add 10 mL of filter-sterilized glucose solution to the glass vials.
3. Transfer the treated mycelia to the glass vials using a 1 mL cut tip (*see* **Note 10**).
4. Mix the contents at 180 rpm between measurements.
5. Dip the pH probe into glass vials for the measurements (*see* **Note 12**).
6. Record the measurements at 0 and 60 min, wash the probe with MQ, and blot dry it between replicates (*see* **Note 13**).
7. Repeat Subheadings 3.2, 3.3, and 3.4 for the biological replicates.

3.5 Determination of the Effective AFA Concentration

1. Obtain Δ pH values for each treatment and control sample by subtracting the final values from the initial values.
2. Create an XY data table. Enter the concentrations of the AFA into X. Enter the average of Δ pH of replicates into Y.
3. From the data table, click Analyze, choose nonlinear regression, choose the panel of equations "Dose-response curves—Inhibition," and then choose the equation "[Inhibitor] vs. response—Variable slope" (*see* **Note 14**).
4. Enter the average Δ pH of negative and positive control as top and bottom constraints ("type: constant equal to") values respectively in the option and plot the dose-response curve (Fig. 3).
5. Obtain the metadata of IC_{50} analysis from the result section (*see* **Note 15**).

3.6 In Vivo Evaluation of Fungal Pathogenicity

The fungal sample post-treatments can be used for *in planta* experiments by employing the blotting paper method [26].

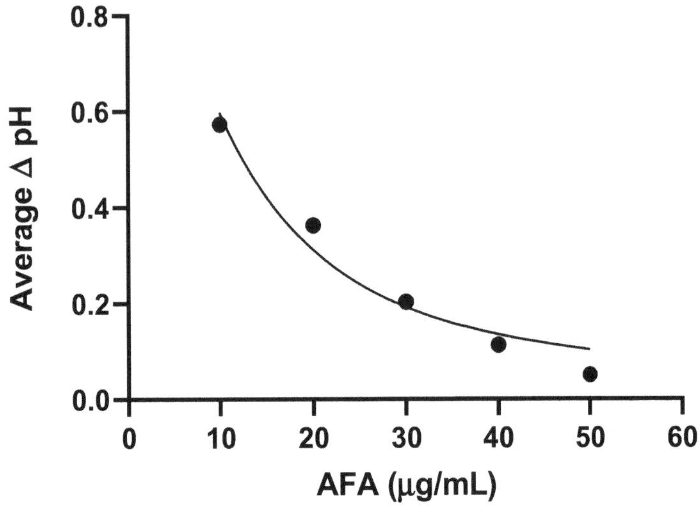

Fig. 3 Dose-response curve. The curve represents the average of the Δ pH initial-final values for each treatment, that is, the extent of acidification as the response to the individual concentrations of AFA used in the assay. The variable slope model used here fits the Hill slope from the data provided, rather than assuming a standard slope. The software output determines the IC_{50} of the AFA, which is the concentration that elicits a response halfway between the maximum response (Top) of the negative control and the maximum inhibition response (Bottom) of the positive control AFA: Antifungal agent; IC_{50}: Inhibitory concentration showing 50% of response

1. Surface-sterilize the chickpea seeds with 2% sodium hypochlorite.
2. Sow seeds in a 15 cm height pot containing soilrite.
3. Uproot 8-day-old seedlings, wash under running tap water, and rinse with sterile water twice.
4. Arrange the plants over the blotter paper in trays and add sufficient water.
5. Transfer an equal number of fungal samples after treatment into each main root of the plants using a 1 mL cut tip.
6. Place the trays with the plants at 28 °C for 8 days under a 16 h photoperiod and 70% relative humidity, moistening the plant daily with sufficient water (*see* **Note 16**).
7. Calculate the disease susceptibility score/disease severity index and analyze the impact of AFA on the pathogenicity of the fungus (*see* **Note 17**).

4 Notes

1. Solvents without inhibitory effect on the fungus can be used to test multiple AFAs. An antifungal protein dissolved in 1 × PBS was used in this chapter for representation.
2. Positive control: A commercially available AFA that shows maximum inhibition response. Negative control: solvent for dissolving AFA.
3. Use a pH meter with a combination electrode probe and a resolution of 0.01 or more.
4. The uniform agar plate preparation will ensure an equal size of the subsequent micro-agar plug inoculum.
5. If the culture contains aerial mycelial growth, remove it using an L-rod, and follow the same step for each subculture.
6. Ensure proper insertion of a single micro-agar plug inside the tip during each transfer.
7. Use a fresh tip whenever there is more than 1 micro-agar plug inside the tip or it is not being released.
8. Tilt the plate 45° and remove the broth from both sides of the mycelia without touching it. Tilt it to the opposite side and remove the excess broth.
9. Use appropriate solvents to wash the positive control, negative control, and test samples.
10. This prevents mycelial damage between transfers. Press the plunger, place the cut tip over the center of the mycelium, and gently release till the entire mycelium is inside the tip. Press the plunger to transfer.
11. Optimize the range of AFA concentrations (ten-fold dilutions for initial screening and two-fold dilutions for accuracy) and number of replicates for the experiment to obtain a reliable dose-response curve.
12. Ensure that the porous reference junction of the probe is dipped inside the test solution along with the glass bulb before measurement.
13. Select the time points based on experiment needs. Avoid rubbing the sensitive glass blub of the probe. Use tissue paper to remove the excess water.
14. The statistical analysis was performed using GraphPad Prism version 10.2.0 for Windows, GraphPad Software, Boston, Massachusetts USA, for further details visit www.graphpad.com.

15. The IC_{50} value denotes the concentration of AFA that elicits 50% of the response. This value can be used to compare multiple AFAs against the pathogen.
16. Choose the method of inoculation and day of plant sample collection based on the phytopathogen under study.
17. The AFA concentration that exhibits the optimum inhibition and simultaneously shows the least infection in the plant can be selected for downstream applications.

Acknowledgments

Projects at M. S. K. lab are funded by the National Institute of Plant Genome Research core funding. We acknowledge DBT-eLibrary and NIPGR library for providing access to e-resources.

References

1. Walker GM, White NA (2017) Introduction to fungal physiology. In: Fungi. Wiley, pp 1–35
2. Casey JR, Grinstein S, Orlowski J (2010) Sensors and regulators of intracellular pH. Nat Rev Mol Cell Biol 11:50–61. https://doi.org/10.1038/nrm2820
3. Kane PM (2016) Proton transport and pH control in fungi. In: Ramos J, Sychrová H, Kschischo M (eds) Yeast membrane transport. Springer International Publishing, Cham, pp 33–68
4. Islam MS, Haque MS, Islam MM et al (2012) Tools to kill: genome of one of the most destructive plant pathogenic fungi Macrophomina phaseolina. BMC Genomics 13:493. https://doi.org/10.1186/1471-2164-13-493
5. Johnston M (1999) Feasting, fasting and fermenting: glucose sensing in yeast and other cells. Trends Genet 15:29–33. https://doi.org/10.1016/S0168-9525(98)01637-0
6. Conrad M, Schothorst J, Kankipati HN et al (2014) Nutrient sensing and signaling in the yeast Saccharomyces cerevisiae. FEMS Microbiol Rev 38:254–299. https://doi.org/10.1111/1574-6976.12065
7. Vankuyk PA, Diderich JA et al (2004) Aspergillus Niger mstA encodes a high-affinity sugar/H+ symporter which is regulated in response to extracellular pH. Biochem J 379: 375–383. https://doi.org/10.1042/bj20030624
8. Vrabl P, Mutschlechner W, Burgstaller W (2008) Characteristics of glucose uptake by glucose- and NH4-limited grown Penicillium ochrochloron at low, medium and high glucose concentration. Fungal Genet Biol 45:1380–1392. https://doi.org/10.1016/j.fgb.2008.07.017
9. Kotyk A, Lapathitis G, Křenková Š (1999) Glucose- and K+-induced acidification in different yeast species. Folia Microbiol 44:295–298. https://doi.org/10.1007/BF02818550
10. Kjellerup L, Gordon S, Cohrt KO et al (2017) Identification of antifungal H+-ATPase inhibitors with effect on plasma membrane potential. Antimicrob Agents Chemother 61. https://doi.org/10.1128/aac.00032-17
11. Vylkova S (2017) Environmental pH modulation by pathogenic fungi as a strategy to conquer the host. PLoS Pathog 13:e1006149. https://doi.org/10.1371/journal.ppat.1006149
12. Liu S, Li J, Zhang Y et al (2020) Fusaric acid instigates the invasion of banana by Fusarium oxysporum f. sp. cubense TR4. New Phytol 225:913–929. https://doi.org/10.1111/nph.16193
13. Marquez N, Giachero ML, Declerck S, Ducasse DA (2021) Macrophomina phaseolina: general characteristics of pathogenicity and methods of control. Front Plant Sci 12. https://doi.org/10.3389/fpls.2021.634397
14. Sinha N, Patra SK, Ghosh S (2022) Secretome analysis of Macrophomina phaseolina identifies an array of putative virulence factors responsible for charcoal rot disease in plants. Front Microbiol 13. https://doi.org/10.3389/fmicb.2022.847832

15. Vázquez-Carrada M, Feldbrügge M, Olicón-Hernández DR et al (2022) Functional analysis of the plasma membrane H+-ATPases of Ustilago maydis. Journal of Fungi 8:550. https://doi.org/10.3390/jof8060550
16. Wu L, Yuan Z, Wang P et al (2022) The plasma membrane H+-ATPase FgPMA1 regulates the development, pathogenicity, and phenamacril sensitivity of Fusarium graminearum by interacting with FgMyo-5 and FgBmh2. Mol Plant Pathol 23:489–502. https://doi.org/10.1111/mpp.13173
17. Sinha R, Irulappan V, Mohan-Raju B et al (2019) Impact of drought stress on simultaneously occurring pathogen infection in field-grown chickpea. Sci Rep 9:5577. https://doi.org/10.1038/s41598-019-41463-z
18. Chilakala AR, Mali KV, Irulappan V et al (2022) Combined drought and heat stress influences the root water relation and determine the dry root rot disease development under field conditions: a study using contrasting chickpea genotypes. Front Plant Sci 13. https://doi.org/10.3389/fpls.2022.890551
19. Irulappan V, Kandpal M, Saini K et al (2022) Drought stress exacerbates fungal colonization and endodermal invasion and dampens defense responses to increase dry root rot in chickpea. MPMI 35:583–591. https://doi.org/10.1094/MPMI-07-21-0195-FI
20. Sharma M, Pande S (2013) Unravelling effects of temperature and soil moisture stress response on development of dry root rot [Rhizoctonia bataticola (Taub.)] Butler in chickpea. Am J Plant Sci 4:584–589. https://doi.org/10.4236/ajps.2013.43076
21. Sinha R, Irulappan V, Patil BS et al (2021) Low soil moisture predisposes field-grown chickpea plants to dry root rot disease: evidence from simulation modeling and correlation analysis. Sci Rep 11:6568. https://doi.org/10.1038/s41598-021-85928-6
22. Chilakala AR, Pandey P, Durgadevi A et al (2023) Drought attenuates plant responses to multiple rhizospheric pathogens: a study on a dry root rot-associated disease complex in chickpea fields. Field Crop Res 298:108965. https://doi.org/10.1016/j.fcr.2023.108965
23. Rai A, Irulappan V, Senthil-Kumar M (2022) Dry root rot of chickpea: a disease favored by drought. Plant Dis 106:346–356. https://doi.org/10.1094/PDIS-07-21-1410-FE
24. Sharma M, Ghosh R, Pande S (2015) Dry root rot (Rhizoctonia bataticola (Taub.) Butler): an emerging disease of chickpea – where do we stand? Arch Phytopathol Plant Protect 48:797–812. https://doi.org/10.1080/03235408.2016.1140564
25. Ghosh P, Roy A, Hess D et al (2015) Deciphering the mode of action of a mutant Allium sativum leaf agglutinin (mASAL), a potent antifungal protein on Rhizoctonia solani. BMC Microbiol 15:237. https://doi.org/10.1186/s12866-015-0549-7
26. Irulappan V, Senthil-Kumar M (2021) Dry root rot disease assays in chickpea: a detailed methodology. JoVE:e61702. https://doi.org/10.3791/61702

Chapter 14

A Guide to Culturing, Maintenance, and Leaf Inoculation Methods for Rapid Screening and Quantification of *Sclerotinia sclerotiorum* Infection in Mustard

Bhanu Malhotra, Ruchi Tiwari, Mohan Varghese, and Naveen C. Bisht

Abstract

Sclerotinia rot poses a significant challenge to the cultivation of oilseed *Brassica* crops, causing the reduction of seed number, weight, and quality leading to >95% yield losses globally. Due to its substantial economic impact on agriculture, *Sclerotinia sclerotiorum* has been extensively studied at the molecular level and has now been adopted as a model to investigate the host-pathogen interactions. Despite various procedures and strategies described in different reports, there exists a considerable disparity in how the pathogen is cultured, maintained, and manipulated across studies.

This chapter provides a comprehensive guide to the fundamental procedures of working with *S. sclerotiorum*. It describes step-by-step methods for its routine culturing and maintenance over artificial media, replenishing and storing laboratory stocks, and conducting detached leaf assays—a robust method used for screening the mustard cultivars. Additionally, we outline the quantification of pathogen load *in planta* through qRT-PCR analysis. The methods presented in the current chapter are reproducible and can be suitably applied when working with different isolates of the pathogen.

Through the present compilation, we aim to address the research gaps in methods involving the pathogen that would not only benefit the new researchers entering the field but also future research endeavors on white mold.

Key words *Brassica* crops, Culture and maintenance, Detached leaf inoculation, Disease screening, Fungal quantitation, Mustard, *Sclerotinia sclerotiorum*, White mold

1 Introduction

The necrotroph *Sclerotinia sclerotiorum* is a devastating fungal pathogen with a dynamically broad host range that infects more than 400 plant species worldwide including many economically important crops like mustard, oilseed rape, sunflower, soybean, tomato, chickpea, and many others [1]. Several factors such as the long-term persistence of the pathogens sclerotia in soil, wider host

Muthappa Senthil-Kumar and Hema Ramanna (eds.), *Plant Biotic Stress: Methods and Protocols*, Methods in Molecular Biology, vol. 2966, https://doi.org/10.1007/978-1-0716-4746-2_14,

range, existence of genetically heterogenous populations in fields, production of airborne ascospores, ineffective fungicidal control due to its soilborne nature, and complex mode of host infection contribute to the success of this notorious pathogen [2]. All the above factors coupled with a lack of adequate host genetic resistance enable it to cause extensive crop damage having a major impact on global agriculture [1, 3]. The agricultural damages inflicted by *S. sclerotiorum* have stimulated extensive research on this necrotroph and its hosts, which is further substantiated by the availability of its high-quality genome sequence and amenability to genetic manipulation [4–6]. Compatible interactions of *S. sclerotiorum* with several different hosts have been studied with pioneering work involving different aspects of pathogen infection on plants such as soybean, lettuce, potato, chickpea, tomato, sunflower, oilseed rape, and mustard [3, 7]. Among them, the occurrence of *Sclerotinia* rot in crops of the genus *Brassica* has received considerable attention and the *Sclerotinia-Brassica* pathosystem is an emerging model for studying the fungal pathogenesis and different aspects of underlying plant immunity. White mold infection has been a major limiting factor in the cultivation of *Brassica* species like mustard and oilseed rape, both in India and worldwide with >90% losses reported across different cultivars along with a qualitative and drastic deterioration of the seed's oil and protein contents [2, 3, 8]. The characteristic white mold symptoms evident during different stages of natural *S. sclerotiorum* infection and those appearing after artificial pathogen inoculation in the Indian mustard fields are illustrated (Fig. 1).

1.1 Sclerotinia *Rot: A Menace of* Brassica *Crops*

The utilization of different cultural and chemical practices like crop rotation, irrigation, and fungicide application to combat white mold enables the reduction in disease severity, but these approaches are hazardous, costly, tedious, time-consuming, and overall ineffective due to the long-term survivability of the pathogens sclerotia in soil [2]. Additionally, changing climatic conditions coupled with modern agricultural practices have further increased the risk of developing *Sclerotinia* rot epidemics by enabling the accumulation of a high pathogen load across fields [9]. Therefore, gaining major significance *Sclerotinia* rot is considered one of the most serious problems of mustard at the global level, with the effective management of this disease still lacking [8, 10].

The most appropriate and promising approach for proper and sustainable management of white mold is developing genetic improvements by obtaining adequate host plant resistance. Although partial resistance to *Sclerotinia* rot has been identified and reported across different *Brassica* crops including various reports in oilseed rape and very few in mustard, overall, resistance to *Sclerotinia* is polygenic, highly complex, and is likely governed by a continuum interplay of several minor effect genes [3, 11,

Fig. 1 *Sclerotinia* rot infection of oilseed mustard. (**a**) Carpogenic germination of sclerotia releases ascospores that after dispersal land on decaying flower petals or senescing plant tissues that are fallen and stuck on the leaf or different plant parts. (**b**) Ascospores-infected petals land on healthy leaves to cause penetration and infection of green tissues. (**c**) Appearance of characteristic necrotic-like symptoms as *S. sclerotiorum* infection progresses in the leaves. (**d**) Infected petals also invade the region of leaf axils from where they spread to the stem. (**e**) As the pathogen grows in the stem, it causes the appearance of characteristic white, cottony-like symptoms often resulting in bleaching of the stem and branches. (**f**) Gradually spreading the fungus girdles through the stem results in wilting of the leaves and the other upper ground parts of the plant. (**g**) Direct myceliogenic germination of *S. sclerotiorum* can also be evidenced under certain environmental conditions to directly produce the mycelium that attacks the ground-level plant tissues. (**h**) The agar plug-based inoculation of *S. sclerotiorum* on the mustard stem in sick plots of the experimental field. (**i**) The white mold disease progression and observance of sclerotia formation on the artificially inoculated stems. (**j**)–(**k**) The *S. sclerotiorum* mycelium found growing and spreading inside the inoculated stem tissue

12]. Thus, at present, there is a complete lack of robust and reliable resistance genotypes of mustard against the white mold pathogen. As a first step in this direction, identification of the elite cultivars and reliable sources of resistance in mustard and other *Brassica* species is the most pressing need of the hour in control of this devastating pathogen. This involves and necessitates the large-scale screening of different germplasms (or cultivars) with different pathotypes/isolates of the pathogen. Therefore, the adoption of standard common laboratory procedures that are reproducible, time-saving, and robust is critical to developing new approaches for future research on *Sclerotinia* rot management. Although various methods of culturing, maintenance, and host inoculation of *S. sclerotiorum* are available and reported, they have not been well-collated and systematically compiled. In the current chapter, utilizing the *S. sclerotiorum* isolate available at our laboratory, we describe detailed stepwise procedures involving the routine cultivation of the pathogen over artificial media, key steps for its laboratory maintenance, revival from the host, preservation, and storage of its laboratory stocks, detached leaf inoculation assay followed by *in planta* fungal quantitation on mustard plants by sharing protocols routinely used by our laboratory.

1.2 Factors Influencing Screening and Identification of Resistant Sources

Different inoculation methods have been used to screen and identify tolerant mustard cultivars to the necrotrophic *S. sclerotiorum* in the past. Commonly used assays include petiole inoculation, cotyledon inoculation, toothpick infection, intact plant inoculation, noninjury screening, and detached leaf inoculation [13–17]. Among these, detached tissue inoculation emerged as one of the robust, fast, and reliable screening approaches that provides a wide applicability in resistance breeding programs against *S. sclerotiorum* and have been extensively utilized to screen the crops of the *Brassica* genus. The identification of white mold resistance in *Brassicas* is challenging and complicated due to the existence of a pathogen population (different pathotypes/isolates) having a varying degree of aggressivity on the host in the field [18]. Therefore, an insight into the variability of the pathogen population is important and needed for the effective screening of the genotypes. Under such circumstances, the tolerance or resistance obtained post-screening with a single isolate may not necessarily reproduce and correlate with the resistance obtained from the field where naturally many pathotypes exist [19, 20]. Therefore, different applications often demand screening of plants against various pathogen isolates whose aggressivity on the test population ranges from low, moderate, to high. The methods described in the chapter can also be suitably and reliably applied to applications involving work with different isolates of *S. sclerotiorum* as desired.

Furthermore, the choice of inoculum from the pathogen is also known to affect the disease severity. For the identification of potential resistant cultivars against *S. sclerotiorum*, the ideal form of inoculum to be used for pathogen bioassays would be ascospores that are responsible for the induction of *Sclerotinia* rot under field conditions [1]. However, the utilization of ascospores as a form of plant inoculum is quite challenging and limited due to the difficulty in routinely producing and maintaining them for longer durations under laboratory conditions. Moreover, an artificial inoculation with ascospores from the pathogen may often lead to inconsistency in infection, and hence alter results across different sets of an experiment [21]. For proper germination ascospores require senescing petals or plant tissue as an appropriate nutrient source which must be carefully ensured and consistently looked at throughout the experiments. Therefore, the *S. sclerotiorum* mycelium has been widely used as the preferred choice of inoculum in different bioassays involving the pathogen and for different applications of plant screening [13, 17].

Scoring the length of lesions obtained after a particular time of pathogen infection is most often used to assess the damage caused by it at the individual plant level since lesion expansion is positively correlated with disease severity, the form of plant damage, the amount of secondary inoculum produced, and with the collapse at the field level. Hence, direct estimation of the lesion sizes on different tissue types is widely utilized as a method to assess quantitative disease resistance against *S. sclerotiorum* [14]. Studies involving the use of other pathogen inoculation methods indicated that the comparison between these methods can be limiting under several instances. Wherein, plant screening with detached leaf assays has shown to possess a wider applicability and comparability not only in *Brassicas* as well as when different *S. sclerotiorum* hosts are considered. Additionally, in studies where a detached assay is applied in conjunction with other screening approaches estimation of lesion sizes after a detached tissue inoculation was reported as the most effective and unbiased screening approach for the identification of host genotypes in terms of resistance and susceptibility to *S. sclerotiorum* [22]. The chapter describes the estimation and evaluation of lesions obtained after detached tissue inoculation involving the leaf tissue of Indian mustard followed by a quantitative real-time PCR assay (qRT-PCR) for the relative quantitation of pathogen load *in planta*. Sharing of these reliable procedures would allow greater accuracy and reproducibility within experiments and result in uniformity within the *Sclerotinia rot* community, aiding the development of appropriate strategies to control the pathogen.

2 Materials

2.1 General Materials and Requirements

1. Absorbent blotting paper sheet: 0.5 mm thickness.
2. Biodegradable saran/cling wrap roll.
3. Black cloth/black paper: for photography.
4. Camera: Canon EOS 750D DSLR series lens 18–55 mm, 24.2 MP for photography (or any suitable camera for photography).
5. CelloVet cellophane sheet: Cut into 3 cm × 3 cm size, soaked in distilled water, and sterilized by autoclaving.

6. Cotton balls/polyethylene bags: to cap detached leaf/stem ends.
7. Cryotubes: 1.8 mL capacity.
8. Drierite desiccant: 4 mesh with indicator.
9. High-capacity cDNA reverse transcription kit: or any suitable kit as desired.
10. Liquid nitrogen.
11. Micro seals for 96-well plates.
12. Mortar and pestle (for use in RNA isolation).
13. Plastic pots: 9 cm (D) × 10 cm (H) round pots for plant cultivation (or larger size pots as needed).
14. Real-time PCR plates: 96-well white semi-skirted clear shell/clear well plates.
15. Round petri plates: 90 mm diameter petri dish radiation sterilized.
16. Silica gel desiccant beads: 2.5–6.0 mm size self-indicating beads.
17. Solerite mix: perlite 1:3, Irish peat moss 1:3, vermiculite 1:3, 25-kg package.
18. Spectrum plant total RNA kit (Catalog No: STRN 50 Sigma Aldrich, Saint Louis, USA): or any suitable kit as desired.
19. Sterile disposable petri plates: 120 mm × 120 mm square-shaped clear and transparent sterile plates.
20. Sterile paper bag: 30 mm × 50 mm for storage of mycelia filter paper stocks.
21. Whatman 3 MM qualitative grade filter paper sheet: cut into 2 × 2 mm size wrapped in aluminum foil and autoclaved before use for mycelial storage.
22. Whatman grade 1 qualitative filter paper sheet: 570 mm (L) × 460 mm (W) cut into 80 mm diameter circle wrapped in aluminum foil and autoclaved before use to fit the 90 mm round petri dish.

2.2 Plant and Pathogen Materials

1. *Brassica juncea* cultivar Varuna, the extensively cultivated mega variety of the Indian mustard was obtained from the University of Delhi South Campus (Department of Genetics), New Delhi, India and utilized for all the experiments.
2. *Sclerotinia sclerotiorum* strain SSD1 (Delhi isolate-1) kindly provided by Prof. Jagreet Kaur (University of Delhi South Campus, New Delhi, India) was maintained and utilized for all the experiments.

2.3 Reagents and Chemicals

1. 0.1% (v/v) diethyl pyrocarbonate solution
2. 0.1% mercuric chloride solution (w/v) for host surface sterilization
3. 1% sodium hypochlorite solution (v/v) for pathogen sclerotia and host surface sterilization
4. 10% glycerol solution (v/v) sterilized by autoclaving for fungal stock preservation
5. 70% ethanol solution (v/v) for decontaminating materials, surfaces, equipment, pathogen sclerotia, and host surface sterilization
6. Agarose.
7. DNA ladders.
8. Ethidium bromide.
9. Ethylenediaminetetraacetic acid disodium salt dehydrate.
10. Formaldehyde.
11. Formamide.
12. Glycerol.
13. iTaq Universal SYBR Green Supermix.
14. Milli-Q water autoclaved for use.
15. Mixed bed resin.
16. MOPS.
17. Nuclease free water autoclaved for use.
18. Other laboratory chemicals for use (as indicated in Subheading 2.4 for Hoagland solution preparation and others).
19. RNaseZAP.
20. Sodium acetate anhydrous.

2.4 Media and Buffer Composition

2.4.1 Plant Growth

1. Hoagland nutrient solution: Prepare individual stocks of the following six components 1 M KNO_3, 1 M $MgSO_4$, 1 M KH_2PO_4 (pH 5.5), 1 M $Ca(NO_3)_2$, 20 mM Fe-EDTA, and the micronutrients (composition mix of 70 mM H_3BO_3, 0.5 mM $CuSO_4$, 14 mM $MnCl_2$, 1 mM $ZnSO_4$, 10 mM NaCl, 0.2 mM $NaMoO_4$, 0.01 mM $CoCl_2$) in glass bottles. Individual stock solutions can be preserved for longer duration by storing at 4 °C. Prepare a 1× strength of the nutrient solution by mixing: 5 mL/L of 1 M KNO_3, 2.5 mL/L each of 1 M KH_2PO_4 pH 5.5 and 20 mM Fe-EDTA, 2 mL/L each of 1 M $Ca(NO_3)_2$, and 1 M $MgSO_4$ and 1 mL/L of micronutrients. Maintain plants by supplementing with a 1/fourth strength of Hoagland nutrient solution twice a week.

2.4.2 Pathogen Cultivation

1. Liquid medium: Potato dextrose broth granulated. Suspend 24.0 g/L in Milli-Q water, autoclave to sterilization, standard pH range: 5.1 ± 0.2.
2. Solid medium: Potato dextrose agar granulated. Suspend 39.0 g/L in Milli-Q water, autoclave to sterilization, standard pH range: 5.6 ± 0.2. PDA plates for culturing *S. sclerotiorum* can either be freshly prepared or preprepared (1 to 2 days before) for laying solid cultures. Preprepared plates can be stored at 4 °C and must be normalized to 22 ± 2 °C before laying pathogen cultures. Generally, 25 mL of PDA media sterilized by autoclaving should be poured into each 90 mm round plate inside a hood and solidified. Additionally, alterations in pathogen growth over solid media can be obtained by preparing thicker and thinner PDA plates by adjusting and pouring different volumes of PDA as desired.

2.4.3 Qualitative Analysis of RNA

1. 5× MOPS buffer: 0.2 M MOPS pH 7.0, 0.05 M sodium acetate, 0.005 M EDTA pH 8.0. For 2 L of MOPS buffer mix, 83.72 g of MOPS, 8.23 g sodium acetate, and 20 mL of DEPC-treated 0.5 M EDTA, adjust the pH to 7.0, and make up the final volume to 2 L with DEPC-treated water. Autoclave the solution, it turns yellow but this does not affect the buffer quality.
2. Deionized formamide: Add mixed-bed resins to formamide and stir them at room temperature for 1 h. Filter twice through the Whatman grade 1 filter paper. Store at −80 °C.
3. RNA sample buffer: Add 10.0 mL of deionized formamide, 3.5 mL of 37% formaldehyde, 2.0 mL of 5× MOPS buffer. Aliquot in 1.5 mL microcentrifuge tubes and store at −20 °C.
4. RNA loading buffer: 50% glycerol, 1 mM EDTA, 0.4% bromophenol blue. Prepare in DEPC-treated water. Aliquot and store at −20 °C.

2.5 Equipment and Other Requirements

1. Bunsen burner/spirit lamp.
2. Centrifuge.
3. Controlled growth chamber/greenhouse/experimental field facility (as required).
4. Cork borer: 4 mm (D) (or larger size as needed).
5. Dissecting forceps: Approximately 5 in. (L).
6. DNA primers for *B. juncea TIPS-41* gene as the endogenous reference control in qRT-PCR: FP RT-5′ TGAAGAGCAGATT GATTTGGCT 3′ and RT RP-5′ ACACTCCATTGTCAGC CAGTT 3′ (as described in Subheading 3.5.5).
7. DNA primers for *S. sclerotiorum Ss Histone 3* (SS1G_09608) gene as the endogenous reference control in qRT-PCR: FP

RT-5′ GGCTCGTACCAAGCAAACTG 3′and RP RT-5′ GA AGTCTTGGGCGATTTCAC 3′ (as described in Subheading 3.5.5 and Table 1).

8. Dry bath/heating block.
9. Electrophoresis gel apparatus and power pack.
10. Filtration unit: benchtop vacuum filter for separation of sclerotia from liquid mycelial cultures.
11. Humidifier: portable for humidity generation in a small room.
12. Incubator 22 °C with platform for plates and shaking facility.
13. Laminar air flow cabinet: with UV, fluorescent lamps, and HEPA filters.
14. Nalgene freezing containers: 1 °C/min Mr. Frosty containers.
15. Nanodrop spectrophotometer machine (for nucleic acid quantitation).
16. Pathogen inoculation workroom.
17. pH meter.
18. Pipettes and tips (all ranges for regular laboratory work and dedicated for RNA isolation).
19. Plastic transparent boxes with lids: rectangular boxes, 23 L capacity airtight container.
20. Real-time PCR machine: or suitable thermal cycler as desired.
21. Scalpel.
22. Sterile disposable transparent needles: 20 mm (W) × 100 mm (H).
23. Surgical blade.
24. Thermocycler machine.
25. Tweezer forceps: approximately 6.2 in. (L).
26. Vacuum desiccator: transparent benchtop chamber.
27. Vortex machine.
28. Water bath.
29. Weighing balance.

3 Methods

The procedures detailed in the present chapter are performed using the SSD1 isolate of *S. sclerotiorum*, which possesses moderate aggressiveness on Indian mustard plants. The described protocols can be reliably utilized and applied as a method for screening and quantifying infection involving different isolates of *S. sclerotiorum*

Table 1
List of *S. sclerotiorum* endogenous reference genes used for qRT-PCR analysis in different studies

S. sclerotiorum study	Endogenous reference gene	Purpose	Gene IDs	Primer details	References
Pathogen transcriptome during infection of *B. napus*	β-tubulin gene	RNAseq validation by qRT-PCR	SS1G_04652	F: ACCGTCGTCGAGCCATATAA R: TAGAGAGCCTCGTTGTCGATAC	[23]
Characterization of a extracellular PGIP-INactivating Effector 1 (SsPINE1)	β-tubulin gene	Biomass determination using genomic DNA quantitative PCR	SS1G_04652	F: ACCTCCATCCAAGAACTC R: GAACTCCATCTCGTCCAT	[24]
Characterization of a small virulence related secreted protein (SsSSVP1)	β-tubulin gene	Candidate gene expression of SsSSVP1 in the wild-type and the mutant strain by qRT-PCR	SS1G_04652	F: TTGGATTTGCTCCTTTGACCAG R: AGCGGCCATCATGTTCTTAGG	[25]
Characterization of a protein elicitor gene (SsCut)	β-tubulin gene	(a) Pathogen biomass quantification in infected plant tissues (b) Expression analysis of plant defense marker genes upon elicitor treatment	SS1G_04652	F: GTCGTCGAGCCATATAACGC R: GGAAACGGAGACAGGTGGTA	[26]
In silico secretome analysis of the pathogen	Actin gene	Candidate virulence related genes expression by qRT-PCR	SS1G_08733	F: TTTCTCCACCACTGCCGAAC R: TCCTTACGAACATCAACATCACA	[27]
Characterization of the p24 family proteins (SsEmp24 & SsErv25)	Actin gene	Candidate genes expression in the wild-type and the mutant strain by qRT-PCR	SS1G_08733	F: GAGCTGTTTTCCCTTCCATTGTC R: GACGACACCGTGCTCGATTGG	[28]
Characterization of a zinc-finger transcription factor (SsNsd1)	Histone gene	Expression profiling of SsNsd1 gene during different developmental stages	SS1G_09608	F: GGCTCGTACCAAGCAAACTG R: GAAGTCTTGGGCGATTTCAC	[29]
Characterization of an isothiocyanate hydrolase gene (*SsSaxA*)	*Histone* gene	(a) Relative quantification of fungal colonization by qRT-PCR (b) Candidate *SsSaxA* genes expression by qRT-PCR	SS1G_09608	F: GGCTCGTACCAAGCAAACTG R: GAAGTCTTGGGCGATTTCAC	[30] & the current study

Gene identifiers as reported from the complete genome sequence of *S. sclerotiorum* [4]
F: Forward primer used for qRT-PCR
R: Reverse primer used for qRT-PCR

on leaf, stem, or other tissues of different *Brassica* crops under a controlled environment.

Detailed procedures involving routine cultivation and maintenance of *S. sclerotiorum* cultures under standard laboratory conditions, regularly replenishing and reviving the pathogen stocks from the host, and detached leaf inoculation method for rapid screening and quantification of fungal infection are described below. The phytopathogen *S. sclerotiorum* is a Biosafety Level 1 pathogenic organism, therefore follow all the basic safety guidelines and regulations for handling, culturing, and conducting experimental work with the pathogen (*see* **Note 1**).

3.1 Routine Cultivation of S. sclerotiorum Cultures Over Artificial Media

Different types of nutrient media are used to cultivate *S. sclerotiorum* under in vitro conditions, where PDA (solid) and PDB (liquid) media have been reported as one of the best media suited for its growth that support the maximum radial mycelial expansion, the highest mycelial dry weight, and the maximum number of sclerotia formation [31, 32]. Herein, routine laboratory cultures of *S. sclerotiorum* were maintained and cultivated on PDA (solid) and PDB (liquid) medium as desired for experiments. Utilization of a different solid or liquid medium for *S. sclerotiorum* cultivation would influence the pathogenic growth depending upon the type of nutrient source used, in which case the growth pattern over the artificial medium should be first examined by the user before laying cultures for experiments.

3.1.1 Surface Sterilization and Cultivation of Sclerotia

1. The *S. sclerotiorum* cultures over the artificial media are usually initiated from the myceliogenic germination of irregularly shaped dormant structures called sclerotia (mycelial aggregates formed by the pathogen that are resistant to physical, biological, and chemical degradation) that possess morphological variabilities among its different isolates.
2. To obtain *S. sclerotiorum* mycelial growth over solid media, a few of its sclerotia are first surface sterilized and then cultivated individually over PDA plates.
3. Collect a few similar-sized sclerotia of the pathogen (3–5 in number and 2 to 5 mm in size for the SSD1 isolate) from storage at 4 °C in a 50 mL Falcon tube having autoclaved MQ, and surface sterilize them inside a laminar airflow (*see* **Note 2**).
4. Perform initial sclerotia washing with 35 mL of autoclaved MQ in a 50 mL Falcon tube, twice for 2 min each.
5. After two initial washes, surface sterilize sclerotia with 35 mL of freshly prepared 70% ethanol solution for 1–1.5 min. Quickly remove the residual ethanol solution from the Falcon tube and proceed to the next step (*see* **Note 3**).

6. After surface sterilization, perform three consecutive washes of sclerotia with 35 mL of autoclaved MQ for 2 min each to remove the residual traces of ethanol.
7. After the final wash, collect the surface-sterilized sclerotia in a new 90 mm sterile petri plate and wait until they appear dry (*see* **Note 4**).
8. After the sclerotia appear dry, cut them equally into small pieces using a presterilized blade (on a scalpel) and place each piece aseptically over the center of preprepared PDA plates with sterile forceps, such that the cut part (appearing whitish) is in direct contact with the solid medium (*see* **Note 5**).
9. Seal all the sclerotia plates with parafilm and place them in an inverted position (upside down) under incubation for 4 days at 22 ± 2 °C in the dark for mycelial colonies to appear (*see* **Note 6**). The morphological characteristics of the cultivation of SSD1 sclerotia after different days of incubation over the PDA medium are depicted (Fig. 2 Panel-1 a–e).

3.1.2 Subculturing the S. sclerotiorum *Sclerotia Culture*

1. To obtain mycelial colonies to be utilized for experiments, the *S. sclerotiorum* is typically passaged and subcultured using a 4-day-old sclerotial plate as desired.
2. For initiating a generation-1 (GEN-1) subculture of *S. sclerotiorum* over the center of new PDA plates, it is advisable to use a single sclerotium plate possessing a fully expanded even mycelial growth at the advancing edge of the agar surface (*see* **Note 7**).
3. Inside the laminar airflow, punch 4 mm diameter mycelial plugs using a sterilized cork borer from the growing edge of a 4-day--old single sclerotium plate to lay a GEN-1 *S. sclerotiorum* culture (*see* **Note 8**).
4. Sequentially lay the GEN-1 culture with the 4 mm mycelial plugs by placing them aseptically either in inverted (mycelial growth on the agar plug is placed downward) or in upright (mycelial growth on the agar plug is placed upward) position with the help of a presterilized needle in the center of new PDA plates individually (*see* **Note 9**).
5. Secure all the GEN-1 culture plates with parafilm and place them in an inverted position under incubation at 22 ± 2 °C in the dark for mycelial growth to appear as desired after 2–3 days.
6. Thus, a typical *S. sclerotiorum* radial agar growth with actively growing mycelial colonies at the growing edge is available for the experiment generally after 7 days of culture initiation (4 days of sclerotial germination and 3 days of its subculture over new plates). Thus, for infecting plants using an agar plug-based inoculation procedure, the mycelial cultures over the

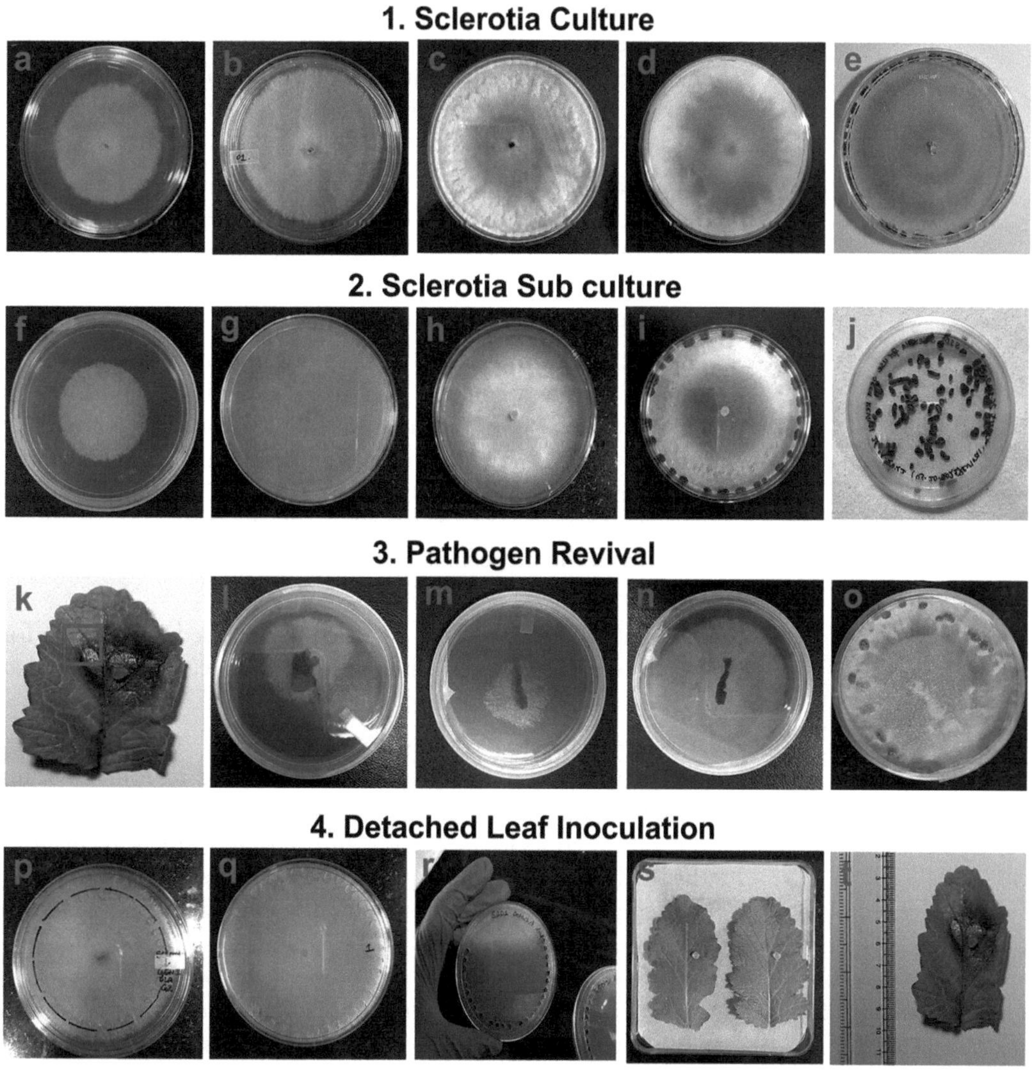

Fig. 2 Schematic representation of the *S. sclerotiorum* cultivation over solid medium, regular laboratory stock revival, and detached leaf inoculation technique. **Panel-1** Cultivation of the SSD1 sclerotia over the PDA medium: (**a**) sclerotia germination and appearance of mycelial colonies after 2 days of incubation; (**b**) a typical sclerotia plate with full mycelial expansion obtained after 4 days of incubation, which is to be utilized for subsequent passage; (**c**) mycelial aggregation indicative of initiation of sclerotia formation from the sclerotia culture; (**d**) melanin pigmentation evidenced from the mycelium growing in the sclerotia culture plate; (**e**) mature sclerotia formation evident at the periphery of the petri plate after approximately 11–12 days of incubation of the sclerotia culture. **Panel-2** Subculturing the SSD1 sclerotia culture; (**f**) emergence of mycelial colonies after 1 day of incubation of a 4 mm size plug taken from the actively growing edge of the sclerotia culture; (**g**) a typical GEN1 culture plate with full mycelial expansion obtained after 3 days of incubation of a 4 mm size plug to be utilized for experiments; (**h**) mycelial overgrowth evidenced on a GEN-1 culture plate after 5 days of incubation of a 4 mm size plug unsuitable for use in the experiment; (**i**) mature sclerotia formation evident at the periphery of the petri plate after approx. 10 days of incubation of the GEN-1 culture; (**j**) mature sclerotia collected from the GEN-1 culture plates dried and stored in sealed petri plates for the next intended use. **Panel-3** Revival of SSD1 from the mustard leaves and stock maintenance; (**k**) typical necrotic lesions apparent on the mustard leaf after 2 days of inoculation with the mycelium agar plug, red box indicates the appropriate leaf area (consisting of both the necrotic and healthy tissue) that must be sampled for fungal revival from the host; (**l**)–(**n**) a number of positive revival plates where the *S. sclerotiorum* mycelium is seen

solid medium can be laid parallelly to fulfill the needs of different experiments as desired (*see* **Note 10**). The morphological characteristics and appearance of mycelial colonies in the SSD1 GEN-1 culture plate during different days of incubation over the PDA medium are depicted (Fig. 2 Panel-2 f–h).

7. Ideally, a 3-day-old GEN-1 culture is the best to be utilized for experiments, but subsequently, one more subculture can be performed from the GEN-1 plates if desired, and the actively growing colonies from the resultant GEN-2 plates can be used for the experiment.
8. The GEN-1 culture plates that are not intended for immediate use can be maintained at the above-described conditions to allow mature sclerotia formation (that are generally completely formed after 10–11 days of plate inoculation) evident at the periphery of the petri plates (Fig. 2 Panel-2 i). The obtained sclerotia from these plates can be collected by gently plucking them from the mycelial mat by using tweezer forceps (*see* **Note 11**).
9. Collected sclerotia should be dried in unsealed petri plates for 1–2 days at 22 °C, after which they must be sealed and stored at 4 °C for the next intended use (Fig. 2 Panel-2 j).
10. It is advisable to replenish and revive the usual laboratory sclerotial stock of *S. sclerotiorum* after three to four passages over the artificial medium as the aggressivity and virulence of the pathogen usually declines with repeated subculturing over the artificial medium. Regular stock revivals are advisable for reducing variability and for obtaining consistent findings from the experiments.

3.1.3 Maintenance of S. sclerotiorum *Cultures in the Liquid Medium*

The growth of the *S. sclerotiorum* mycelium in the liquid medium is often required for numerous types of downstream applications. For example, mycelial growth in the liquid broth is often required to obtain a sufficient amount of fungal mycelium that can be utilized for genomic DNA, RNA, or for protein extraction from the pathogen. Moreover, in addition to the agar plug-based inoculation

Fig. 2 (continued) growing and advancing over the PDA medium from the symptomatic plant tissue; (**o**) mature sclerotia formation evidenced from the revival plate that can be collected, dried, and preserved for the long-term or used as required. **Panel-4** Detached inoculation of the mustard leaves with SSD1; (**p**) a 4-day-old sclerotia culture plate used to excise 4 mm mycelial agar plugs from the uniformly growing region at the edge (black mark) to obtain a subsequent GEN-1 culture; (**q**) actively growing mycelial colonies obtained after 3 days of incubation of a typical GEN-1 culture plate from which 4 mm plugs were punched for plant inoculation; (**r**) and (**s**) agar plugs excised from a GEN-1 culture were placed in an inverted position on one or the other side of the mid-vein over the adaxial surface of the leaves arranged in square petri plates lined with moist blotting paper; (**t**) appearance of clear necrotic symptoms on a mustard leaf photographed after 48 h of *S. sclerotiorum* inoculation for measurement of the lesions with the ImageJ tool

method, plant infection can also be achieved by using a liquid inoculum preparation of *S. sclerotiorum* in which the concentration of the fungal inoculum can be adjusted by delivering the desired mycelial fragments for infection. To prepare the macerated mycelial suspension of *S. sclerotiorum*, it is first cultured in the PDB (liquid) medium, to obtain its growing mycelium that is subsequently crushed, filtered, and used to obtain the desired hyphal fragment count for plant infection, as per the method described earlier [13]. Evaluation of the antifungal activity of different compounds, toxicity induced by defense compounds, and fungal growth inhibitors or toxins may also require the cultivation of the mycelium in broth. To cultivate *S. sclerotiorum* in the liquid medium, initiate its culture over the solid media by first germinating the pathogen sclerotia and subsequently subculturing the mycelial agar plugs to obtain a GEN-1 plate, as described in Subheadings 3.1.1 and 3.1.2.

1. Inside the hood, use a fully expanded GEN-1 culture plate to excise mycelial agar plugs from the actively growing PDA culture to inoculate the PDB medium flasks and cultivate the liquid broth for 3 days at 22 ± 2 °C, under a rotation of 100–120 rpm (*see* **Note 12**).
2. The *S. sclerotiorum* actively growing mycelium from the liquid broth would be available for the experiment generally after 10 days of the culture initiation (4 days of sclerotial germination, 3 days of its subculture on GEN-1 plates, and 3 days of its cultivation in the liquid medium). Therefore, to use the macerated suspension-based plant inoculation method, the cultures can be laid parallelly to fulfill the need of the experiment as desired (*see* **Note 13**).
3. After 3 days of cultivation in the liquid broth, the mycelial colonies can be harvested and blended to prepare the fungal inoculum as described before [13].
4. Additionally, according to the needs of different downstream applications, the user can cultivate the *S. sclerotiorum* mycelium in the PDB medium by adjusting the volume of the media to be inoculated with the respective days of culture incubation and harvesting the mycelia as needed.
5. The *S. sclerotiorum* liquid cultures that are not intended for further use can be maintained at the above-described conditions to allow mature sclerotia formation in the liquid medium that can be separated after filtration from mycelial mats, then rinsed, and dried at 22 °C before being stored at −20 °C or 4 °C as described before.

3.2 Reviving the *S. sclerotiorum* Cultures (Stock Maintenance)

Repeated subculturing over artificial media often leads to the attenuation of pathogenic virulence. To maintain/preserve the infectivity of the isolate after subsequent passages (subcultures) over the artificial media, the fungi should be regularly revived from suitable host plants and maintained thereafter.

1. To revive *S. sclerotiorum*, punch mycelial agar plugs (4 mm diameter) from the advancing edge of a 3-days old GEN-1 culture plate (as described above in Subheading 3.1.2) to infect healthy plant leaves. Leaves of the mustard plants were utilized for reviving the pathogen (*see* **Note 14**).
2. Harvest approximately 15–20 detached leaves from the host using scissors in a ziplock bag (maintain over ice) shortly before performing pathogen inoculation, and if desired, rinse the leaves one or two times with autoclaved MQ and place over trays or square petri plates (*see* **Note 15**).
3. Proceed for the inoculation step only after the water on the leaf surface appears dry (*see* **Note 16**).
4. Pick a 4 mm mycelial plug from a fully grown GEN-1 plate with a presterilized needle and place it in an inverted position by gently pressing it over the adaxial surface of the leaf tissue on either side of the midrib. Further, similarly proceed to inoculate at least 15–20 leaves individually and ensure that the mycelial agar plug adheres to the plant tissue and does not fall off.
5. After inoculation, maintain the leaf samples under high humidity in a humid chamber (where RH > 90%) at 22 °C suitably under dark (initially a 12 h dark period) to facilitate infection (*see* **Note 17**).
6. Two days postinfection, the spreading necrotic sector can be visualized with the naked eye on the host tissue, at this stage, the leaf area with necrosis should be sampled from the necrotic margin and harvested in such a way that the sections contain both the necrotic (diseased) and the normal (healthy) leaf tissue (Fig. 2 Panel-3 k) (*see* **Note 18**).
7. After 48 h, sample the symptomatic leaf area into 5–10 mm size pieces with a wiped scissor and harvest at least 15–20 infected leaf sections in a sterile 50 mL Falcon tube containing autoclaved MQ (*see* **Note 19**).
8. After sampling all the subsequent revival steps must be conducted in a hood near the flame from a Bunsen burner to ensure an aseptic environment.
9. Wash the infected leaf samples twice with 35 mL of autoclaved MQ by pouring it into the 50 mL Falcon tube for 1–2 min each.
10. After two initial water washes, surface-sterilize the harvested leaf sections with 35 mL of freshly prepared 70% ethanol solution for exactly 1 min. Quickly decant the residual ethanol

solution from the Falcon tube and proceed to the next step (*see* **Note 20**).

11. Quickly rinse the sterilized leaf sections with 35 mL of autoclaved MQ for 30 s, to get rid of the residual ethanol if any from the surface. Decant as much liquid as one can from this step (*see* **Note 21**).
12. Subsequently, perform one to two more washes with 35 mL of autoclaved MQ for 1–2 min each, and after the last wash collect the leaf sections in a 90 mm petri dish that is lined with a sterile filter paper.
13. After 5–7 min as the residual water is absorbed by the filter paper and leaf sections appear dry, each of the infected leaf is ready to be placed over a nutrient medium (*see* **Note 22**).
14. Each leaf section should be placed individually over a PDA plate such that the necrotic area is in direct contact with the solid medium. To ensure contact, gently press the necrotic sector with forceps such that it touches the solid surface of the media. This contact with the nutrient medium is necessary and allows the pathogen to subsequently passage to the solid medium from the infected area.
15. Plate approximately 12–15 leaf sections individually over 12–15 preprepared 90 mm PDA plates. After plating all the sections, the petri plates should be parafilmed, stacked, and wrapped in a saran wrap.
16. Place all the PDA plates in an inverted position under optimal incubation at 22 ± 2 °C in dark for 2–3 days.
17. As the pathogen grows from the symptomatic necrotic plant tissue over solid media, the initials of mycelial growth can be visualized with the naked eye after 3 days of incubation over the nutrient medium (Fig. 2 Panel-3 l and m).
18. After 5–6 days as sufficient mycelial expansion is visible over the PDA plates, remove any plates without growth or with any contamination (Fig. 2 Panel-3 n) (*see* **Note 23**).
19. Pure cultures showing mycelial expansion from the necrotic area of the leaf to the nutrient medium should be retained and placed undisturbed to allow them to form sclerotia over the PDA medium (*see* **Note 24**).
20. All the plates should be suitably maintained in an inverted position under incubation at 22 ± 2 °C in the dark until complete sclerotia formation, which is usually evident during 10–14 days of plating depending upon the type of isolate being revived (Fig. 2 Panel-3 o).
21. Once sclerotia have been completely formed, they can be collected from all the plates using sterile forceps in 90 mm petri

dishes and should be dried in unsealed petri plates for 1–2 days at 22 °C. Completely dried sclerotia would produce a rattling sound when shaken in petri dishes.

22. Sclerotia revived from the host tissue can be preserved for the long term by storing at −20 °C or kept at 4 °C for routine use as required by the user.

3.3 Storage of S. sclerotiorum Laboratory Stocks

The regular laboratory stock culture of *S. sclerotiorum* should be maintained with the aim that the stock is preserved in a viable state without effects or alterations in its physiological, morphological, and genetic characteristics such that it can be utilized as a principal source of material for subsequent research works. After re-isolation of the pathogen from symptomatic tissues of a suitable host, the pathogen stocks can be maintained by either preserving its mycelia or its sclerotia.

3.3.1 Mycelia Preservation

1. The *S. sclerotiorum* mycelia stocks can be stored either over the filter paper or over cellophane squares after drying for long-term storage at −20 °C.
2. Take the Whatman 3 MM filter paper or cellophane sheet and cut them into squares of sizes 2–3 mm^2, and autoclave the cut pieces in a suitable glass container (*see* **Note 25**).
3. Punch a 4–5-mm size agar plug from the actively growing *S. sclerotiorum* culture that needs to be preserved and transfer it into the center of a fresh PDA plate.
4. With a sterile forceps, place 10–12 pieces of filter paper squares (2 mm^2 each) around the rim of the 90 mm petri dish inoculated with the mycelial agar plug. The size of the Whatman filter paper can generally vary, but usually smaller size pieces should be preferred such that the fungal mycelium can grow through it easily.
5. Postinoculation, parafilm the plates and incubate them as indicated above at 22 ± 2 °C in an inverted position in the dark to obtain mycelial growth over the placed filter papers. As the culture is mature with time, proceed to collect the filter papers inside a hood (*see* **Note 26**).
6. Use a sterilized forceps to pick the mycelial colonized filter papers from PDA petri dishes into new sterile petri plates and allow them to dry for 48 h in a suitable vacuum desiccator. Preservation of even slightly humid filter paper stocks highly diminishes the culture viability; hence, the user should ensure that the mycelia-colonized filter papers are completely dry.
7. Alternatively, the filter papers can also be dried in rubber containers (that are dry with an airtight seal) having silica gel at the bottom to absorb moisture.

8. Take the completely desiccated mycelial filter papers using a sterile forceps and place them in a flat autoclave bag or sterile envelopes having 1–4 cm^3 of silica gel. If stocks are prepared with great caution, the dried filter paper mycelium can be maintained for sufficiently longer durations and recovered as required.
9. Store mycelial stocks as desiccated hyphae-colonized filter papers by keeping the bags in airtight plastic containers at –20 °C storage.
10. *S. sclerotiorum* mutants or genetically modified isolates that do not form sclerotia (are impaired in sclerotia formation or produce deformed sclerotia that are not viable), even after prolonged incubation, can be preserved and stored as dried-mycelium colonized over filter paper at −20 °C, a method that provides a standard long-term preservation technique of such strains (*see* **Note 27**).
11. Additionally, the mycelial cultures can also be preserved by preparing glycerol stocks from 4 to 5 mm diameter agar plugs taken from an actively growing PDA culture in cryotubes having a 10% glycerol solution after slow freezing in suitable containers before being stored at −80 °C.

3.3.2 Sclerotia Preservation

1. Storage of pathogen sclerotia stocks at −20 or – 80 °C is also a robust long-term preservation method that allows maximum viability in terms of its germination and growth, without the reduction in its pathogenicity [33].
2. Dried pathogen sclerotia can be stored and preserved at 4 °C (for regular use) at −20 °C (for long term) by placing them in sterile vials with a suitable drierite desiccant at the bottom.
3. Completely dried sclerotia (that rattle when shaken in petri plates) can alternatively be stored in a 10% glycerol solution as a cryoprotectant in 1.8 mL sterile cryotubes after slow freezing at 1 °C/min in Nalgene freezing containers before being placed at −80 °C.
4. Frozen cultures stored at −80 °C storage in cryovials should be thawed carefully for recovery by pipetting out the culture in sterile water or directly thawing the cryotubes by transferring them to a water bath at 37 °C until the ice is melted.
5. It should be ensured that the surfaces are properly disinfected with 70% ethanol prior to opening the tubes for recovery. The cultures should be carefully incubated at ambient growth conditions after being transferred to either broth or agar medium as desired (*see* **Note 28**).

3.4 Detached Leaf Inoculation Assays

Detached tissue inoculation is a rapid way to screen, identify, and assess tolerance/resistance or susceptibility of a large number of cultivars under test against the necrotrophic *S. sclerotiorum*, with the advantage of using either plant tissue type: the leaves or the stem. In the present chapter, a stepwise protocol for performing detached inoculation involving the leaf tissues of mustard crops is described; however, the method can also be reliably used with the detached stem tissues as desired.

3.4.1 Plant Material Production and Preparation

The plant material set to be evaluated against the pathogen should include a test population (such as different genotypes as required) along with appropriate controls and the suitable replicates of each to be utilized within an experiment must be devised prior to conducting it (*see* **Note 29**).

1. Mustard plants can be cultivated under either the experimental field condition, in a greenhouse compartment, or under a controlled growth chamber to be utilized for the detached tissue inoculation experiment (*see* **Note 30**).
2. For conducting unbiased experiments, if the accessions to be tested against *S. sclerotiorum* were cultivated in the experimental field it is advisable to follow a randomized complete block design within different replications [34]. Plants selected randomly from each plot should be used to sample tissues that are subjected to the disease resistance assessment [35].
3. The detached tissue inoculation in this protocol is performed using the leaf tissues of various cultivars of the oilseed mustard crop *Brassica juncea*, by cultivating them under a controlled growth chamber facility. However, the method can be used reliably for screening different tissue types of mustard against the pathogen (*see* **Note 31**).
4. All the plants to be utilized for screening within an experiment should be cultivated under similar conditions to reduce the influence of external factors in causing experimental variabilities. For consistency, it should be taken care that the perturbations caused due to external abiotic and biotic factors on plants being evaluated in an experiment should be kept at a minimum.
5. Sow the seeds of the respective cultivars in pots of diameter 9 cm or larger (as desired and assembled in a tray) filled with an autoclaved mix of 3 parts of solerite mix: 1 part of sand for cultivation.
6. The leaf tissues to be used for the detached leaf inoculation experiment can be harvested from plants at the stage when plants are 3–4-week old, keeping in note that an appropriate size leaf is available for inoculation (*see* **Note 32**).

7. Sow seeds of the respective plant genotypes to be screened by making two digs/cells in each pot. Two to four seeds of a genotype should be sown in each cell and 1 week after the emergence of seedlings, plants should be thinned to either one or two seedlings per pot and cultivated thereafter as needed (*see* **Note 33**).
8. Post sowing the *B. juncea* plants were maintained under a short-day regime at 24 °C (day)/18 °C (night) temperature, with 70–80% relative humidity, and a 10 h light (day) with 400 mmol m^{-2} s^{-1} balanced light spectrum using fluorescent (70-W) and halogen incandescent (100 W) lamps and 14 h dark (night) photoperiod in the plant growth chamber facility (PGF) at the NIPGR. Slight variations in the plant growth conditions may be adopted by the user as per the available place for plant cultivation (such as a greenhouse compartment or a growth chamber) and the experimental requirement.
9. Cultivate plants by maintaining them over a 1/fourth strength Hoagland solution twice a week for proper growth (*see* **Note 34**).
10. On the day of the experiment when parallel pathogen cultures are available, plant tissues can be harvested for inoculation.
11. Six to eight fully expanded leaves of an accession type should be sampled using independent plants of the genotype at the 3–4-week-old stage with scissors, placed in labeled ziplock bags, and kept over ice until transportation to the laboratory (*see* **Note 35**).
12. The harvesting of either the leaf or the alternative plant tissue should be done carefully as required and as quickly as possible, such that the harvested samples are transported fresh to the laboratory for the experiment (*see* **Note 36**).
13. Post-harvesting, if required, all the leaves (or other tissues) may be washed one or two times with sterile water and wiped with a tissue to remove excess water droplets on their surface until their placement in petri dishes or glass plates (*see* **Note 37**).
14. Arrange all the harvested leaves in square petri plates that are lidded, labeled, and lined with blotting paper moistened with sterile water before proceeding with pathogen inoculation. Typically, one, two, or more leaves can be kept in a single square petri dish depending upon the leaf size. Alternatively, tissues can also be placed and arranged in glass plates or platform prepared with moist paper towels or filter paper to allow sufficient humidity for plant infection.

3.4.2 Obtaining the S. sclerotiorum Cultures for Inoculation

1. For agar plug-based plant inoculation, the cultures of the pathogen should be initiated (in parallel to the host cultivation) before 7 days of the date of conducting the experiment. Pathogen cultures should be initiated when the host plants are almost 3 weeks old, such that the actively growing colonies are available for inoculation of the host at the 4-week-old stages (approximately 1-month old).
2. To obtain *S. sclerotiorum* for plant infection, the cultures should be initiated after surface sterilization of pathogen sclerotia by individually cultivating them over PDA plates and incubating them under the desired conditions, as described above in the Subheading 3.1.1 (*see* **Note 38**).
3. The *S. sclerotiorum* GEN-1 cultures should be suitably used for an agar plug-based mycelial inoculation of the host. Superfluous passage of the *S. sclerotiorum* solid culture can reduce pathogenic virulence, the GEN-1 subculture is generally the best to be used for plant inoculation and to reduce the possibility of experimental variabilities. *S. sclerotiorum* solid cultures were passaged no more than two times prior to being used in our experiments.
4. Fresh mycelial colonies over the solid PDA medium for plant infection should be obtained after subculturing the mycelial agar plugs from the growing edge of the best sclerotia plate (with uniformly distributed mycelial growth at the rim) following incubation for 2 to 3 days at 22 ± 2 °C under dark as described in Subheading 3.1.2 (Fig. 2 Panel-4 p and q) (*see* **Note 39**).

3.4.3 Plant Inoculation

1. After the leaves have been arranged in labeled square petri plates, pathogen inoculation can be performed at the laboratory workbench (or in a dedicated pathogen inoculation workroom) by using young pure cultures of either a single *S. sclerotiorum* isolate or more than one isolate as desired.
2. In a detached tissue inoculation experiment, different applications can involve the use of more than one isolate of *S. sclerotiorum* for plant screening, or alternatively, the aggressiveness of different isolates may be evaluated prior to the experiment to select the most suitable isolate for the experiment.
3. Use a sterile cork borer to punch mycelial agar plugs of 4 mm diameter each from the periphery of the appropriately selected fully grown 7 days old *S. sclerotiorum* culture (4 days of sclerotial germination and 3 days of subsequent subculture) and sequentially pick and inoculate the mycelial agar plugs at the region on one or the other side of the midvein over the adaxial surface of the leaf (Fig. 2 Panel-4 r and s) (*see* **Note 40**).

4. Mycelial agar plugs from the growing *S. sclerotiorum* culture should be placed carefully over each leaf such that the hyphal side is down and centered between the main leaf vein and the leaf edge and must be slightly pressed to ensure adequate contact with the leaf surface. Ensure that the agar plugs stick to the surface of the leaf and do not fall (*see* **Note 41**).
5. The inoculation of a small set of leaves can be easily completed by using plugs from a single *S. sclerotiorum* GEN-1 culture plate or alternatively, if a large number of test samples have to be inoculated in an experiment more than one GEN-1 culture plate can be used.
6. Around 35–40 agar plugs of 4 mm diameter can be easily punched and are available for inoculation from a single 90 mm plate that has full mycelial growth till the edge. Inoculation with plugs obtained from a single GEN-1 plate would reduce variabilities among different inoculated samples. In case a large number of samples have to be tested in an experiment, more GEN-1 plates would be required. In the latter case, the user can reduce experimental variabilities in samples by ensuring that the control and test samples are inoculated from the agar plugs coming from the same GEN-1 plates.
7. Negative control must be included in the experiment by inoculation with the agar plug of the same size that is devoid of the mycelium.
8. To obtain consistent findings from the assay, detached inoculation should be performed by using fully expanded true leaves that are nearly of uniform size (similar-sized), ensuring that the amount of inoculum is uniform and its placement is constant throughout samples.
9. Postinoculation, square petri plates should be secured with lids and kept over the platform in trays or boxes that are filled with water at the bottom and sealed from the top to ensure humid conditions conducive for infection. Alternatively, plants can also be maintained in a suitable chamber with a humidifier under high RH conditions >90% without the requirement of trays or boxes. The infected samples should be placed at 22 ± 2 °C under dark for 48 h for infection symptoms to develop. Both relative humidity and temperature play an important role in *S. sclerotiorum* disease establishment; therefore, it is necessary to consider these factors while screening plants against the pathogen [2, 14–16].
10. The assay should be conducted in different biological repeats (performing the tests at independent times), and each independent repeat must include a sufficient number of leaves (n) before drawing any conclusions about the genotypes under evaluation.

3.4.4 Disease Assessment and Data Analyses

1. Different ways to assess and estimate the outputs of the lesions developed after the detached tissue inoculation experiment have been used in the past, which can be suitably adopted or modified as per the experiment [15, 36–39].
2. Water-soaked infection is a characteristic symptom of the early infection phase of *S. sclerotiorum*, while tissue necrosis results during the phase when the disease is established. It is therefore advisable to estimate the experimental results after the development of brown necrotic lesions rather than the appearance of water-soaked infection over the plant tissue surface.
3. Characteristic necrotic lesions over mustard plants can be generally visualized, photographed, and quantified at a 48-h time point; however, this may vary (Fig. 2 Panel-4 t) (*see* **Note 42**).
4. For analyzing results, each inoculated leaf should be first photographed by using a suitable camera (Canon EOS 750D series with a lens 18–55 mm, 24.2 MP) and placing a scale at its corner of the leaf for quantification (used for marking the known distance and converting it into pixels) with appropriate labeling of the genotype code. Place a white sheet over a black cloth during photography of the samples such that both the scale markings and the leaf samples are clearly visible and ensure that shadows are avoided while photographing the leaves to enable accurate measurements with downstream tools (Fig. 2 Panel-4 t). It should be taken care that the photos are taken at approximately the same distance from the leaf by keeping the position of the camera static throughout the photography. If in an experiment the position of the scale and camera are undisturbed throughout the photography, the results of the entire samples of an experiment can be estimated by setting the scale globally at once for all the measurements and thus saving time.
5. After the leaves have been photographed, the appearance of clear measurable lesions can be estimated in different ways by using the ImageJ tool (Version: 1.53t) or any other suitable software such as the leaf doctor (Version: 1.1) that estimates the disease severity by estimating the area of the plant tissue that is diseased and converting it into pixels [40–42].
6. Depending upon different experimental needs the obtained lesions can be measured and estimated as required by the user (Fig. 3). Either the long (L) and the short diameter (W) of the lesion on each leaf can be measured and the lesion sizes (in cm) can be estimated as $(L + W)/2$. Alternatively, the entire lesion areas (A) and leaf areas (B) can be estimated (in cm^2) to calculate the percentage of lesion per unit leaf as $(A \times 100)/B$ (in %) or direct quantification of the lesion diameter (in cm) or lesion areas of an ellipse (in cm^2) can also be done by using the free hand selection tool used to mark the entire boundary of

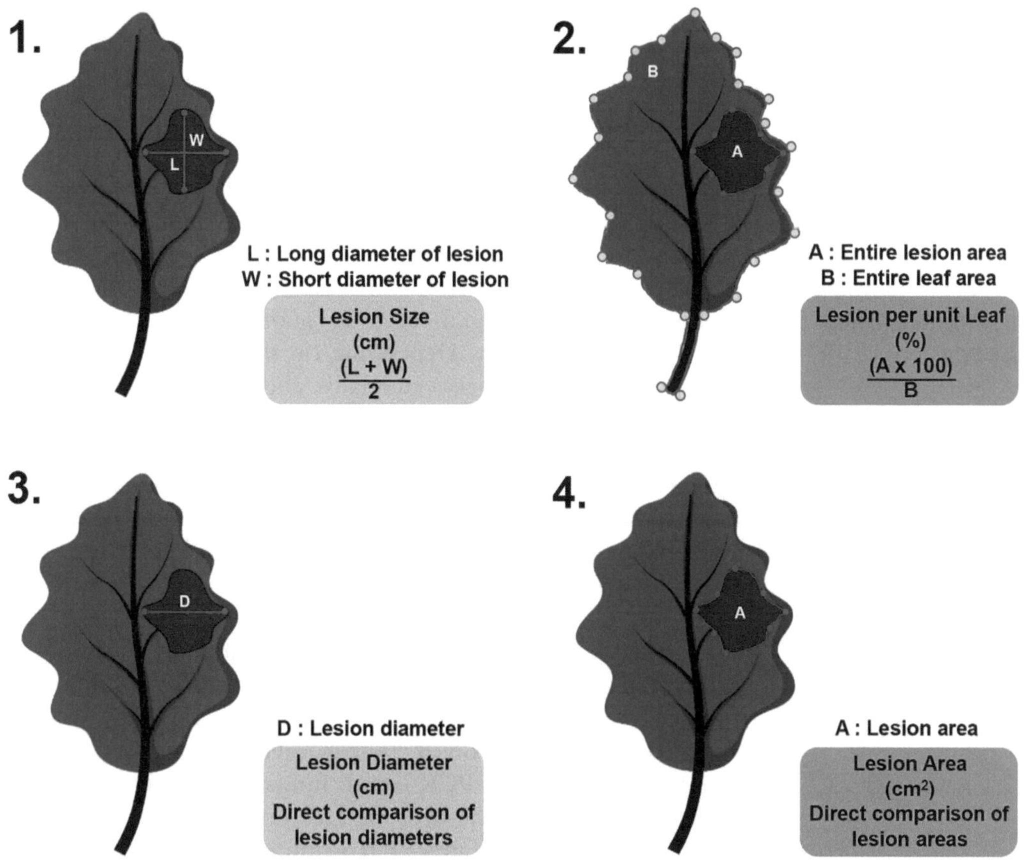

Fig. 3 Suitable approaches available for the measurement and estimation of lesions obtained after the detached leaf inoculation assay. For comparison of the genotypes under test, the necrotic sector apparent after the detached leaf inoculation assay can be measured, analyzed, and represented in various ways as needed. (**1**) The lesion sizes (in cm) can be estimated as (L + W)/2 after measurement of the long (L) and the short diameter (W) of the lesions using a straight-line tool available in the image processing software. (**2**) The percentage of lesion area per unit leaf area can be estimated as (A × 100)/B (in %) after measuring the entire lesion areas (A) and leaf areas (B) using a free hand selection tool. (**3**) The lesion diameters (D) can be estimated with a straight-line tool for direct comparison of lesion length (in cm) between the genotypes. (**4**) The lesion areas (A) of an infected leaf (in cm^2) can be measured by using the free hand selection tool to mark the boundary of the entire necrotic region over the leaf for comparison of different genotypes whose leaf areas are variable

the lesion for comparison of different genotypes [28, 38, 39, 43–45] (*see* **Note 43**).

7. The data can be evaluated by presenting the result of the samples under tests in the form of mean ± standard error by preparing graphs with the Prism 9 software (GraphPad Version: 9.5.1, https://www.graphpad.com/features) or as desired and including a suitable statistical analysis depending upon the number of sample sets to be compared in an experiment (such as the student's *t-test* or one-way or two-way analysis of

variance to compare the differences in means) to evaluate the significant differences amongst them at the desired *p* value.

8. Conducting the experiment in various replications would provide a sufficient number of observations (n) for each genotype under test and would also increase the power of the statistical tool used to interpret the results, and thus enable the user to draw appropriate conclusions from the experiment.

3.5 In Planta *Fungal Quantification Using qRT-PCR*

Real-time PCR method is regularly used for the quantification of the fungal plant pathogen levels *in planta* because of its high sensitivity and specificity. Depending on the biological questions, it is important to select between relative and absolute quantification needs [46]. For relative quantification measurements, the expression of the fungal target gene is compared with an appropriate reference gene of the host. The results are expressed as a target/reference ratio. Reference genes of each organism must be selected appropriately and their stability must be evaluated under different conditions prior to use.

3.5.1 Plant Inoculation and Sample Harvesting

1. Infected plants (as described in Subheading 3.4.3) should be harvested at the requisite time points for assessment of the relative levels of fungal colonization. Herein, the relative SSD1 colonization on *B. juncea* leaves was assessed at 24-h postinfection. qRT-PCR-based fungal quantitation at later time points when the necrotic area on the sample has spread over the leaf (such as after 48 h postinfection) often leads to the decreased copy number of plant housekeeping genes that causes increased variability in the data sets often leading to inaccuracy in quantitation. If qRT-PCR is to be performed for later time points of infection, the necrotic area containing the plug should be carefully removed and the sample size must be increased by pooling leaf samples from independent plants.
2. Harvest the SSD1-infected samples at the desired time point by removing the necrotic area containing the plug from the infected leaf with scissors, place them in sterile 50 mL Falcon tubes, and freeze the samples immediately in liquid nitrogen. The most common source for RNase contamination is hand, so always wear gloves while handling the samples. The samples that are not required for immediate use should be stored at −80 °C until analysis.

3.5.2 Isolation of Total RNA

A sterile work environment is a must for isolating RNA from the samples. Leaf samples should be kept in liquid nitrogen while performing RNA isolation. Care must be taken to prevent the thawing of the samples and to avoid poor RNA yields and quality. Since RNases are ubiquitous, extreme care must be taken while handling the samples to prevent RNase contamination and degradation of samples (*see* **Note 44**).

Autoclaving does not ensure the complete elimination of RNases; therefore, diethylpyrocarbonate (DEPC) treatment must be done for all the plasticware, glassware, and materials to inactivate RNases before RNA isolation. Make a 0.1% (v/v) DEPC solution in Milli-Q water and mix well over the stirrer. Dip all the plasticware, glassware, and other materials to be decontaminated in 0.1% DEPC solution and leave overnight for stirring in dark. After the treatment, dry all the materials in the incubator (at 60–80 °C), and pack and autoclave them before use. Additionally, a 0.1% DEPC solution may be aliquoted in a separate glass bottle and autoclaved for use in eluting RNA, in RNA loading buffer preparation, or in other requirements. Mortar, pestle, and other required materials should be covered with aluminum foil and baked at 180 °C overnight before use.

1. Prechill a forceps and a 1.5 mL microcentrifuge tube with liquid nitrogen. Take out the leaf samples in liquid nitrogen for performing RNA isolation and weigh 100 mg of leaf by removing the tissue from either side of the midvein with forceps (*see* **Note 45**).
2. RNA isolation can either be done manually or with any of the available RNA isolation kits as desired. Herein, RNA from the infected samples was isolated with the Spectrum™ Plant Total RNA kit (Sigma Aldrich) following the manufacturer's instructions. RNA isolation with the above kit is conducted at room temperature by using a lysis buffer and purification over a silica membrane column to obtain high quality of pure RNA for downstream use.

3.5.3 Qualitative and Quantitative Analyses of RNA

The quantity and quality of the isolated RNA should be assessed by checking the optical density with a Nanodrop spectrophotometer. A 260/280 ratio of ~2.0 is accepted as pure for RNA. Additionally, the integrity of total RNA should be assessed by visualizing it over the agarose formaldehyde gels as described.

1. Prepare a 1% (w/v) agarose gel by heating 1 g of agarose in 20 mL of 5× MOPS buffer and 72 mL of DEPC-treated water in a microwave. Cool it and add 17.6 mL 37% formaldehyde and 5 μL of 10 mg/mL ethidium bromide to the mixture. Pour the gel and allow it to solidify.
2. Mix one part of the RNA sample with two parts of the RNA sample buffer to a total volume of 10–30 μL. Heat all the samples to 65 °C for 5 min and then cool them on ice for 2 min. Add 2 μL of RNA loading buffer to the samples.
3. Pre-run the gel for approximately 10 min in 1× MOPS buffer. Load the RNA samples in the wells. It is generally

recommended to run the RNA gels at low voltage to avoid heat denaturation of the samples. The RNA integrity must be checked by visualization of intact RNA bands in the gels before proceeding further.

3.5.4 cDNA Synthesis

Approximately 2 μg RNA from each sample should be used for corresponding cDNA synthesis with the available cDNA synthesis kits. Herein, 2 μg RNA from each sample was used to synthesize cDNA by reverse transcription with a high-capacity cDNA reverse transcription kit (Applied Biosystems) as per the manufacturer's instructions.

3.5.5 Primer Design and Evaluation of Specificity for qRT-PCR

1. Appropriate selection of both the fungal and the plant endogenous reference gene is critical for reliable and accurate quantitation for avoiding variations in the qRT-PCR analysis. qRT primers must be designed from the conserved region of the gene preferably spanning an intron with a small amplicon size. Care must be taken to design primers by following all the standard and recommended guidelines for qRT-PCR analysis [47].
2. Different endogenous reference controls for qRT-PCR analysis have been previously used for *S. sclerotiorum* for various applications involving different isolates of the pathogen (Table 1). From these, both the *Ss β-tubulin* (SS1G_04652) and *Ss Histone 3* (SS1G_09608) genes have been used for the quantitation of fungal load *in planta* [26, 30]. The *Ss Histone 3* (SS1G_09608) was used as a target gene for relative quantitation of fungal colonization in *B. juncea* samples, which has been previously used for quantitation of pathogen load both in *Arabidopsis thaliana* and previously by our laboratory in *B. juncea* samples [30, 48] (*see* **Note 46**).
3. Primers for *Ss Histone 3*:

   FP RT: 5′ GGCTCGTACCAAGCAAACTG 3′.

   RP RT: 5′ GAAGTCTTGGGCGATTTCAC 3′.
4. Agarose gel analysis revealed that the PCR amplification with the above primers resulted in a single band of size 235 bp from the *Histone 3* gene of *S. sclerotiorum* strain SSD1.
5. Different traditional housekeeping genes from *B. juncea* were previously designed and optimized for use in our laboratory [49]. *B. juncea tonoplastic intrinsic proteins 41* (*TIPS-41*) gene primers had the most stable expression under SSD1 treatment and were therefore utilized for the relative quantification measurements [48] (*see* **Note 47**).

6. Primers for *TIPS-41*:

   FP RT: 5′ TGAAGAGCAGATTGATTTGGCT 3′.

   RT RP: 5′ ACACTCCATTGTCAGCCAGTT 3′.

7. Primer specificity must be verified after amplification with a conventional PCR and evaluation of samples on the gel. The primer pairs must amplify a single copy gene of the expected size and should possess a single peak after sequencing and melt curve analysis.

3.5.6 Relative Quantification of *S. sclerotiorum* in *B. juncea* Samples by qRT PCR

1. Prepare 50 μL of 1:20 diluted cDNA for each of the infected and control leaf samples. For each biological replicate (each independent experiment), three technical replicates should be prepared for both the target and reference gene.
2. Prepare a master mix of SYBR Green with the cDNA of the respective samples. For each 20 μL reaction, the master mix should contain 16 μL of 1× iTaq Universal SYBR Green Supermix and 2 μL of 1:20 diluted respective cDNA sample. A no template control (NTC) should also be included in a few wells of the plate having sterile water instead of a template (e.g., Control samples 1 & 2 and Infected samples 1 & 2 cDNA SYBR mix) (*see* **Note 48**).
3. Prepare a primer mix with the respective FP and RP in 1.5 mL tubes to obtain 200 nM final concentration of each primer in a total 20 μL reaction volume (e.g., *Ss Histone 3* Mix and *TIPS-41* mix).
4. Add 2 μL of the respective primer mix in each well individually. Aliquot 18 μL of each cDNA SYBR Green master mix (from step 2) and add in the respective wells with a pipette. Carefully seal the plates with a microseal adhesive cover for a 96-well plate. Briefly spin the samples before setting up a qRT-PCR reaction.
5. Place the PCR plate containing the samples into the real-time machine. Use the Bio-Rad CFX96 software to set up the plate by indicating the desired protocol, the position of samples, target names, and NTCs on a plate.
6. Adjust the Bio-Rad CFX96 program on the SYBR Green I manufacturer template to include a 2 min 50 °C than denaturation step—2 min 95 °C—followed by 40 cycles of 95 °C for 15 s and 60 °C for 1 min (single data acquisition point), followed by the default melt curve analysis of heating to 95 °C, cooling to 60 °C for 15 s, and then heating to 95 °C at a rate of 1 °C/5 s (continuous data acquisition). Give the command and initiate the run.

3.5.7 Data and Statistical Analysis

1. After the completion of the run, a data file is generated. Analyze the amplification curves using the CFX Maestro Software bundled with the real-time PCR thermocycler. Analyze the melt curve profile of the qRT-PCR reaction for any genomic DNA contamination, formation of primer dimers, and nonspecific products. Export the results to the excel file.
2. The results are reported from the machine as the mean C_q value of the PCR replicates per sample (standard deviation is also determined), for both the target and the reference genes.
3. The results should be exported to Microsoft Excel and relative gene expression can be quantified using the $2^{-\Delta\Delta C_T}$ method [50].
4. For the relative quantification of fungal colonization in the infected samples, the expression of the fungal target gene should be compared with the host reference gene expression. The results are expressed as the target/reference ratio of *Ss Histone 3* gene expression to the *TIPS-41* gene expression (Table 2).
5. Analyze the obtained results with suitable statistical analysis. Depending on the sample sets in an experiment, the means of the two groups can be compared by the student's *t-test,* while three or more groups by one-way analysis of the variance using an available statistical module (e.g., Microsoft Excel or SPSS Version: 23.0 IBM, New York https://www.ibm.com/spss), followed by suitable post hoc tests to evaluate the significant differences among the data sets at the desired *p* value.

Table 2
Relative quantitation of *S. sclerotiorum* colonization on *B. juncea* samples after 24 h postinfection using qRT-PCR analysis

Target genes	Samples	Mean C_T	SD	*Ss Histone 3/TIPS-41*
TIPS-41	Control sample 1	29.15	0.64	–
TIPS-41	Control sample 2	29.42	0.34	–
TIPS-41	Infected sample 1	29.05	0.33	–
TIPS-41	Infected sample 2	29.13	0.27	–
Ss Histone 3	Control sample 1	NA	NA	–
Ss Histone 3	Control sample 2	NA	NA	–
Ss Histone 3	Infected sample 1	28.10	0.43	1.93
Ss Histone 3	Infected sample 2	29.15	0.39	1.01

The expression of *S. sclerotiorum Histone 3* mRNA was normalized against the constitutive *B. juncea TIPS-41* gene expression level

4 Notes

1. It is advisable to use UV light before and after working with the pathogen. Daily decontamination and disinfection of all the work surfaces should necessarily be done. Contaminating cultures or materials should be disinfected prior to disposing, generally by autoclaving them. The user should follow a strict aseptic environment to avoid any contamination in pure cultures of *S. sclerotiorum*.
2. Each sclerotium is generally halved (cut into two pieces) and plated individually on PDA plates. To initiate a routine *S. sclerotiorum* culture, typically 4–5 plates of sclerotia are germinated, which can be used for subsequent subculture as needed. The quantity of sclerotia germination is solely at the user's discretion based on the experimental requirements.
3. Ethanol generally penetrates the sclerotium and can inhibit its growth. Exceeding the sterilization step beyond 1.5 min can lead to a delay in growth or inhibition of sclerotial germination. Alternatively, sclerotia can also be sterilized using a 0.5–1% (v/v) sodium hypochlorite solution for 1–1.5 min instead of a 70% ethanol solution.
4. After the final water wash, sclerotia can be collected in a sterile petri dish and left aside to dry for 2–3 min. Alternatively, they can be collected and dried on sterilized filter paper before plating.
5. Each sclerotium can be cut into two or three pieces, depending on its size. Each piece should be gently pressed onto the surface of the solid medium to adhere over it and prevent falling off. To ensure an aseptic environment during culturing, the scalpel and the forceps should be sterilized both before and after contact with the fungus.
6. The time required to obtain fully expanded myceliogenic growth over the agar surface in a 90 mm petri dish from sclerotial germination may vary with sclerotia from different isolates of *S. sclerotiorum*. Additionally, it depends on the size of each piece placed on the PDA medium. If a different isolate is used, the user should examine the timing of obtaining fully expanded mycelial growth over the PDA medium. This examination will enable them to establish subsequent subcultures accordingly.
7. All the routine *S. sclerotiorum* GEN-1 plates for the experiment should be laid from a single sclerotial plate that has uniform mycelial growth at the growing edge, which would prevent any uneven growth and difference in timing of obtaining a full plate from the downstream cultures.

8. Inoculation can also be performed by using mycelial agar plugs of varying diameter sizes that would enable the user to alter the radial mycelial growth of the subsequent subculture as desired at an earlier or later time. Additionally, the quantity of GEN-1 culture plates is solely at the user's discretion as per the need of the experiment.
9. If the mycelial agar plug is placed in an inverted position on the surface of a new PDA plate, the resultant fully expanded *S. sclerotiorum* plate would be available after 2.5 days and if the plug is placed in the upright direction, the resultant plate with a full mycelial expansion would be obtained after 3 days of incubation for the experiment. The GEN-1 culture can be laid in either position as required in the downstream experiment by the user. Ensure that the agar plug in either position is gently pressed over the surface of the solid medium such that it adheres and does not fall off.
10. Different characteristics of *S. sclerotiorum* mycelial growth over artificial media, such as colony color, growth type, and radial expansion (in mm), vary among its isolates. The growth pattern described above aligns with the SSD1 isolate of *S. sclerotiorum*. However, if a different isolate is used, the user should monitor and examine the growth pattern over artificial media before conducting the experiment.
11. Timing of complete sclerotia formation, the number of sclerotia in plates, and various morphological parameters—such as size, shape, growth pattern, average weight, and other sclerotial features—can differ among different isolates of *S. sclerotiorum*. When using a different isolate, users should examine these parameters by establishing in vitro cultures in replicates.
12. Usually, 4–5 mm diameter plugs can be excised to inoculate a 100 mL PDB culture in a 500 mL flask and cultivated for 3 to 4 days to prepare the mycelial suspension. The number of agar plugs, their size (in mm), and the volume of the liquid medium to be inoculated should be accordingly adjusted by the user as per the requirement of the experiment.
13. The pattern and timing of obtaining a growing mycelium culture can differ slightly or considerably if a different *S. sclerotiorum* isolate is being used, in such a case the pattern of its growth over both the solid and liquid medium should be first examined by the user before conducting the experiment.
14. The pathogen can be passaged typically from any suitable host tissue such as the young leaves, stems, or pods. Additionally, fungal revival can be obtained using either intact plants (cultivated in pots) or detached tissues like leaves or stems as per the availability and convenience of the user. The timing of

obtaining the necrotic disease symptoms would differ if a different host plant is used for pathogen revival.

15. If plants being used for pathogen revival were cultivated under a controlled environment, then the washing step may be omitted; however, if the tissue is harvested from field-cultivated plants ensure that the washing step is performed to remove any soil or dirt over the surface.
16. Over-wetness on the surface of the plant tissue quickly flares infection and often leads to extensive necrosis of the tissue, which makes difficult to sample the sections for revival.
17. If detached leaves or stem tissues are used, the cut samples can be maintained in square petri plates lined with wet blotting paper. If intact plants are used for pathogen revival, they can be maintained in a plastic box, having some water at the bottom and covered with a lid.
18. If only the necrotic area of the diseased tissue is cut and sampled, subsequent sterilization and washing steps often result in the disintegration of the leaf sections, making them challenging to plate. Additionally, users should refrain from sampling the mycelial plug used for leaf inoculation and must ensure that the samples are harvested and cut from the margin of the necrotic region on the leaf sections—specifically, away from the plug—where the *S. sclerotiorum* mycelium is growing in the host.
19. With plug-based inoculation, the initial necrotic symptoms on mustard leaves are usually evident after 24 h; however, harvesting of the leaf sections for revival should be done after 48 h of inoculation such that a clear spreading necrotic lesion is available for sampling. If a different host is utilized for revival, timing of obtaining the necrotic lesions may differ slightly or considerably. The leaf sections drawn should be harvested in water to prevent drying of the leaves after sampling and the user should quickly proceed to perform the subsequent steps.
20. Ethanol generally penetrates the fungal mycelium and can prevent fungal growth, so the sterilization step should be performed for not more than 1 min, as exceeding the time can lead to a notable growth inhibition of the pathogen, subsequently making its revival difficult. Additionally, surface sterilization can also be performed using 1% (v/v) sodium hypochlorite or a 0.1% (v/v) mercuric chloride solution for 1 min instead of a 70% ethanol solution.
21. *S. sclerotiorum* can be frequently killed during the usual procedures of its surface sterilization; therefore, infected tissues should be always washed with sterile MQ before plating them.

22. Start plating the leaf sections as the water appears dried, do not leave them over the filter paper for long as this often quickly dehydrates and disintegrates the sections before plating.
23. At this stage, plates growing with any contamination can be discarded and only pure cultures should be retained. Unfortunately, if all the plates after revival possess some contamination, the hyphal tip isolation method can be used to isolate mycelial colonies and obtain a pure culture of *S. sclerotiorum* over new PDA plates.
24. At any stage till sclerotia formation, if contamination is observed, the emerging *S. sclerotiorum* mycelium colonies can be punched with a cork borer from the uncontaminated region and can be subsequently subcultured over new PDA plates.
25. To prevent the sticking of cellophane squares onto each other, they can be soaked in sterile water for some time until their placement over PDA plates.
26. Do not leave the cultures for >2 weeks long before collection of the filter papers, as longer incubation times can result in the reduction of viability during storage. Additionally, the growth time may vary with different isolates and can be extended depending on the strain being used.
27. If the pathogenic mutants fail to be recovered from the dried mycelium colonized over filter paper, they can alternatively be preserved for extended durations (more than a year or so) as a 3–4-day-old growing mycelial culture over the PDA medium at 4 °C.
28. During initial recovery from the storage conditions ensure that the recommended medium and growth conditions specific to the strain being recovered are used at first before performing any downstream subculture to ensure optimal recovery.
29. If transgenic crops or GMOs are being utilized in the experiment, it should be ensured that they must be strictly maintained, propagated, and used for R&D activities as per the standard regulatory requirements and guidelines for their deployment.
30. The decision to cultivate plants under either condition is solely at the user's discretion based on the availability of resources, space, and convenience. When using plants from the experimental field or a greenhouse compartment for screening against the pathogen, users should ensure that the harvested plants (and tissues) are free from other abiotic or biotic constraints that might affect crop performance in the test. Additionally, any spraying or application of chemicals or fungicides before harvesting and conducting the assay could impact the performance of the tested cultivars.

31. Tissues such as stems and branches can be used for pathogen inoculation as needed. Plants must be cultivated until the mature plant stage to obtain robust stems for pathogen inoculation.

32. The size of the true leaf is a crucial parameter in the detached leaf inoculation experiment. Inoculating smaller leaves with the mycelial agar plug can often lead to the expansion of necrotic lesions to the leaf periphery, introducing inaccuracies in lesion size estimation. Therefore, appropriately sized true leaves should be used for pathogen inoculation. Typically, 3–4-week-old *B. juncea* plants that have fully expanded true leaves are best suited for experiments. However, for different needs, the detached leaf experiment can also be reliably performed with leaves from plants as young as half a month old or older. If detached inoculation involves a different tissue type, plants must be cultivated until the appropriate stage for harvesting such tissues for the experiment as needed.

33. The number of seedlings of a genotype to be sown in a pot should be in accordance with the size of the pot used, the stage for which plants must be maintained and cultivated for the experiment, and the other experimental requirements. If plants have to be maintained until the 1-month-old stage, it is best to cultivate and grow one seedling per pot, to allow appropriate plant growth.

34. Plants cultivated in the greenhouse compartment or controlled growth chamber can be maintained under nutrient that is favorable for appropriate plant growth (such as for plant height, leaf width expansion, and stem base diameter). Alternatively, plants cultivated in the field should be grown in the presence of appropriately irrigated sandy loamy soil that received fertilizers with standard agronomic practices.

35. The experimental design framework including the requirement of the number of independent plants of a genotype, the number of leaves of an accession type to be inoculated (n) within an experiment, and the appropriate controls to be included is at the user's discretion. These parameters should be decided prior to conducting the experiment. It is important to sample various leaves of a genotype from independent plants of the accession to draw appropriate conclusions from an unbiased experiment.

36. If inoculation is to be performed using stems or branches, main stems ranging from 15 to 30 cm in size and its first two branches can be harvested from a region 10–12 cm above the ground or as desired with their ends wrapped in suitable polyethylene bags to retain freshness. Alternatively, the detached

part of the leaf base may be capped with a moist cotton pad to retain its freshness.

37. If tissues being used for detached inoculation were harvested from plants cultivated under a controlled environment, then the washing step may be omitted; however, if the tissue is harvested from field-cultivated plants ensure that the washing step is performed to remove any soil or dirt over the surface. Ensure that the tissue surface is dry before performing inoculation, as over-wetness of the plant tissue quickly flares infection and often leads to extensive necrosis, which can subsequently alter the conclusions being drawn from the experimental results.
38. The number of sclerotial plates to be laid must be according to the experimental requirement. Usually, four to five sclerotia plates are sufficient to obtain at least one or two plates with uniform mycelial expansion at the periphery, which can be used to obtain even growth of the subsequent GEN-1 cultures.
39. The agar plug should be placed either in the inverted or in the upright position or few plates in both, adjusted for obtaining them according to the desired time of plant inoculation. The number of GEN-1 culture plates to be inoculated is at the user's discretion as per the need of the experiment, keeping in note of both obtaining number of plates with uniform growth at the edges and the number of plant leaves (or other tissue) to be inoculated in an experiment.
40. Mycelial agar plugs can be punched at the laboratory workbench, but one should ensure that the cork borer and the needle used to inoculate are flame-sterilized both before and after use with the pathogen. The size of the mycelial plug used for inoculation can be moderated by the user as per their requirement. Typically, enough GEN-1 plates should always be available for the detached leaf inoculation experiment such that the best plates with uniform and even mycelial expansion at the border can be selected to inoculate plants.
41. If a different tissue, such as the stem or branch, is used for inoculation, a region over them can be wounded artificially with an appropriately sized cork borer to place the mycelial agar plugs. The 30 cm long stems can be inoculated at two positions, with a gap of 10 cm each, while the 10 cm sections can be inoculated with a single plug at the middle point or as desired in the experiment.
42. Development of the necrotic lesions over the host tissue may vary slightly or considerably from laboratory to laboratory depending upon different factors and variations adopted in the protocol such as the type of isolate used for infection, the type of host and the tissue inoculated, the experimental

conditions, and several others. For obtaining consistent results, it is generally advisable to estimate and quantify the results of the experiment by photographing them only after a sufficient and clear spreading necrotic sector is apparent.

43. If a different tissue type, like stems or branches, is used in the assay lesion length (in cm) and perimeter of stem and branch (in cm) can be recorded to present the results as desired.
44. The working area dedicated to RNA isolation must be decontaminated with RNaseZAP. Always wear gloves while handling the samples. Pipettes, chemicals, and other equipment required for RNA isolation can be reserved for RNA application only and kept separate from other uses.
45. Repetitive freeze-thaw often leads to poor quality of RNA. So, work quickly to prevent the thawing of the samples.
46. If the use involves the detection and quantitation of other isolates of *S. sclerotiorum* by qRT-PCR analysis, the selection of an appropriate reference gene must be done and the primer set must be validated before use.
47. Traditional *B. juncea* housekeeping gene such as *ACT* is frequently used as a reference gene for qRT-PCR analysis. However, variation of *ACT* is exhibited under SSD1 treatment (data not shown). Both the *B. juncea UBQ9* and *TIPS-41* genes possess the most stable expression during various stress conditions.
48. SYBR Green is light-sensitive; therefore, lights must be switched off while preparing of master mix and the use of SYBR Green.

Acknowledgments

NCB acknowledges the funding support by the Newton-Bhabha Fund (BT/UK/PORI/03/AKP/2018-19), Science and Engineering Board, India (grant no. EMR/2016/006433), and a core grant from the DBT-NIPGR. Financial Assistance to BM (CSIR, India), RT (DBT, India) and MV (CSIR, India) is sincerely acknowledged. The authors declare no conflict of interest.

References

1. Bolton MD, Thomma BP, Nelson BD (2006) *Sclerotinia sclerotiorum* (Lib.) de Bary: biology and molecular traits of a cosmopolitan pathogen. Mol Plant Pathol 7(1):1–16
2. Derbyshire MC, Denton-Giles M (2016) The control of *Sclerotinia* stem rot on oilseed rape (*Brassica napus*): current practices and future opportunities. Plant Pathol 65(6):859–877
3. Wang Z, Ma LY, Cao J et al (2019) Recent advances in mechanisms of plant defense to *Sclerotinia sclerotiorum*. Front Plant Sci 10: 1314

4. Amselem J, Cuomo CA, van Kan JA et al (2011) Genomic analysis of the necrotrophic fungal pathogens *Sclerotinia sclerotiorum* and *Botrytis cinerea*. PLoS Genet 7(8):e1002230
5. Derbyshire M, Denton-Giles M, Hegedus D et al (2017) The complete genome sequence of the phytopathogenic fungus *Sclerotinia sclerotiorum* reveals insights into the genome architecture of broad host range pathogens. Genome Biol Evol 9(3):593–618
6. Rollins JA (2003) The *Sclerotinia sclerotiorum pac1* gene is required for sclerotial development and virulence. Mol Plant-Microbe Interact 16(9):785–795
7. Xu L, Li G, Jiang D et al (2018) *Sclerotinia sclerotiorum*: an evaluation of virulence theories. Annu Rev Phytopathol 56:311–338
8. Sharma P, Meena PD, Verma PR et al (2016) *Sclerotinia sclerotiorum* (Lib) de Bary causing *Sclerotinia* rot in oilseed *Brassicas*: a review. J Oilseed Brassica 1(2):1–44
9. He DC, Zhan JS, Xie LH (2016) Problems, challenges and future of plant disease management: from an ecological point of view. J Integr Agric 15(4):705–715
10. Yadav MS, Saroj S, Nasim A et al (2012) *Sclerotinia* rot: a serious problem of rapeseed-mustard in Rajasthan and Haryana, Technical bulletin 15. NCIPM, New Delhi
11. Singh KP, Kumari P, Rai PK (2021) Current status of the disease-resistant gene (s)/QTLs, and strategies for improvement in *Brassica juncea*. Front Plant Sci 12:617405
12. Wei L, Jian H, Lu K et al (2016) Genome-wide association analysis and differential expression analysis of resistance to *Sclerotinia* stem rot in *Brassica napus*. Plant Biotechnol J 14(6): 1368–1380
13. Garg H, Sivasithamparam K, Banga SS et al (2008) Cotyledon assay as a rapid and reliable method of screening for resistance against *Sclerotinia sclerotiorum* in *Brassica napus* genotypes. Australas Plant Pathol 37:106–111
14. Gupta NC, Sharma P, Rao M et al (2020) Evaluation of non-injury inoculation technique for assessing *Sclerotinia* stem rot (*Sclerotinia sclerotiorum*) in oilseed *Brassica*. J Microbiol Methods 175:105983
15. Mei J, Wei D, Disi JO et al (2012) Screening resistance against *Sclerotinia sclerotiorum* in *Brassica* crops with use of detached stem assay under controlled environment. Eur J Plant Pathol 134:599–604
16. Uloth MB, You MP, Cawthray G et al (2015) Temperature adaptation in isolates of *Sclerotinia sclerotiorum* affects their ability to infect *Brassica carinata*. Plant Pathol 64(5): 1140–1148
17. Zhao J, Peltier AJ, Meng J et al (2004) Evaluation of *Sclerotinia* stem rot resistance in oilseed *Brassica napus* using a petiole inoculation technique under greenhouse conditions. Plant Dis 88(9):1033–1039
18. Ge XT, Li YP, Wan ZJ et al (2012) Delineation of *Sclerotinia sclerotiorum* pathotypes using differential resistance responses on *Brassica napus* and *B. juncea* genotypes enables identification of resistance to prevailing pathotypes. Field Crop Res 127:248–258
19. Taylor A, Coventry E, Jones JE et al (2015) Resistance to a highly aggressive isolate of *Sclerotinia sclerotiorum* in a *Brassica napus* diversity set. Plant Pathol 64(4):932–940
20. Uloth MB, You MP, Finnegan PM et al (2013) New sources of resistance to *Sclerotinia sclerotiorum* for crucifer crops. Field Crop Res 154: 40–52
21. Bradley CA, Henson RA, Porter PM et al (2006) Response of canola cultivars to *Sclerotinia sclerotiorum* in controlled and field environments. Plant Dis 90(2):215–219
22. Hüller GDC, Jaccoud Filho DDS, Pierre MLC et al (2016) Different methods of assessing susceptibility of soybean genotypes to white mold. J Biosci 32(2):389–402
23. Seifbarghi S, Borhan MH, Wei Y et al (2017) Changes in the *Sclerotinia sclerotiorum* transcriptome during infection of *Brassica napus*. BMC Genomics 18:1–37
24. Wei W, Xu L, Peng H et al (2022) A fungal extracellular effector inactivates plant polygalacturonase-inhibiting protein. Nat Commun 13(1):2213
25. Lyu X, Shen C, Fu Y et al (2016) A small secreted virulence-related protein is essential for the necrotrophic interactions of *Sclerotinia sclerotiorum* with its host plants. PLoS Pathog 12(2):e1005435
26. Zhang H, Wu Q, Cao S et al (2014) A novel protein elicitor (SsCut) from *Sclerotinia sclerotiorum* induces multiple defense responses in plants. Plant Mol Biol 86:495–511
27. Guyon K, Balagué C, Roby D et al (2014) Secretome analysis reveals effector candidates associated with broad host range necrotrophy in the fungal plant pathogen *Sclerotinia sclerotiorum*. BMC Genomics 15:1–19
28. Xie C, Shang Q, Mo C et al (2021) Early secretory pathway-associated proteins SsEmp24 and SsErv25 are involved in morphogenesis and pathogenicity in a filamentous phytopathogenic fungus. MBio 12(6):e03173-21

29. Li J, Mu W, Veluchamy S et al (2018) The GATA-type IVb zinc-finger transcription factor *SsNsd1* regulates asexual–sexual development and appressoria formation in *Sclerotinia sclerotiorum*. Mol Plant Pathol 19(7):1679–1689
30. Chen J, Ullah C, Reichelt M et al (2020) The phytopathogenic fungus *Sclerotinia sclerotiorum* detoxifies plant glucosinolate hydrolysis products via an isothiocyanate hydrolase. Nat Commun 11(1):3090
31. Hossain MS, Khatun F, Islam MM (2018) Evaluation of culture media, temperature and pH for maximum growth of *Sclerotinia sclerotiorum* (lib.) De bary causing white mold of rapeseed mustard. Bangladesh. J Plant Pathol 34(1&2):1–4
32. Prajapati S, Kumar N, Godika S et al (2022) Assessment of different media on growth and sclerotia formation of *Sclerotinia sclerotiorum* causing *Sclerotinia* Rot of Chilli under laboratory condition. Biol Forum 14(1):249–253
33. Pottinger BM, Ridgway HJ, Bourdôt GW et al (2008) Influence of storage method on the viability, mycelial growth and pathogenicity of *Sclerotinia sclerotiorum* sclerotia. Australas Plant Pathol 37:609–614
34. Khan M, Hasija RC, Hooda BK et al (2017) Relative efficiency of experimental designs in relation to various size and shape of plots and blocks in Indian mustard (*Brassica juncea* L.) crop. Int J Agric Stat Sci 13(1):253–258
35. Rana K, Atri C, Akhatar J et al (2019) Detection of first marker trait associations for resistance against *Sclerotinia sclerotiorum* in *Brassica juncea–Erucastrum cardaminoides* introgression lines. Front Plant Sci 10:1015
36. Davidson AL, Blahut-Beatty L, Itaya A et al (2016) Histopathology of *Sclerotinia sclerotiorum* infection and oxalic acid function in susceptible and resistant soybean. Plant Pathol 65(6):878–887
37. Dura S, Lujan P, Puppala N et al (2020) Screening US peanut mini-core accessions for resistance to *Sclerotinia* blight caused by *Sclerotinia sclerotiorum*. Can J Plant Sci 101(1): 53–60
38. Kull LS, Vuong TD, Powers KS et al (2003) Evaluation of resistance screening methods for *Sclerotinia* stem rot of soybean and dry bean. Plant Dis 87(12):1471–1476
39. Prova A, Akanda AM, Islam S et al (2018) Characterization of *Sclerotinia sclerotiorum*, an emerging fungal pathogen causing blight in hyacinth bean (*Lablab purpureus*). Plant Pathol J 34(5):367
40. Pethybridge SJ, Nelson SC (2015) Leaf doctor: a new portable application for quantifying plant disease severity. Plant Dis 99(10):1310–1316
41. Rasband WS (2018) ImageJ. US National Institutes of Health, Bethesda
42. Schneider CA, Rasband WS, Eliceiri KW (2012) NIH image to ImageJ: 25 years of image analysis. Nat Methods 9(7):671–675
43. Verma R, Kaur J (2021) Expression of barley oxalate oxidase confers resistance against *Sclerotinia sclerotiorum* in transgenic *Brassica juncea* cv Varuna. Transgenic Res 30:143–154
44. Xiao X, Xie J, Cheng J et al (2014) Novel secretory protein Ss-Caf1 of the plant-pathogenic fungus *Sclerotinia sclerotiorum* is required for host penetration and normal sclerotial development. Mol Plant-Microbe Interact 27(1):40–55
45. Arya GC, Tiwari R, Bisht NC (2021) A complex interplay of Gβ and Gγ proteins regulates plant growth and defence traits in the allotetraploid *Brassica juncea*. Plant Mol Biol 106:505–520
46. Chini V, Foka A, Dimitracopoulos G et al (2007) Absolute and relative real-time PCR in the quantification of *tst* gene expression among methicillin-resistant *Staphylococcus aureus*: evaluation by two mathematical models. Lett Appl Microbiol 45(5):479–484
47. Thornton B, Basu C (2011) Real-time PCR (qPCR) primer design using free online software. Biochem Mol Biol Educ 39(2):145–154
48. Tiwari R, Kaur J, Bisht NC (2021) Extra-large G-proteins influence plant response to *Sclerotinia sclerotiorum* by regulating glucosinolate metabolism in *Brassica juncea*. Mol Plant Pathol 22(10):1180–1194
49. Chandna R, Augustine R, Bisht NC (2012) Evaluation of candidate reference genes for gene expression normalization in *Brassica juncea* using real time quantitative RT-PCR. PLoS One 7(5):e36918
50. Livak KJ, Schmittgen TD (2001) Analysis of relative gene expression data using real-time quantitative PCR and the $2^{-\Delta\Delta C_T}$ method. Methods 25(4):402–408

Chapter 15

Differential-Display Reverse Transcription-PCR for Unraveling the Molecular Mechanisms in Foot Rot (*Phytophthora capsici* sp. nov.) Tolerance in Black Pepper*Piper nigrum* L.

Basil Babu Paul and Deepu Mathew

Abstract

Foot rot, caused by the oomycete *Phytophthora capsici* is the most devastating disease in black pepper. Understanding the genes and pathways involved in the field tolerance to foot rot is important in deciding the breeding, gene editing, and transgenesis strategies. Differential-display reverse transcription-PCR (DDRT-PCR) is a reverse genetic approach for gene finding. The protocol involves isolating the foot rot pathogen from infected plant material, preparing the inoculum, inoculatingboth tolerant and susceptible plants, collecting leaf samples at regular intervals, isolating RNA, generating the first-strand complementary DNA by reverse transcription, amplifying double-stranded DNA fragments using independent PCR reactions with common anchored and different arbitrary primers, resolving them through electrophoresis in denaturing urea polyacrylamide gel, developing the gel by silver staining, identifying the differentially expressed bands, eluting DNA from those bands, and sequencing to identify the genes related to foot rot tolerance in black pepper. This chapter details the materials and procedures involved from the selection of tolerant and susceptible cultivar to the identification of differentially expressed genes. Compared to the widely used RNA-Seq, DDRT-PCR is a less technology-intensive but cost-effective strategy, which can be practiced in a medium-class laboratory for identifying the major genes and pathways contributing to a specific trait.

Key words DDRT-PCR, Fungal resistance, Gene expression, Molecular techniques, Pathway analysis, Urea PAGE

1 Introduction

Black pepper (*Piper nigrum* L.), the king of spices, with a global production of 0.57 million tons, accounts for a 7.1 billion dollar world trade. India has an annual production of 0.12 million tons from an area of 0.26 million ha with a productivity of 426 kg/ha [1]. The world's largest exporters are Vietnam, India, and Brazil, having the export shares of 35.3%, 15.0%, and 7.0%, respectively,

Muthappa Senthil-Kumar and Hema Ramanna (eds.), *Plant Biotic Stress: Methods and Protocols*, Methods in Molecular Biology, vol. 2966, https://doi.org/10.1007/978-1-0716-4746-2_15,

followed by Indonesia, Malaysia, and Sri Lanka, which together account for more than 20% of the exports. The United States, followed by the Netherlands, Germany, the United Kingdom, and Japan, account for more than 50% of global black pepper imports. Since 2016, European imports have grown by 5% annually and reached 0.077 million tons in 2020 [2].

Foot rot, also known as quick wilt, caused by the oomycete pathogen *Phytophthora capsici*, is the most serious disease of black pepper. This fast-spreading disease causes devastation in the plantations, leading to 60–80% crop losses. All parts of the plants are prone to the infection by this soilborne oomycete. Dried remnants and debris in soil are the primary sources of infection [3]. Even though large-scale screening of black pepper germplasm has been performed by research groups around the world, a truly resistant or immune line is yet to be obtained [4]. Two field-tolerant/ moderately resistant cultivars, IISR-Shakthi and IISR-Thevam, have been developed and released for commercial cultivation by ICAR-Indian Institute of Spices Research (IISR), Kozhikode, India [5]. Besides this, Panniyur-8, Vijay and Panniyur-9 varieties released by Kerala Agricultural University, India, are also field-tolerant to this disease.

Even though the roles of few genes, *hydroxymethyl glutaryl CoA reductase* (*Hmgr*) [6], *β-1,3-glucanase* [7], *chitinase* [8], *phenylalanine ammonia lyase* [9], *phenylalanine ammonia lyase*, *peroxidase* [10], *osmotin* [11], and *cytosolic ascorbate peroxidase* (*cAPX*) [12], in foot rot tolerance is shown and some of the pathways are understood through transcriptome analyses [13, 14], a thorough understanding of the molecular mechanisms is yet to be drawn. Differential display reverse transcription PCR (DDRT-PCR) methodology to study the differential gene expression under any specific situation [15]. Even with the current availability of robust gene expression analysis strategies, DDRT-PCR remains as a preferred strategy for the identification of the differentially expressed genes, due to its low cost, simplicity, and medium throughput. This chapter details the stepwise procedures involved in finding the genes and pathways involved in foot rot tolerance in black pepper through DDRT-PCR.

2 Materials

2.1 Plant Material

The root cuttings of two black pepper cultivars, one with field tolerance to foot rot and other susceptible, has to be used in this study. IISR-Shakthi and IISR-Shubhakara, released for commercial cultivation by ICAR-IISR have field tolerance and susceptibility to this disease, respectively. Their seedlings have to be planted and maintained under glasshouse facility.

2.2 Pathogen Isolation, Inoculum Preparation, and Inoculation

1. Infected leaf samples.
2. Distilled water.
3. Petri plates.
4. Sterile scalpel.
5. Carrot agar medium.
6. Cork borer.
7. Inoculation loop.
8. Microscope.
9. Laminar airflow (LAF) hood.

2.3 Total RNA Extraction

1. TRIzol Reagent.
2. Chloroform.
3. Ice-cold isopropanol (100%).
4. Polyvinylpyrrolidone (PVP).
5. β-mercaptoethanol (β-ME).
6. Diethyl pyrocarbonate (DEPC).
7. 75% ethanol (DEPC-treated).
8. Autoclaved DEPC-treated water.
9. Autoclaved DEPC-treated pestle and mortar.
10. 1.5 mL microcentrifuge tubes.
11. Micropipettes.

2.4 Analysis of Quality and Quantity of Total RNA

1. Agarose.
2. 10× MOPS (3-(N-morpholino) propane sulfonic acid) (pH 7.0).
3. 6× gel RNA loading dye.
4. *Hind*III/*Eco*RI double-digest ladder/100 bp DNA ladder.
5. Ethidium bromide solution (0.5μg/mL).
6. Eletrophoresis unit.
7. NanoDrop® ND-1000 (Thermo Fisher Scientific, USA).

2.5 Differential-Display Reverse Transcription Polymerase Chain Reaction (DDRT-PCR)

2.5.1 First-Strand cDNA Synthesis

1. RNA sample.
2. Oligo (dT)18 primer (500μg/mL).
3. First-strand buffer (5×).
4. 10 mM dNTP mix (2.5 mM each).
5. DTT (0.1 M).
6. M-MuLV reverse transcriptase (200 units/μL).
7. RNaseOUT Recombinant Ribonuclease Inhibitor (40 units/μL).

8. Autoclaved DEPC-treated water.
9. Thermal cycler.

2.5.2 Second-Strand cDNA Synthesis

1. 10× reaction buffer A.
2. dNTP mix (25μM),
3. T11C anchored primer (2μM).
4. AP (1–8) arbitrary primer (2μM).
5. cDNA (25 ng).
6. Taq DNA polymerase (5 U/μL).
7. Nuclease-free water.
8. Thermal cycler.

2.5.3 Primers Used in cDNA Second-Strand Synthesis

T11C with the sequence AAGCTTTTTTTTTTTT (16 bases), AP1 with the sequence AAGCTTGATTGCC (13 bases), AP2 with the sequence AAGCTTCGACTGT (13 bases), AP3 with the sequence AAGCTTTGGTCAG (13 bases), AP4 with the sequence AAGCTTCTCAACG (13 bases), AP5 with the sequence AAGCTTAGTAGGC (13 bases), AP6 with the sequence AAGCTTGCACCAT (13 bases), AP7 with the sequence AAGCTTAACGAGG (13 bases), and AP8 with the sequence AAGCTTTTACCGC (13 bases).

2.5.4 Materials for Control PCR Reaction (Total Volume: 50μL)

2.0μL of newly synthesized first-strand cDNA (1:1000 dilution), 5.0μL of 10× PCR buffer, 1.0μL of 10 mM dNTP mix, 3.0μL of 25 mM $MgCl_2$, 1.5μL of forward actin primer (ActF: 5′-TCCATCTTGGCATCTCTCAG-3′), 1.5μL of reverse actin primer (ActR: 5′-GTACCCGCATCAGGCATCTG-3′), 0.5μL of Taq DNA polymerase (5 U/μL), and 35.5μL of nuclease-free water.

2.5.5 Temperature Profile for Control Thermal Cycling

The initial denaturation is at 94 °C for 3 min (1 cycle). Denaturation occurs at 94 °C for 30 s, repeated for 35 cycles. Annealing takes place at 58 °C for 30 s, followed by extension at 72 °C for 45 s.

2.5.6 Thermal Cycling for Second-Strand cDNA Synthesis (Total Volume: 20μL)

2.0μL of 10× reaction buffer A, 1.6μL of dNTP mix (25μM), 2.0μL of T11C anchored primer (2μM), 2.0μL of AP (1–8) Arbitrary primer (2μM), 2.0μL of cDNA (25 ng), 0.2μL of Taq DNA polymerase (5 U/μL), and 10.2μL of nuclease-free water.

2.5.7 Temperature Profile for PCR for Second-Strand cDNA Synthesis

The initial denaturation is at 94 °C for 2 min. Denaturation occurs at 94 °C for 30 s. Annealing takes place at 40 °C for 2 min, followed by extension at 72 °C for 1 min. From denaturation to annealing, steps are repeated for 40 cycles. The final extension is at 72 °C for 10 min, and the reaction is held at 4 °C for 10 min.

2.6 Urea PAGE and Silver Staining

2.6.1 Equipment used

1. Sequi-Gen GT sequencing cell (BioRad Laboratories Inc., USA) along with PowerPac HV 5000.
2. Reciprocating shaker.

2.6.2 Chemicals used

1. Urea PAGE Bind-Silane:995μL of ethanol, 5μL Sigmacote of Bind-Silane, and 5μL of acetic acid.
2. Urea PAGE Repel-Silane 995μL of ethanol, 500μL Sigmacote of Repel-Silane, and 5μL of acetic acid.
3. **Monomer solution (40%):** 38 g acrylamide, 2 g of Bis-acrylamide, and 100 mL of distilled water.
4. **Ammonium persulphate (10%) (APS):** 0.1 g of APS (0.1 g) and 1 mL of distilled water.
5. **Urea acrylamide solution (6%):** 25.2 g of urea, 6 mL of 10× TBE, 9 mL of 40% monomer solution, double-autoclaved distilled water to make up volume to 60 mL, 600μL of APS, and 60μL of TEMED.
6. **10× TBE (pH 8.0):** 27 g of Tris base, 13.75 g of boric acid, 2.323 g of EDTA, and double-autoclaved distilled water to make up volume to 250 mL.
7. **0.5× TBE (1 L):** 50 mL of 10× TBE, double-autoclaved distilled water to make up volume to 1 L.
8. **Formamide dye (10 mL):** 9.5 mL of formamide, 2.5 mg of bromophenol blue, 2.5 mg of xylene cyanol, and 100μL of 0.5 M EDTA (pH 8.0).
9. **NaOH (10%) for silver staining.**
10. **Fixer solution:** 150 mL of absolute ethanol, 7.5 mL of acetic acid, and autoclaved double-distilled water to make up volume to 1.5 L.
11. **Stainer solution:** 2.25 g of silver nitrate, 2.25 mL of 37% formaldehyde, and autoclaved double-distilled water to make up volume to 1.5 L.
12. **Developer solution:** 22.5 g of sodium hydroxide, 2.25 mL of 37% formaldehyde, and autoclaved double-distilled water to make up volume to 1.5 L.

2.7 Isolation of Differential cDNA Bands from Polyacrylamide Gel

1. Distilled water.
2. Microcentrifuge tubes.
3. Sterile scalpel.
4. UV transilluminator (Wealtec Bioscience Co. Ltd, Taiwan).

2.8 Elution of DNA from the Sliced Gel Fragments

1. GeneJET gel extraction kit including DEA buffer, DEB buffer, wash buffer, and desalting buffer (Thermo Scientific).
2. Vortexer (Tarson Spinwin MC-00, India).

3. 1.5 mL microcentrifuge tubes.
4. Agarose gel electrophoresis units.
5. Thermal cycler.
6. Incubator.

3 Methods

3.1 Collection, Isolation, and Maintenance of Pathogen

1. Collect black pepper plants showing typical symptoms of *Phytophthora* foot rot disease.
2. Wash the infected leaves thoroughly with mild detergent (Tween 20) and running tap water, air-dry, and blot with sterile tissue paper.
3. Cut the infected leaf into small bits with a sterile blade and incubate in sterile distilled water for 24 h in an air-conditioned chamber for the development of sporangia. Observe the leaf bits under microscope for hyphae and lemon-shaped sporangia.
4. Inoculate the leaf bits showing the development of mycelia and sporangia, in Petri plates with carrot agar medium (CAM), aseptically in laminar airflow (LAF) hood, and incubate at low temperature (24 ± 2 °C).
5. Examine the Petri plates for the mycelial growth from the next day onward.
6. The oomycete culture has to be further purified by using hyphal tip method.

3.2 Inoculum Preparation and Artificial Inoculation on Plants

1. Using a cork borer, make 5 mm diameter discs from 7-day-old pathogen cultures on CAM and place the discs in Petri plates containing autoclaved sterile water.
2. Wipe the tender leaves of the black pepper plants raised in pots with distilled water, and remove the dust particles.
3. Using sterile inoculation needle, give a pinprick on the abaxial side of the leaves of both susceptible and tolerant plants, place the culture disc over the prick, and cover the prick with thin layer of sterile wet cotton.
4. Water the plants and cover individual whole plants with polyethylene bags to maintain high humidity (Fig. 1).
5. Maintain a low temperature for the development of symptoms.

3.3 Collection of Leaf Sample

1. Collect the leaf samples from control plants, susceptible plants, and tolerant plants at time intervals of 0, 2, 4, 6, 12, 24, and 48 h after pathogen inoculation.
2. Label and pack the samples in aluminum foil in icebox, and proceed immediately with the RNA extraction.

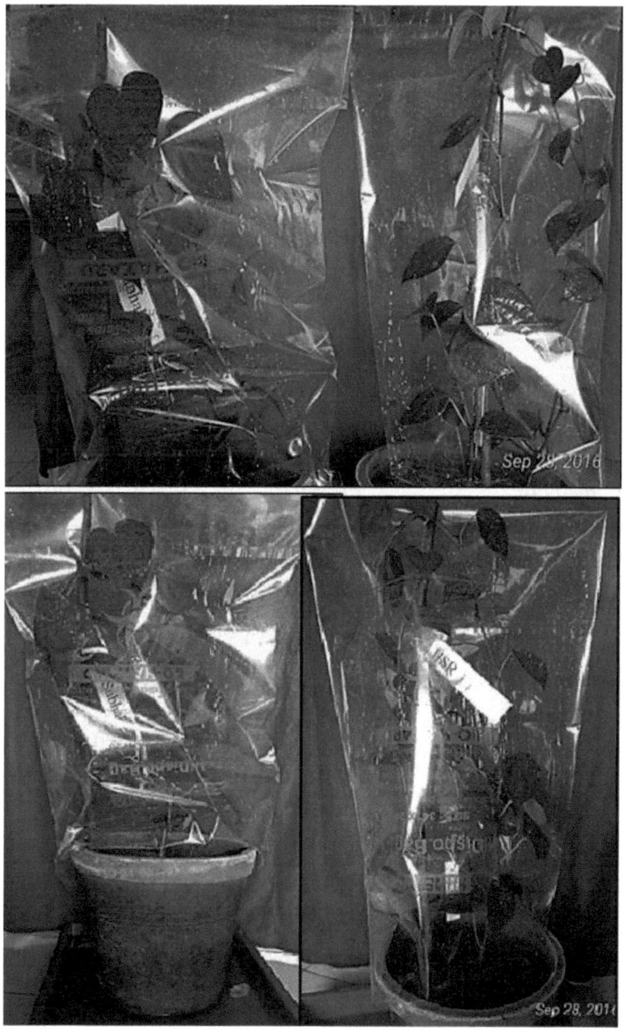

Fig. 1 IISR-Subhakara (top) and IISR-Shakthi (bottom) plants artificially inoculated with *Phytopthora capsici* culture discs and maintained at 25 ± 2 °C and high relative humidity (>90%) by covering with polybag and spraying distilled water

3.4 Total RNA Extraction

1. Pulverize the collected leaf samples from control, susceptible, and tolerant plants of varying time intervals in liquid nitrogen using DEPC-treated mortar and pestle.
2. Add 1 mL of TRIzol reagent and transfer the homogenate to a 2 mL centrifuge tube.
3. Incubate the homogenate at room temperature till the solution becomes brown, and centrifuge the content at 12,000×*g for 10 min at 4 °C.

4. Transfer the supernatant to a 2 mL centrifuge tube, add 200μL of chloroform, shake the content vigorously for 15 s, and keep at room temperature for 10 min.
5. Centrifuge the contents at 12,000×*g* for 10 min at 4 °C, transfer the supernatant into 1.5 mL microcentrifuge tube, and add 500μL of ice-cold isopropanol.
6. Mix the contents by slight inversion, incubate in ice for 10 min, and centrifuge at 12,000×*g* for 10 min at 4 °C.
7. Discard the supernatant and wash the pellet with 1 mL of 75% ethanol by centrifuging at 5000×*g* for 2 min at 4 °C.
8. Discard supernatant, dry the pellet in LAF, and dissolve in 20μL of DEPC-treated water.

3.5 Analysis of Quality and Quantity of Isolated RNA

3.5.1 Quality Analysis by Formaldehyde-Agarose Gel Electrophoresis

1. Prepare 1 L 1× MOPS buffer by diluting 100 mL of 10× MOPS buffer in 900 mL of double-autoclaved DEPC-treated water. This buffer is used for gel preparation and for filling the electrophoresis tank.
2. Seal the open end of casting tray with cello tape, keep it on a horizontal platform, and place the comb at the one end.
3. Dissolve agarose (0.5 g) in 5 mL of 10× MOPS and 47.3 mL of autoclaved DEPC-treated water by boiling in a microwave oven, and allow it to cool for a few minutes at room temperature.
4. Add 3μL of ethidium bromide, followed by 2.7 mL of formaldehyde to the molten gel. Transfer to the tray and allow the gel to solidify at room temperature.
5. Submerge the gel to a depth of 1 cm in 1× MOPS buffer in the electrophoresis tank, with wells facing the cathode.
6. Denature 5μL of total RNA in a thermal cycler at 96 °C for 3 min and immediately cool on ice.
7. Mix the denatured RNA with 1μL of 6× RNA loading dye and load onto the gel. Also, load 3μL of DNA ladder in the nearby well to mark the molecular weight.
8. Run the gel at 50 V and document the image.

3.5.2 Spectrophotometric Analysis of Isolated RNA

1. Wipe the paddle stand and load 1μL of double-autoclaved DEPC-treated water as blank to set the instrument to zero (*see* **Note 1**).
2. Load 1μL of sample and measure the absorbance ratios OD_{260}/OD_{280} and OD_{260}/OD_{230}.
3. Samples with OD_{260}/OD_{280} ratio in the range of 1.8–2.0 and OD_{260}/OD_{230} ratio >1.0 are considered as good quality RNA with no DNA or protein contamination.

3.6 Differential-Display Reverse Transcription PCR

3.6.1 First-Strand cDNA Synthesis

1. Incubate 3μL RNA (~1μg), 1μL oligo-dT primer, and 8μL nuclease-free water at 65 °C for 5 min in a 0.2 mL microcentrifuge tube in a thermal cycler, and then quick-chill on ice. Spin the mixture briefly to collect the content at the bottom of the tube (*see* **Note 2**).
2. Add 4μL 5× first-strand buffer, 2μL 10 mM dNTPs, 1μL RNAseOUT Recombinant Ribonuclease Inhibitor, and 1μL M-MuLV reverse transcriptase enzyme to the tube, one by one, in the order.
3. Gently mix the contents, incubate at 42 °C for 10 min, inactivate the reaction by incubating 70 °C for 5 min, and store at − 80 °C.

3.6.2 Control PCR Amplification

1. Dilute the cDNA generated with the control first-strand cDNA reaction to 1:1000 in nuclease-free water (*see* **Note 3**).
2. Composition of the reaction mixture and the thermal cycling program are shown in Subheadings 2.5.4 and 2.5.5, respectively.
3. Load 20μL of the reverse transcription-PCR product on 1% agarose gel and observe for the 450 bp band, representing the *actin* gene.

3.6.3 Second-Strand cDNA Synthesis

1. Composition of the PCR reaction mixture and thermal cycling program for second-strand synthesis are presented in Subheadings 2.5.6 and 2.5.7, respectively. For each arbitrary primer combination, separate PCR reaction is done (*see* **Note 4**).
2. Resolve the PCR product from each primer combination independently in 6% Urea polyacrylamide gel electrophoresis (PAGE) system and visualize by silver staining.

3.7 Urea PAGE

1. Wipe the IP (Integral plate) and bind plate thoroughly with distilled water and allow to dry.
2. Further wipe with plates with absolute alcohol, followed by 10% NaOH and dry for 30 min.
3. Apply Bind-Silane and Repel-Silane onto the dried bind plate and integral plate, respectively.
4. Drop a few microliters of distilled water on both sides of the spacer sheets, place them on the sides of the IP, and cover with bind plate with Bind-Silane-applied surface facing the IP.
5. Place the clamping unit along the sides of the plates and lock by folding the lever toward the plates.
6. Position the entire unit vertically in the bottom frame and lock using the knots on both sides in the bottom frame.
7. Maintain the entire unit parallel to the work area and perform level-adjust using level balancer.

8. Inject acrylamide-bis-acrylamide monomer solution slowly through the cavity in the bottom frame, with no bubble formation.
9. Place the comb at the top of the gel in inverted position and allow solidification.
10. After solidification, remove the bottom frame and transfer the plates to electrophoresis tank containing 0.5× TBE buffer. Fill the inner chamber of IP with 0.5× TBE buffer, revert the comb with shark tooth toward the gel, and purge the wells with a pipette.
11. Load the wells with 1.0μL of formamide dye, connect the power supply unit (BioRad PowerPac HV), and allow for a pre-run at 1200 V for 30 min. Switch off the power and purge the wells again to avoid air bubbles.

3.8 Sample Preparation and Electrophoresis

1. Prepare the samples by mixing 4.0μL of the sample with 2.0μL loading dye (10.0μL formamide, 0.5 M EDTA, pH 8.0, 10.0 mg xylene cyanol, and 10.0 mg bromophenol blue).
2. Denature the samples at 96 °C for 5 min, immediately dip in ice and load into the well.
3. Restart the power supply and run at 1200 V for ~3 h, till three-fourth of the gel is covered.
4. Switch off the power, dismantle the plates from the unit, apply ice on the plates, and allow the plates to cool. Separate the plates with the help of spatula.

3.9 Silver Staining

1. Immerse urea gel bind plate in the fixer solution and maintain under shaking condition for 10 min.
2. Wash the plate twice in distilled water, transfer to a tray containing 0.1% silver nitrate solution, and shake intermittently for 10 min.
3. Wash the plate quickly in distilled water, transfer to the developer solution, and shake at 50 rpm till the bands develop.
4. Transfer the gel to 10% acetic acid and maintain with no shaking for 3 min.
5. Keep the plate in distilled water for 5 min and transfer to the sequencing gel casting stand for drying (Fig. 2).

3.10 Isolation of Differential cDNA Bands from Polyacrylamide Gel

1. Rehydrate each of the differentially expressed cDNA fragments with 10μL distilled water for 20 min.
2. Place the gel on a transilluminator and scoop out each fragment with a sterile surgical blade, avoiding much exposure to UV.
3. Incubate each gel slice in 40μL of distilled water for 2 h at room temperature.

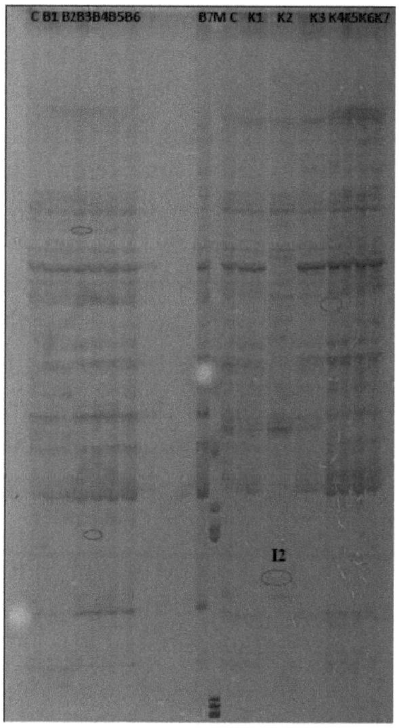

Fig. 2 Profile resolved on denaturing PAGE on double-stranded DNA fragments amplified in a thermal cycler using anchored primer T11C and arbitrary primer AP5, from single DNA strands reverse transcribed from the mRNA transcripts. B1-B7 and K1-K7 are fragments generated from the RNA samples collected 0, 2, 4, 6, 12, 24, and 48 h after pathogen inoculation in Subhakara and Shakthi cultivars. *C* Control treatment, *M* marker. Differentially displayed bands, which were further eluted from the gel and sequenced are shown in red circles

3.11 Elution of DNA from Gel Fragments

Elute cDNA fragments from the gel slices using GeneJET Gel Extraction Kit, following manufacturer's guidelines:

1. Weigh the gel slice in 1.5 mL microcentrifuge tube and add 3× gel volume of gel solubilization buffer (DEA buffer).
2. Resuspend the gel in gel solubilization buffer by vortexing and heat the contents at 75 °C, with intermittent vortexing in every 2 to 3 min, until the gel is completely dissolved.
3. Add 0.5× gel solubilization volume of binding buffer (DEB buffer) and mix thoroughly.
4. Place the spin column in a 2 mL collection tube, transfer the solubilized gel into the spin column, and centrifuge at 12,000×*g* for 1 min.
5. Discard the filtrate and add 500μL of wash buffer (W1) to the spin column and centrifuge at 12,000×*g* for 30 s.

6. Discard the filtrate, add 700μL of desalting buffer (W2) to the spin column, and centrifuge at 12,000×*g* for 30 s. Repeat this step to ensure the complete removal of salts.
7. Discard the filtrate, place the spin column in the collection tube, and centrifuge the column at 12,000×*g* for 1 min to remove the residual buffer.
8. Transfer the spin column to a fresh 1.5 mL microcentrifuge tube.
9. Pre-warm the eluent at 65 °C to improve the elution efficiency and add 50μL of eluent to the center of the spin column.
10. Allow the column to stand for 1 min at room temperature and centrifuge at 12,000×*g* for 1 min.
11. Check the eluted cDNA fragments on 1.0% agarose gel and store at −20 °C.
12. Use 10 μL of eluent for selective PCR amplification with modified dNTP concentration of 2 mM and with the same anchored [T11C] and arbitrary primers [AP (1–8)].
13. Electrophorese the product in 1.5% agarose gel, elute DNA from the band, and perform Sanger sequencing.

3.12 Sequencing and Sequence Analysis

The forward and reverse primers in the selective PCR amplification reaction will be used for sequencing on Sanger platform.

1. Generate a contig from the forward and reverse sequences, resulting from the paired-end sequencing of DNA eluted from each band, using online tool 'Emboss Merger' (http://www.bioinfo.nhri.org.tw/gui/) or CAP3 sequence assembly program (https://doua.prabi.fr/software/cap3).
2. Identify the differentially expressed genes through BLASTn of each contig sequence.
3. Obtain the respective AGI codes for the sequence of cDNA or the transcript-derived fragments using the translated BLAST analysis at (www.ncbi.nlm.nih.gov/Blastx).
4. Perform eNorthern analysis using Botany Array Resource (http://bbc.botany.utoronto.ca/) by selecting expression browser option from the BAR homepage.

4 Notes

1. The samples showing three intact bands in the formaldehyde agarose gel can be further analyzed to determine the quality and quantity of the isolated total RNA using a microvolume spectrophotometer.

2. The total RNA predominantly consists of rRNA, tRNA, and mRNA. Isolation of intact mRNA is essential for successful first-strand synthesis and PCR amplification. The mature mRNA molecules carry poly (A) tail at their 3′ termini, to which oligo dT primers used in the first step in cDNA synthesis will anneal. Reverse transcriptase enzyme can synthesize cDNA from the isolated mRNA under favorable conditions.
3. In order to verify the result of first-strand cDNA synthesis, a control PCR reaction is carried out using the primers of the housekeeping gene *Actin* (ActF 5′-TCCATCTTGGCATCT CTCAG-3′) and (ActR 5′-GTACCCGCATCAGGCATCTG-3′), generate the band at 450 bps.
4. Maintain a negative control is to ensure that no nonspecific amplification occurs due to contamination in any reagents. Use special thin-walled tubes to ensure rapid and equal temperature changes throughout the reaction mixture.

Acknowledgments

This work was financially supported by Department of Biotechnology (DBT), Ministry of Science and Technology, Government of India.

References

1. Sabu SS (2022) Implications of trade agreements on India's trade in black pepper and its products. Thesis submitted to Kerala Agricultural University, India, p 162
2. Permata R, Ichsan I (2023) An analysis of pepper exports and imports on economic growth in Indonesia. J Malikussaleh Public Econ 5(2): 26–34
3. Rajan PP, Sarma YR, Anandaraj M (2002) Management of foot rot disease of black pepper with *Trichoderma* spp. Indian Phytopathol 55(1):34–38
4. Bhai RS, Anandaraj M, Sarma YR, Veena SS, Saji KV (2007) Screening of black pepper (*Piper nigrum* L.) germplasm for resistance to foot rot disease caused by *Phytophthora capsici* Leonian. J Spices Aromat Crops 16(2): 115–117
5. Anonymous (2004) ICAR-Indian Institute of spices research, research highlights 2003–04. IISR, Kozhikode, p 5
6. Girija D, Beena PS, Nazeem PA, Purushothama MG (2007) Molecular cloning of a cDNA fragment encoding *hydroxy methyl glutaryl CoA reductase* in *Piper colubrinum*. In: Keshavachandran R, Nazeem PA, Girija D, John PS, Peter KV (eds) Recent trends in horticultural biotechnology, new India publishing agency, New Delhi, pp 663–669. ISBN:9788189422592
7. Vijesh Kumar IP, Johnson GK, Anandaraj M (2016) Real-time quantitative RT-PCR of some defense response genes in *Piper colubrinum* challenge inoculated with *Phytophthora capsici*. Int J Agric Sci Res 6(1):69–78
8. Varma RS, George KJ, Balaji S, Parthasarathy VA (2009) Differential induction of chitinase in *Piper colubrinum* in response to inoculation with *Phytophthora capsici*, the cause of foot rot in black pepper. Saudi J Biol Sci 16(1):11–16
9. Anh TT, Linh TB, Phong NV, Bien TLT, Tram THN, Don LD (2018) Expression of proteins related to *Phytophthora capsici* tolerance in black pepper (*Piper nigrum* L.). Int J Agric Innov Res 6(4):2319–1473
10. Hao C, Xia Z, Fan R, Tan L, Hu L, Wu B, Wu H (2016) De novo transcriptome sequencing of black pepper (*Piper nigrum* L.) and an analysis of genes involved in phenylpropanoid

metabolism in response to *Phytophthora capsici*. BMC Genomics 17:822

11. Anu K, Jessymol KK, Chidambareswaren M, Gayathri GS, Manjula S (2015) Downregulation of *osmotin* (PR5) gene by virus-induced gene silencing (VIGS) leads to susceptibility of resistant *Piper colubrinum* Link. to the oomycete pathogen *Phytophthora capsici* Leonian. Indian J Exp Biol 53(6):329–334
12. Vandana VV, Bhai RS (2018) Differential expression of PR genes in response to *Phytophthora capsici* inoculation in resistant and susceptible black pepper (*Piper nigrum* L.) lines. Eur J Plant Pathol 150(3):713–724
13. Wang P, Liu X, Guo J, Liu C, Fu N, Shen H (2015) Identification and expression analysis of candidate genes associated with defense responses to *Phytophthora capsici* in pepper line "PI 201234". Int J Mol Sci 16:11417–11438
14. Mahadevan C, Krishnan A, Saraswathy GG, Surendran A, Jaleel A, Sakuntala M (2016) Transcriptome-assisted label-free quantitative proteomics analysis reveals novel insights into *Piper nigrum-Phytophthora capsici* phytopathosystem. Front Plant Sci 7:785
15. Bauer D, Warthoe P, Rohde M, Strauss M (1994) Detection and differential display of expressed genes by DDRT-PCR. Genome Res 4:S97–S108

Chapter 16

Estimation of Foliar Volatiles Emitted by Mycorrhizal Colonized Blackgram (*Vigna mungo* L) Infested with *Spodoptera litura*

Selvaraj Anandakumar, Dhandapani Senthamilselvi, and Thangavel Kalaiselvi

Abstract

Plant-emitted volatile organic compounds (VOCs) play a significant role in signaling and tolerance to biotic stressors, including insect pest infestation. Mycorrhizae, a symbiotic fungus, improves the tolerance of blackgram plants to *Spodoptera litura* by altering the profiling of foliage VOCs. Recently, gas chromatography-mass spectrometry (GC-MS) coupled with headspace (HS) trapping of VOCs is the most frequently used analytical technique to understand the metabolic process and responses of plants to biotic stresses. This method performs four steps such as (i) trapping of HS-VOCs, (ii) concentration and enrichment of VOCs, (iii) transfer of VOCs from air sample to analytical device, and (iv) detection and identification of compounds. HS volatiles is trapped using fiber polymers like Tenax TA and desorbed thermally in GC-MS with TD autosampler and thermal desorption (TD). The identification of VOCs compounds is performed by searching mass spectral peaks against NIST mass spectral library. This chapter provides the detailed procedure for the estimation of plant-produced VOCs using HS sapling coupled with gas chromatography-mass spectrometry (TD–GC/MS) with TD autosampler and thermal desorption method.

Key words Blackgram, Mycorrhizae, *Spodoptera litura*, Volatiles, Thermal desorption (TD), GC-MS, NIST library

1 Introduction

Blackgram, *Vigna mungo* (L.), is an important leguminous pulse crop and a major contributing source of protein to the world's growing population. It gets colonized symbiotically by mycorrhizae in addition to *Rhizobium,* which fixes atmospheric nitrogen by forming root nodules and supplies nitrogen to the plants [1]. Mycorrhizae symbiosis improves plant nutrition and growth by mobilizing unavailable inorganic soil phosphate from soil to plants and improves plant tolerance to both biotic and abiotic

Muthappa Senthil-Kumar and Hema Ramanna (eds.), *Plant Biotic Stress: Methods and Protocols*, Methods in Molecular Biology, vol. 2966, https://doi.org/10.1007/978-1-0716-4746-2_16,

stresses, including herbivorous insect infestation [2, 3]. *Spodoptera litura* is a polyphagous pest that feeds several crops including blackgram, which causes intense damage to plants and leads to economic yield loss [4]. Mycorrhizae colonization improves plant defense against herbivorous insects through various direct and indirect mechanisms including regulation of plant volatile organic compounds (VOCs) production, which act as a cue for repellent pests and attraction predators to particular crop pests [5]. Even though, the exploration of plant-released VOCs is required to understand the metabolic activity of plants during various levels of biotrophic interaction and its significant role in plant defense against different types of stressors including herbivorous insect infestation.

Plants produce blends of low-molecular-weight VOCs to build communication with other organisms and the surrounding environment. They are very specific to plant parts such as roots, leaves, flowers, and fruits. The majority of VOCs belong to the category of terpenes, phenolics, benzenoids, nitrogen, and sulfur-containing compounds, indole, as well as fatty acid derivatives [6]. Plant-derived VOCs have a diver's application and biological functions include improving resistance and priming of plant defense to herbivore attack [7]. In recent advanced research, we could understand the responses of plant systems to biotic stress by studying the plant's released VOCs using various methods. However, extraordinary instrumentation and procedure are essential to trap the VOCs and analyze them with adequate resolution and sensitivity.

Generally, plant VOCs have been analyzed by using gas chromatography-mass spectrometry (GC-MS) coupled with volatile trapping. Various principle steps are involved in the analysis of plant VOCs including (i) collection or trapping of sample, (ii) concentrating and enriching of sample, (iii) transfer of VOCs from air sample to analytical device, and (iv) detection and identification of compounds. Based on the necessity various methods are used to collect the headspace VOCs from plants such as conventional distillation or solvent extraction and advanced solvent-less headspace (HS) sampling methods. HS sampling of plant-released VOCs can be achieved by using two different methods, such as dynamic and static sampling through an active and passive processes [8].

HS sampling coupled with GC-MS analysis is the most frequently used analytical technique for the trapping, extraction, and concentration of VOCs. Based on specific physicochemical characteristics of the volatile compounds, different fiber polymers are currently used to trap the HS gaseous VOCs, such as Tenax TA, polydimethylsiloxane, divinyl benzene, carboxen, polyacrylate, and polyethylene glycol [7, 8]. It is based on a successive transfer of VOCs from plants to the headspace and from the headspace to a polymer fiber coating. The adsorbed VOCs are then released in the injection port of the analytical instrument by thermal desorption. Vaporized compounds are carried out to the stationary phase of the column-containing oven by gas dissolved in a mobile phase and separation of each compound from the mixture based on the molecular mass (mz) at its appropriate retention time (RT). The extracted compound comes out from the gas chromatograph and then enters the detector system. Based on the registered signal given by the compounds in the detector system, spectral peaks of the chromatogram are formed, and the concentration is determined based on the RT and area of the signal produced by the compounds. Finally, the identification of plant-produced VOCs by comparing obtained MS spectral peaks with the library [7, 8].

Static HS sampling/trapping of VOCs using a Tenax TA fiber-containing column coupled with GC-MS analysis is a novel, efficient, and reliable method for analyzing VOCs from plants and microbial systems [7, 8]. As compared to conventional techniques, HS sampling GC-MS analysis has its advantages and disadvantages: (i) the absence of extraction steps with solvents is to eliminate the high cost and excessive preparation time; (ii) simple and fast modern tool; (iii) more employed in many analytical fields such as agriculture, environmental, microbial, and food as well as clinical analysis; and (iv) it combines extraction, concentration, and cleanup into a single step. However, this technique presents some shortcomings, such as high temperatures can cause degradation or chemical modifications of volatile constituents [7, 8].

2 Materials

2.1 Planting Material and Mycorrhizal Inoculum

1. Blackgram (CO6) seeds (*see* **Note 1**).
2. The liquid-based arbuscular mycorrhizal (AM) inoculum contains 100 spores of *Glomus intraradices* per mL of inoculum (*see* **Note 2**).

2.2 Seed Surface Sterilization

1. 10% sodium hypochlorite (NaOCl).
2. 70% ethanol.
3. Sterile distilled water.
4. Sterile petri plates and forceps.

2.3 Plant Growth Medium and Intermittent Sterilization

1. Pot mixture: prepare by mixing of Garden land red soil and sand at the ratio of 2:1 (w/w) (*see* **Notes 3** and **4**).
2. Autoclave (*see* **Note 5**).
3. 2 kg plastic pots.

Table 1
Component of modified Hoagland nutrient solution for nutrients supplements to plants

Chemicals	Concentration of stock solution	Volume of stock solution per liter of the final solution (mL)
Potassium nitrate (KNO_3)	1 M	6.0
Magnesium sulfate ($MgSO_4.7H_2O$)	1 M	1.0
Calcium nitrate tetrahydrate ($Ca(NO_3)_2.4H_2O$)	1 M	4.0
Ammonium dihydrogen phosphate ($NH_4H_2PO_4$)	1 M	2.0
Potassium chloride (KCl)	25.0 mM	2
Boric acid (H_3BO_3)	12.5 mM	2
Manganese chloride ($MnCl_2{\cdot}4H_2O$)	1.0 mM	2
Zinc sulfate ($ZnSO_4.7H_2O$)	1 mM	2
Copper sulfate ($CuSO_4.5H_2O$)	0.25 mM	2
Sodium molybdate ($Na_2MoO_4{\cdot}2H_2O$)	0.25 mM	2
Iron EDTA	50 mM	2
MES (10×)		2
pH		6.8

2.4 Nutrient Solution

1. Modified Hoagland nutrient solution needs for growing plants [9].
2. Add the appropriate volume of all the stock solution (Table 1) of prepared nutrients to 1 L beaker and make up the final volume to 1 L using a volumetric flask after adjusting the final pH of the solution to 6.8 using NaOH or HCl (*see* **Note 6**).

2.5 Mycorrhizal Colonization Assessment

1. 2% HCl.
2. 10% KOH.
3. 0.05% trypan blue staining solution is prepared by adding 0.05 g of trypan blue in a mixture of lactic acid: glycerol: water (2:2:1).
4. Destaining solution: mix lactic acid, glycerol, and water at the ratio of 2:2:1 and use as a destaining solution.
5. Test tubes.
6. Forceps.
7. Microscopic slides.
8. Light microscope.

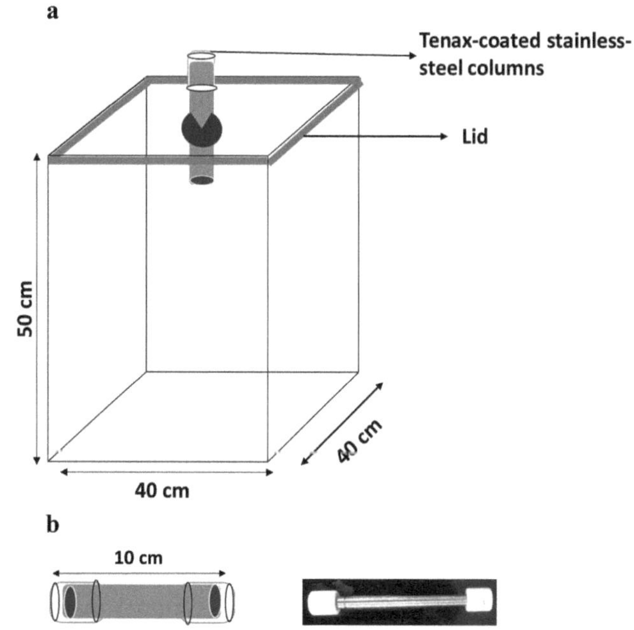

Fig. 1 Schematic diagram of (**a**) headspace (HS) volatile trapping glass column with Tenax-coated stainless-steel columns and (**b**) Tenax-coated stainless-steel columns

2.6 Insect Material and Infestation

1. The eggs of *Spodoptera litura* are purchased from the National Bureau of Agricultural Insect Resources (NBAIR), Bangalore, or collected from the field. Place the eggs in plastic containers with castor leaves (*Ricinus communis*) as feed and incubate them at high relative humidity (about 80%). Then transfer the 100 numbers of first instar neonates to the new plastic containers with castor leaves till they reach third instar stage and use them as a source of insect material.
2. Stainless steel mesh cage (40 × 40 × 60 cm).

2.7 Volatile Collection

1. Headspace collection box (40 × 40 × 50 cm) made up of glass with a movable lid (Fig. 1).
2. Tenax-coated stainless steel columns.

2.8 Volatile Detection and Identification

1. GC-MS autosampler with a TurboMatrix 150 thermal desorber (TD) and 30 m × 250 μm capillary column coated with 5% phenyl methyl siloxane.
2. NIST library 14 (*see* **Note 7**).

3 Methods

3.1 Surface Sterilization of Seeds

1. Immerse the blackgram (CO6) seeds into the 10% sodium hypochlorite (NaOCl) solution for 10 min.
2. Then transfer the seeds to a sterile petri dish containing 70% ethanol and incubate it for 2 min.
3. Finally, wash the seeds with sterile distilled water four times to remove the sterilant residuals from the seed surface (*see* **Note 8**).

3.2 Seed Biotization

1. Take 1 mL of liquid suspension containing 200 numbers of *G. intraradices* (AM fungal) spores in a sterile petri dish.
2. Add 0.1 mL sterile 2% carboxyl methyl cellulose (CMC) sticker solution to the same petri dish containing AM fungal spore inoculum.
3. Then add 40 numbers of surface-sterilize blackgram seeds and mix well to coat approximately 5 AM fungal spores per seed (*see* **Note 9**).
4. Finally, transfer one biotized seed to each of the three replicates of a 2 mL microcentrifuge tube containing 1 mL of sterile distilled water and mix well. Then transfer the suspension to a sterile petri dish and observe under a stereo zoom microscope to examine the approximate number of AM fungal spores adhered on the seed surface during seed biotization.

3.3 Seed Sowing and Plant Growth Condition

1. Intermittently steam-sterilize the red soil and sand mixture (2:1) at 121 °C for 30 min with 15 psi, which kills native AM fungal spores in the soil along with other microorganisms (*see* **Note 10**).
2. Fill the 2 kg of plastic pots with 1.5 kg of sterile soil mixture and transfer the pots to the greenhouse.
3. Sow the biotized seeds by dibbling three seeds per plastic pot filled with a sterile soil mixture.
4. Thinning the seedlings after 5 days of seed germination and maintaining the single seedling per pot.
5. The pots are equally irrigated with 100 mL Hoagland nutrient solution (HNS) and 100 mL of tap water alternatively at two-day intervals.
6. Grow the plants under greenhouse conditions with the optimum temperature of 25 °C to 30 °C and light intensity of 80 $\mu Em^{-2} s^{-1}$ for 30 to 35 days.

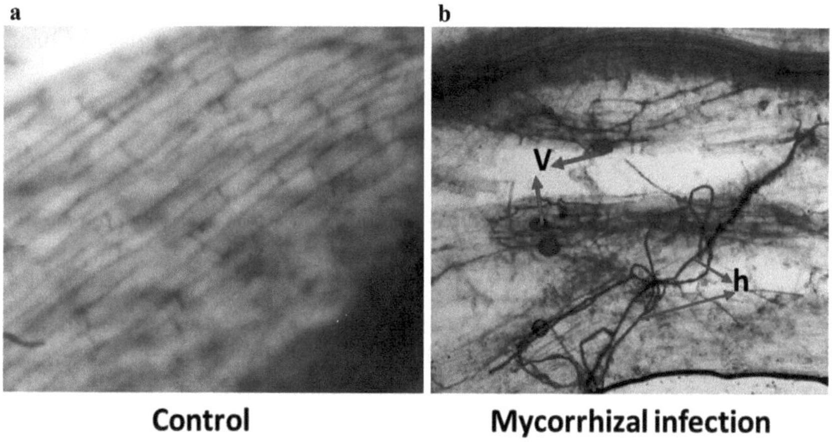

Fig. 2 Presence of mycorrhizal infection in blackgram. (**a**) control (**b**) mycorrhizal infection with intraradical vesicles (v) and hyphal threads (h)

3.4 Examination of Root Mycorrhizal Colonization

1. Uproot the plant with a minimum distribution of roots after 30 to 35 days of sowing.
2. Then, roots are gently washed with running tap water to remove adhering soil particles on the root surface and cut into 1–2 cm bits.
3. Immerse the root bits in 10% KOH solution and keep it in a boiling water bath for 30 min.
4. Transfer the root bits into 2% HCl solution and incubate it for 2 min.
5. Then, discard the HCl solution and wash it with sterile distilled water [10].
6. After washing root bits are stained with 0.05% trypan blue for 12 h and observe AMF infection under a bright field microscope (Fig. 2; *see* **Note 11**).
7. Find out the percent root mycorrhizal colonization by following the formula

$$\text{Mycorrhizal root colonization }(\%) = \frac{\text{No.of infected root bits}}{\text{Total no.of root bits observed} \times 100}$$

3.5 Insect Treatments

1. After 35 days of sowing, release the third instar *S. litura* larvae on the blackgram plant at the rate of 3 numbers of *S. litura* larvae per plant (*see* **Note 12**).
2. Then cover the plants individually in a stainless steel cage to prevent the escape of *S. litura* larvae and allow the setup for 24 h after induction of third instar *S. litura* larvae on blackgram plants.

3.6 Collection of Volatiles

1. After 24 h of *S. litura* larvae infestation, transfer the culturing pot with the insect-infested plant into a glass VOCs collection column with a movable top lid.
2. Then, close the VOCs collection glass column airtightly by using the glass-made top lid of the column; the airtight glass column with one point on the top lid of the column, which is the site of the headspace VOCs collection.
3. Insert the one-end opened stainless steel thermal desorption (TD) tube with Tenax TA into the opening point on the top lid with a stopper of the glass column by manually penetrating the septum and sealing it tightly, while the other end of the TD tube close with cap (*see* **Note 13**).
4. Incubate this setup for 2 to 3 h to trap the shoot volatiles by exposing Tenax TA to the headspace.
5. Remove the tubes after incubation aseptically and close the opening end of the tube with polypropylene caps.
6. Then, immediately place the tubes into the GC/MS-TD autosampler.

3.7 GC-MS Analysis/Detection of Volatiles

1. Insert the Tenax TA TD tube into the GC/MS autosampler with TurboMatrix 150 thermal desorber (TD) to desorb VOCs from the Tenax trap thermally by heating at a rate of 12 °C s^{-1} from 35 °C to 280 °C.
2. The separation of components after injection into the GC/MS–TD analysis is attained by the following conditions: 10:1 split, Helium gas carrier at 1.2 mL min^{-1}, and 20 Psi. The initial oven temperature of 35 °C is held at 2 min and further enhancement to 140 °C at 2 °C min^{-1}. The final trapping temperature is set at 280 °C, attained by increasing 10 °C min^{-1} and held at this temperature for 10 min. The component from the GC capillary column is split 1:1 (v/v) between the MS detector with electron impact mass spectra at 70 eV, and positive ion mode of compounds with a m/z scan range from 40 to 500.
3. The GC-MS-generated chromatogram is used further for analysis and VOCs identification (Fig. 3).

3.8 Identification of Volatiles

1. The GC/MS spectra are used further for analyses and identification of VOCs by searching/comparing the obtained mass spectra with NIST/EPA/NIH Mass Spectral Library, NIST 14, installed in the GC-MS equipment based on most probable hits.
2. The volatile compounds that exhibited more than 90% similarity with mass spectra in the NIST library are shortlisted as tentative compounds of interest.

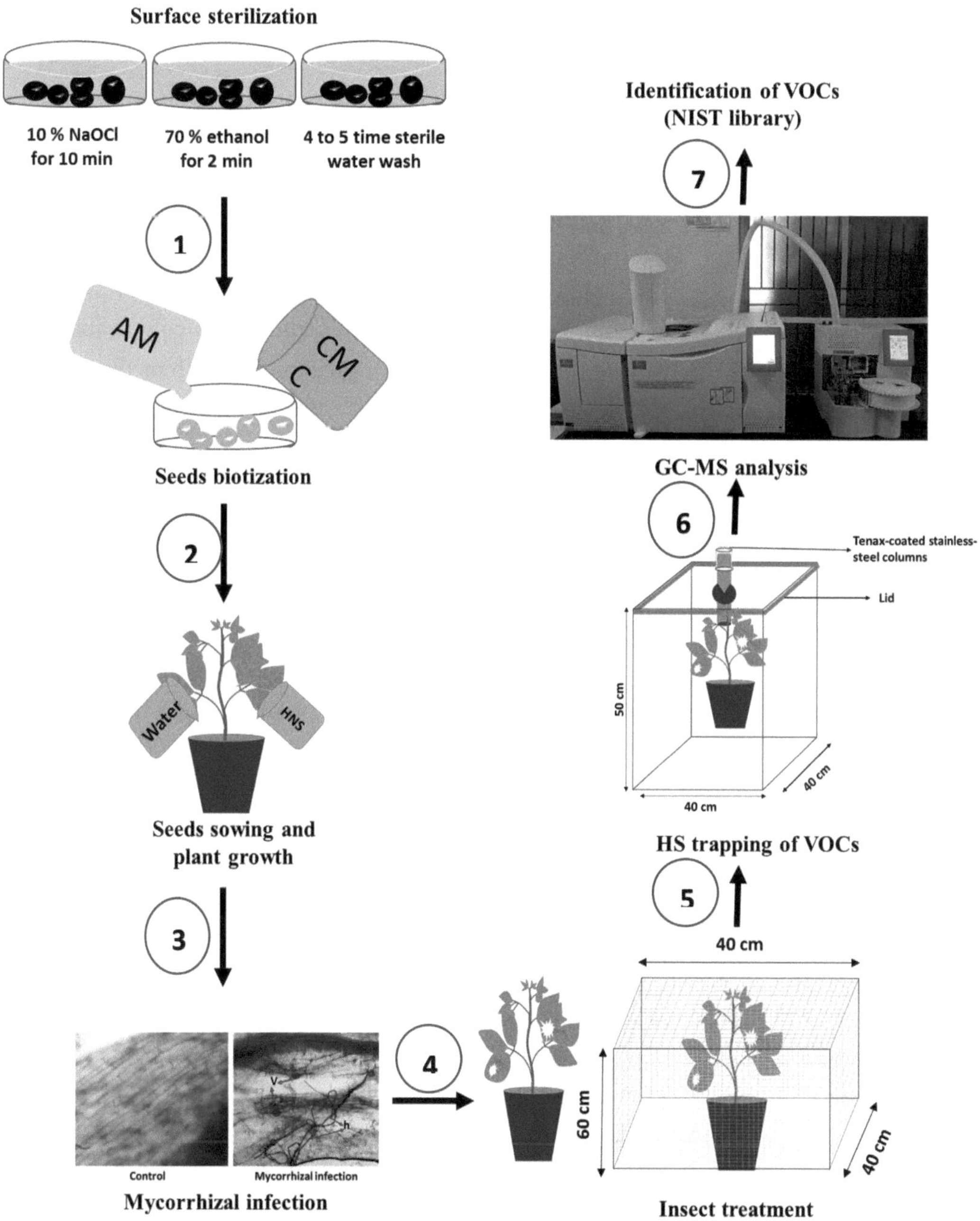

Fig. 3 Schematic representation of GC-MS analysis of VOCs coupled with headspace (HS) trapping and NIST identification of VOCs

4 Notes

1. Blackgram variety CO 6 is a short duration high yielding determinate variety that matures in 60–65 days suitable for Kharif and Rabi season. It is moderately resistant to mungbean yellow mosaic virus (MYMV), stem necrosis, and root rot diseases and tolerant to aphids, stem flies, and spotted pod borer infestation.
2. Mycorrhizal viable spore inoculum load is a more important deciding factor for percent mycorrhizal root colonization.
3. Mixing the soil with sand for preparing growth media is important for improving soil texture, and drainage, and helping easy root growth.
4. Intermittent sterilization of soil is achieved by alternating heating and 24 h incubation at room temperature for 3 days to kill all resistance spores germinated during incubation followed by heating.
5. The heating of soil is to archive by autoclaving with the condition at 121 °C for 30 min with 15 psi.
6. All the nutrient stock solution required for preparing modified Hoagland nutrient solutions are prepared using distilled water and autoclaving it to eliminate microbial contamination.
7. NIST library is the mass spectral reference library widely used to identify unknown compounds by library searching of MS spectra.
8. Seed surface sterilization eliminates pathogens from seeds surface.
9. Optimization of the mycorrhizal spore load necessary for effective mycorrhizal colonization in plant roots is the most important need to find out.
10. The physiochemical properties of plant growth media need to examine before taking sowing seeds.
11. To find out the mycorrhizal colonization percentage needs to observe 50–100 stained root bits under a microscope.
12. The population of *S. litura* larvae introduction per plant is fixed based on the economic threshold level.
13. Before sampling, each fiber was reconditioned for 5 min in the GC injector port at 230 °C.

Acknowledgments

This work was supported by the Department of Agricultural Microbiology, Tamil Nadu Agricultural University, Coimbatore 641 003, India for providing required facilities and GOI-MHRD (FAST)—Microbes to Feed the World (grant F.No. 5-5/2014-TS-VII dt. 04.09.2014) for providing students research fellowship.

References

1. Marimuthu S, Gnanachitra M, Prabu Kumar G, Surendran U (2023) Effect of organic and inorganic sources of phosphorus for enhancing productivity and phosphorus use efficiency in blackgram under acid soils. J Plant Nutr 46:1845–1855
2. Varghese R, Ray JG (2023) Sustainable agriculture of tropical spices: arbuscular mycorrhizal fungi as an ecotechnological tool-a critical review. J Appl Res Med Aromat Plants 36: 100507
3. George NP, Ray JG (2023) The inevitability of arbuscular mycorrhiza for sustainability in organic agriculture—a critical review. Front Sustain Food Syst 7:1124688
4. Ganguly S, Srivastava CP, Premkumari S, Meena RS, Singh RN, Srivastava K, Prakash P (2023) Has insecticidal pressure influenced *Spodoptera litura* (Fabricius, 1775) population genetic structure and genetic diversity in India? Biologia 78:837–849
5. Meier AR, Hunter MD (2019) Mycorrhizae alter constitutive and herbivore-induced volatile emissions by milkweeds. J Chem Ecol 45:610–625
6. Niinemets Ü, Kännaste A, Copolovici L (2013) Quantitative patterns between plant volatile emissions induced by biotic stresses and the degree of damage. Front Plant Sci 4: 262
7. Tholl D, Hossain O, Weinhold A, Röse US, Wei Q (2021) Trends and applications in plant volatile sampling and analysis. Plant J 106:314–325
8. Alborn HT, Bruton RG, Beck JJ (2021) Sampling of volatiles in closed systems: a controlled comparison of three solvent less volatile collection methods. J Chem Ecol 47:930–940
9. Taiz L, Zeiger E (2002) Photosynthesis: physiological and ecological considerations. Plant Physiol 9:172–174
10. Phillips JM, Hayman DS (1970) Improved procedures for clearing roots and staining parasitic and vesicular-arbuscular mycorrhizal fungi for rapid assessment of infection. TBMS 55:158–161

Chapter 17

Development of an *Agrobacterium tumefaciens*-Mediated *In Planta* Transformation Strategy in Pigeonpea for Resistance to *Helicoverpa armigera*

Maniraj Rathinam and Rohini Sreevathsa

Abstract

In planta transformation strategies aid in the development of transgenics in difficult-to-regenerate or recalcitrant plant species. Many economically important crops including the legume pigeonpea fall under the group of plants that are less amenable to tissue culture. The strategy essentially involves *in planta* inoculation of apical meristems of germinating seeds and allowing them to grow into seedlings *ex vitro*. Since T_0 plants are chimeric in nature, transformants are selected in the T_1 generation. In this chapter, we present a shoot apical meristem-targeted *in planta* transformation in pigeonpea for resistance to pod borer, *Helicoverpa armigera*. This protocol can be extrapolated to other plant species as well as for the transformation of genes governing other traits.

Key words Pigeonpea, *Agrobacterium tumefaciens*, Shoot apical meristem, Transformation

1 Introduction

Transgenic technology has emerged as a powerful tool in plant research and used by scientists globally. Indirect DNA transfer methodology, which is the most commonly used method for obtaining transgenic plants, involves the introduction of T-DNA, harboring gene(s) of interest into plant cells by a pathogenic bacterium *Agrobacterium tumefaciens*. So far, a number of successful protocols are available both in crop plants and tree species. In plant species that are amenable to transformation by *A. tumefaciens* and regeneration T-DNA-directed gene transfer remains the method of choice because of its ease, efficient transgene delivery, and integration into the host genome. Based on the utility of tissue culture and plant regeneration, *A. tumefaciens*-mediated genetic transformation is further segregated into (a) plant transformation methods that exploit plant regeneration and *in vitro* tissue culture techniques; (b) transformation methods that avoid tissue culture

Muthappa Senthil-Kumar and Hema Ramanna (eds.), *Plant Biotic Stress: Methods and Protocols*, Methods in Molecular Biology, vol. 2966, https://doi.org/10.1007/978-1-0716-4746-2_17,

regeneration steps. Hence, they are referred to as tissue culture-dependent (*in vitro*) and independent (*in planta*) *A. tumefaciens*-mediated plant transformation strategies, respectively. The major disadvantage of tissue culture-based transformation methods is that they are time-consuming and can yield somaclonal variations in the regenerated plants. Moreover, not all the plants are equally amenable to *in vitro* regeneration. In this regard, scientists have come up with transformation protocols that totally avoid or reduce tissue culture steps [1, 2]. These are called as non-tissue culture-based or *in planta* transformation strategies. Over the decades of establishment of transgenic technology in various plant biotechnology laboratories, *in planta* transformation protocols directing the T-DNA to various plant parts, like flowers, seeds, and apical meristem, have been established [3–5]. Development of a large number of uniform plants *in vitro* in short time, with less labor efforts and minimal reagent requirements are the major advantages of *in planta* transformation protocols. These methods have been commonly preferred and used for the transformation of genes into recalcitrant plant species. This article deals with a shoot apical meristem-targeted *Agrobacterium*-mediated *in planta* transformation strategy in pigeonpea. We present development of pigeonpea transgenics with *cry2Aa* and the following high-throughput analyses for the identification of transformants. Here, *Agrobacterium* is directed toward the plumule, cotyledonary node, and surrounding regions of aseptically germinated 2-day-old young seedlings and allowed to grow, with the characterization for T-DNA integration happening in the T1 generation [3].

2 Materials

Prepare all chemicals, hormones, growth media, etc. in recommended solutions (use only double-distilled autoclaved water and molecular grade solvents) and follow recommended storage conditions.

2.1 In planta *Transformation*

1. Freshly revived culture of *Agrobacterium tumefaciens* (EHA 105) harboring pBINAR-*cry2Aa* (*see* **Note 1**).
2. Freshly prepared LB broth.
3. AB minimal media:

   AB glucose: Dissolve 2.25 g glucose in 450 mL of distilled water and take 45 mL for each bottle and autoclave.

   AB salt: Dissolve 2 g NH_4Cl, 0.6 g $MgSO_4.7H_2O$, 0.3 g KCl, 0.03 g $CaCl_2$, and 0.005 g $FeSO_4.7H_2O$ into 100 mL of distilled water. Autoclave and store it under room temperature.

AB buffer (pH 7.0): Dissolve 6 g K_2HPO_4 and 2 g NaH_2PO_4 in 80 mL of distilled water, set pH, and make up to 100 mL. Autoclave and store it under room temperature.

Take 45 mL AB glucose, add 2.5 mL AB salt, 2.5 mL AB buffer, and 3 mL of starter culture.

4. Antibiotic stock solutions [50 mg/mL kanamycin and 10 mg/mL rifampicin].
5. Petri plates containing the bacterial colonies.
6. Autoclaved double-distilled water.
7. Washed and autoclaved glassware.
8. Filter paper discs cut in the size of petri plates.
9. Soilritc.
10. Pots with soil.

2.2 Identification of Putative Transformants

1. Seeds of primary transformants (T0) and wild type.
2. Prepare kanamycin solution at different concentrations (10, 20, 40, 60, 80, and 100 mg/L) in autoclaved distilled water.
3. Soilrite or quartz sand.
4. Pots.

2.3 Molecular Characterization of the Putative Transformants

2.3.1 Plant DNA Isolation by CTAB Method

1. DNA extraction buffer composition: 2% CTAB, 1.4 M NaCl, 100 mM Tris.cl, 200 mM EDTA, and 0.2% β-mercaptoethanol in autoclaved double-distilled water
2. Polyvinylpyrrolidone (PVP)
3. Chloroform: isoamylalchol (24:1)
4. Isopropanol
5. 70% ethanol
6. 95% ethanol
7. RNase (10 mg/mL)
8. Proteinase K (1 mg/mL)
9. 3 M sodium acetate (pH 5.2)
10. 1× TE buffer: 10 mM Tris and 1 mM EDTA; autoclave and syringe filter before storing it at room temperature

2.3.2 PCR Analysis

1. Genomic DNA of transgenic and wild-type pigeonpea plants
2. Forward and reverse primers (10 mM stock)
3. Taq DNA polymerase
4. 10× PCR reaction buffer (10 mM pH 9.0 Tris–HCl, 50 mM KCl, 1.5 mM $MgCl_2$, 0.01% gelatin)
5. dNTPs stock (10 mM each)

6. PCR tubes
7. Agarose
8. 1× TAE buffer: 1 L buffer contains 40 mM Tris base, 20 mM glacial acetic acid and 1 mM EDTA. pH should be around 8.6 (do not adjust); autoclave and store it under room temperature
9. PCR reaction mixture (total volume: 50 μL): 42.57 μL of nuclease-free water, 5.0 μL of 10× Taq buffer, 0.1 μL of 100 mM dNTPs, 0.5 μL of 10 mM forward primer, 0.5 μL of 10 mM reverse primer, 0.33 μL of Taq polymerase enzyme, and 1.0 μL of template cDNA

2.3.3 Genomic Southern Analysis

1. Genomic DNA of transgenic and wild-type pigeonpea plants
2. Specific restriction enzymes
3. 1× TAE buffer
4. 20× SSC: 3 M NaCl and 300 mM $Na_3C_6H_5O_7$, pH 7.0; autoclave and store it under room temperature
5. Denaturing solution: 1.5 M NaCl and 0.5 M NaOH; autoclave and store it under room temperature
6. Neutralizing solution: 0.5 M Tris, 1.5 M NaCl, and 1 mM EDTA, pH 7.5; autoclave and store it under room temperature
7. Autoclaved double-distilled water
8. Whatman 3 filter paper
9. Nylon membrane
10. DIG PCR labeling kit
11. Hybridization buffer.
12. DIG-labeled T-DNA-specific DNA probe
13. Low stringency buffer: prepare 2× SSC and 0.1% SDS in autoclaved double-distilled water
14. High stringency wash: prepare 0.5× SSC and 0.1% SDS autoclaved double-distilled water
15. Detection kit
16. Maleic acid buffer (pH-7.5): 0.1 M maleic acid and 0.15 M NaCl in double-distilled water; autoclave and store it under room temperature
17. Maleic acid washing solution: 0.3% tween in Maleic acid buffer
18. Detection buffer (pH 9.5): 0.1 M Tris-HCl and 0.1 M NaCl in double-distilled water; autoclave and store it under room temperature
19. Films, developer, and fixer

2.3.4 Gene Expression Analyses

1. Total RNA isolated from transgenic and wild-type plants
2. cDNA synthesis kit
3. PCR reagents as mentioned above
4. SYBR green master PCR mix

2.3.5 Protein Expression Analysis by Western Blot

1. Extraction buffer: contains 50 mM sodium phosphate (pH 7.0), 150 mM NaCl, 1 mM PMSF, 10 mM β-mercaptoethanol, and 0.1% Tween 20
2. Anti-Cry2Aa protein antibody
3. NAP (Non Animal Protein) blocker
4. Secondary antibody (goat anti-rabbit IgG)
5. TMB substrate (3,3′,5,5′-tetramethylbenzidine)
6. PVDF membrane

2.4 Bioefficacy Studies Against Helicoverpa armigera

1. Transgenic and wild-type plants in vegetative phase
2. Laboratory-reared second instar *H. armigera* larvae
3. Brushes, filter paper discs, Petri plates, and polythene sheets to cover plants

3 Methods

3.1 Shoot Apical Meristem-Targeted In planta Transformation in Pigeon Pea

3.1.1 Seed Sterilization

1. Sterilize Pigeonpea seeds by dipping them in 1% bavastin solution for 10 min and then with 0.1% mercuric chloride solution for few seconds.
2. After treatment with each sterilant, wash the seeds thoroughly with double-distilled autoclaved water.
3. Then finally wash with distilled water and submerge in double-distilled autoclaved water for 5 min.

3.1.2 Seed Germination

1. Allow the sterilized seeds to soak overnight in glass bottles in double-distilled autoclaved water.
2. Next day (DAY-2) transfer the seeds to Petri plates with filter paper discs soaked in water.
3. Place the Petri plates in dark for 24 h at 30 °C. The plumule emerges out and seeds begin to germinate (*see* **Note 2**).

3.1.3 Preparation of the Inoculum

1. Choose the *Agrobacterium tumefaciens* strain with the respective binary vector (pBINAR:*cry2Aa*) to be transformed.
2. Inoculate a single colony into 5 mL LB broth added with respective antibiotics aseptically in laminar airflow chamber. Incubate at 28 °C, 220 rpm for 24 h on the first day (starter culture), that is, the day the seeds are sterilized.

3. On the second day, inoculate 3 mL of starter culture into 50 mL AB minimal media [6] aseptically in laminar hood, and incubate at 28 °C and 220 rpm for 16–18 h.
4. Simultaneously, crush mature tobacco leaves (yellowish-green), grind using an autoclaved mortar and pestle in 10–15 mL autoclaved double-distilled water and allow to stand overnight in dark (*see* **Note 3**).
5. On the third day, add the leaf extract to the culture in AB medium and incubate at 220 rpm, 28 °C for 4–5 h.
6. Use the culture for infection.

3.1.4 Transformation and Recovery of Transformants

1. Take the germinating embryos and remove the seed coat. Prick the embryo at the region of the meristem. Prick each seed 4–6 times with a sterile needle (Fig. 1).
2. Transfer the pricked seeds to the culture containing the leaf extract and incubate at 28 °C, 50 rpm for 1 h (*see* **Note 4**).
3. Transfer the embryos to soilrite and grow in culture room for 7–10 days at 26–28 °C, 35 μmol m^{-2} s^{-1} of light intensity, with a photo period of 14 h under florescent light.
4. Transfer the plantlets to the transgenic glasshouse.

1. Prepare kanamycin solutions of different concentrations in double-distilled water and pour into glass flasks.
2. Drop 1-day old (*see* **Note 5**) wild-type seeds (in double-distilled water) into them. Keep the bottles in incubator with constant rocking at 50 rpm for 5 h at 32 °C.
3. Transfer the seeds to trays filled with autoclaved quartz sand (*see* **Note 6**) and maintain under net house conditions for about 8–10 days.

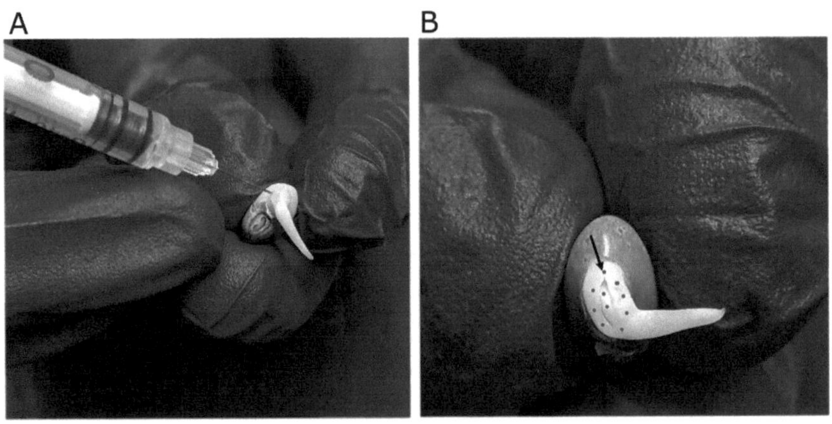

Fig. 1 Targeting T-DNA to the apical meristem of pigeonpea seedlings by pricking with an insulin syringe (**a**); Red dots signify pricking sites (**b**)

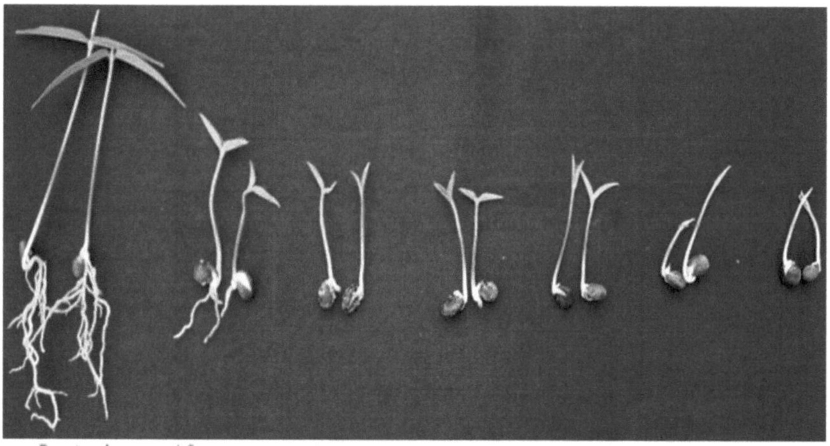

Fig. 2 Standardization of kanamycin screening with wild-type seedlings

3.2 Selection Marker-Based Screening for the Identification of Putative Transformants

3.2.1 Standardization of Kanamycin Concentration for Screening

4. Take equal number of seeds devoid of antibiotic treatment as untreated control, plant in a separate tray and grow along with the remaining treatment trays.
5. Observe the plants for necrotic symptoms and lack of root growth in different concentrations (Fig. 2).

3.2.2 Identification of Putative Transformants Resistant to Kanamycin

1. Drop 1-day old seeds of different T0 plants into 70 mg/L kanamycin, and incubate at 50 rpm for 5 h at 32 °C.
2. Transfer the seeds to trays filled with autoclaved quartz sand (*see* **Note 6**) and allow them to grow in the net house. Maintain both untreated control (wild-type seeds without kanamycin treatment) and treated control (wild-type seeds treated with kanamycin) in different trays.
3. After 8–10 days, select the plants that could tolerate kanamycin and grew normally and transfer to pots supplemented with ¼th strength Hoagland solution.

3.3 Molecular Characterization of Putative Transformants

3.3.1 Plant DNA Isolation by CTAB Method

1. Grind 500 mg of leaf sample (*see* **Note 7**) in a prechilled mortar and pestle; add 50 mg PVP for each sample and transfer to sterile 50 mL Falcon tubes.
2. Add 5 mL extraction buffer (preheated at 60 °C) and mix tubes by inversion and incubate at 60 °C for 1 h with intermediate shaking.

3. Bring the samples to room temperature and extract with chloroform: isoamylalchol (24:1), mix by inversion till emulsion is formed.
4. Centrifuge tubes at 3000 rpm for 20 min and collect the supernatant into a fresh tube.
5. Repeat the step if cloudiness is found.
6. To the supernatant, add 2/3 volume of isopropanol and leave the tubes in −20 °C freezer for 1 h or overnight.
7. Spin down at 3000 rpm for 10 min at 25 °C.
8. Wash the pellet with 70% ethanol (5 mL) and dry at 37 °C.
9. Dissolve pellet in 300 μL of 1X TE buffer, and keep it in 4 °C for overnight.
10. Transfer the dissolved DNA into 1.5 mL tube, add 3 μL of RNase, and incubate for 1 h at 37 °C.
11. Add 3 μL of Proteinase K and incubate for 15 min at 37 °C.
12. Extract with equal volume of chloroform: isoamylalchol (24: 1), mix by inversion till emulsion is formed.
13. Spin at 10,000 rpm for 10 min.
14. To the supernatant, add 1/10th volume of 3 M sodium acetate, pH 5.2, and 2.5 times of 95% ethanol.
15. Leave the tubes overnight in −80 °C freezer.
16. Spin at 12,000 rpm for 15 min at 4 °C.
17. Wash pellets twice with 70% ethanol.
18. Dry the pellet and dissolve in 100 μL of 1X TE buffer.
19. Allow DNA to dissolve in 4 °C overnight.
20. Quantify the genomic DNA by loading the samples on 0.8% agarose gel and use 100–150 ng of genomic DNA to perform PCR amplifications and 15 μg of DNA to perform southern blotting.

3.3.2 PCR Analysis

1. PCR analysis is carried out with the genomic DNA of putative transgenic and wild-type plants. The PCR reaction mixture (25 μL) containing 1 U Taq DNA polymerase, 1× assay buffer, 150 μM of each dNTP, 0.5 μL each of appropriate forward and reverse primer at a final concentration of 0.25 μM, and 100 ng of template DNA is used to amplify the transgene.
2. PCR amplification is set in a thermal cycler programmed with a hot start of initial denaturation at 95 °C for 5 min. Set the PCR reaction profile comprising 30 cycles, with strand separation at 94 °C for 1 min, annealing based on respective primers and extension at 72 °C for 1 min. Extend the program at 72 °C for 10 min.

3. Visualize the PCR products by electrophoresis on a 0.8% agarose gel, stain with ethidium bromide, and observe under ultraviolet light.

3.3.3 Genomic Southern Analysis

Southern hybridization is performed to confirm the physical integration of transgene into the plant genome along with its copy number [7].

1. Restriction digestion: Take approximately 15 μg of genomic DNA from each plant sample (putative transgenic plants and wild type) and digest overnight at 37 °C with the respective high-fidelity restriction enzyme (20 U/μL). For the genomic DNA restriction reaction (total volume: 50 μL), include A μL of plant genomic DNA (equal to 15 μg), B μL of nuclease-free water (to make up the total volume of 50 μL), 5.0 μL of restriction enzyme, and 5.0 μL of the respective reaction buffer.
2. Mix the digested DNA sample (50 μl) with 8 μL of 6× DNA loading dye and separate on 0.8% agarose TAE gel (without ethidium bromide) by running overnight at 40 V at room temperature in freshly prepared 1× TAE buffer.
3. Next day morning, stain the gel ethidium bromide (EtBr) (10 mg/mL) for 20 min in dark and document on a UV illuminator for checking the seperation of the restricted DNA samples.
4. Rinse the gel with double-distilled autoclaved water for removing excess EtBr.
5. Treat the gel firstly with depurination solution (0.25 N HCl) in a glass tray. Use about 300–400 mL depurination solution to submerge the gel and place for gentle shaking for 20 min.
6. After depurination, rinse the gel with autoclaved double-distilled water twice and then treat with denaturation solution (0.4 N NaOH) for 30 min. Change the buffer after 15 min and add fresh denaturation solution.
7. After 30 min, rinse the gel again with autoclaved double distilled water and treat with neutralization solution for 1 h to reduce the pH of the gel from 13 (during denaturation) to 9, which is essential for proper binding of DNA to the nylon membrane during capillary transfer.
8. Equilibrate the gel in 20× SSC solution for 10 min before transfer.
9. Keep a gel-casting tray in inverted position on a tray reservoir containing 20× SSC. Place a wick made up of 3 mm Whatman filter paper on the casting tray drenched with 20× SSC and remove air bubbles, if any, by moving glass rod over it.

10. Place the treated gel upside down on the wick and remove air bubbles by gently rolling glass rod over the gel.
11. Cut a positively charged nylon membrane, exactly to the gel size and place firstly in autoclaved double-distilled water and then in 20× SSC. Later, place the membrane over the gel, remove air bubbles by rolling a glass rod. Place 3 mm Whatman filter paper (3–4 pieces presoaked in 20× SSC) of size of the gel over the nylon membrane followed by a stacking with blotting papers till a height of 8 cm above the gel.
12. Finally, place a glass plate along with some weight on the stack of blotting papers and for 18–20 h for capillary blotting.
13. Disassemble the arrangement carefully after transfer and remove the membrane carefully, wash in 2× SSC solution, dry by placing it on 1 mm Whatman filter paper.
14. UV cross-link the membrane using a cross-linker at 22 J/cm for 2 min and store at +4 °C until further use.
15. Label a fragment of the respective transgene as a probe with DIG-dUTP during PCR using gene-specific primers and DIG PCR probe synthesis kit.
16. For detection of desired DNA fragments on the membrane, perform pre-hybridization, hybridization, and detection by following the instructions provided by DIG Luminescent Detection Kit.
17. Pre-hybridization is performed for 3 h using Easy Hyb buffer at 42 °C, followed by hybridization with the desired probe for 16–18 h at 42 °C.
18. After this, the membrane is proceeded with washing steps.
19. Proceed with low-stringency wash for 2× 15 min at 35 °C.
20. Proceed with high-stringency wash for 2× 5 min at 65 °C.
21. After the stringency washes, treat the membrane with maleic acid washing solution for 10 min.
22. Add 150 mL of 1× blocking solution (component of detection kit) to the membrane and keep it for 60 min on a shaker.
23. Discard the blocking solution and incubate with 50 mL of antibody solution for 45–50 min (5 μL of antibody to 50 mL of blocking solution).
24. Wash 2× 15 min with maleic acid washing solution.
25. Equilibrate the membrane with 25 mL of detection buffer for 3–5 min (discard it thereafter).
26. Add 4 mL of detection solution containing 40 μL of substrate and incubate at room temperature (no shaking) for 1 h.
27. Drain the substrate after 1 h and wrap the membrane and keep in dark overnight.

28. Wrap the hybridized membrane first with saran wrap and then place in an intensifying cassette for exposure to X-ray film for 1 h in dark. Develop the exposed X-ray film by treatment with developer and fixer to visualize the results.

3.3.4 Gene Expression Analyses

Total RNA is isolated from both transgenic and wild-type plants using total RNA isolation kit as per the manufacturer's instructions. This kit has been standardized for pigeonpea samples. cDNA is synthesized from 2.5 μg total RNA according to the manufacturer's instructions and used for transcript quantification.

1. To evaluate transcript accumulation, 1 μL of the cDNA mix was used as a template for amplification of internal control and transgenic fragments.
2. The semiquantitative RT-PCR reaction setup and PCR amplification conditions as given below. Load the amplified products on 1–1.5% agarose gel by electrophoresis at constant voltage of 60 V.
3. RT-PCR reaction conditions: The RT-PCR reaction conditions are as follows: initial denaturation at 95 °C for 4 min (1 cycle), denaturation at 94 °C for 1 min, annealing temperature 30 s, extension step at 72 °C for 30 s, final extension at 72 °C for 7 min (1 cycle).
4. For qRT-PCR analyses, 1 μL of diluted cDNA template was taken from both transgenics and wild-type plants.
5. Set the reaction according to SYBR Green qPCR kit (Agilent). PCR program is of 5 min initial denaturation at 95 °C, followed by 40 cycles of denaturation at 95 °C for 20 s, annealing based on the specific transgene for 20 s, and extension at 72 °C for 15 s.
6. A final melting curve consisting of 95 °C for 30 s, 65 °C for 30 s, and 95 °C for 30 s one cycle for ascertaining the variations of transgene expression among transgenic events *vis-à-vis* wild type.
7. Normalize the Ct values of samples with the Ct value of internal control to calculate ΔCt value.

3.3.5 Protein Expression Analyses

1. Total protein is isolated from 100 mg leaf sample of both transgenic and wild-type plants by finely grinding with liquid nitrogen.
2. Add 500 μL of extraction buffer.
3. Allow the tubes to stand on ice for 30 min and centrifuge the samples at 10,000 rpm for 5 min at 4 °C.
4. Collect the supernatant into fresh tubes and centrifuge again to remove leftover cell debris.

5. Collect the final supernatant into fresh tubes and store in −20 °C freezer.
6. Quantify the isolated protein by Bradford's assay.
7. For western blot analysis, prepare 20–25 μg of total protein samples by adding 6× protein loading dye, denature at 95 °C for 10 min, and snap-chill on ice.
8. Load the samples onto an SDS-PAGE as per the standard protocol [8]. Later, assemble the gel on the western transfer apparatus as per the standard protocol [8] and blot onto nitrocellulose membrane at a constant voltage (40 V) for 3 h in transfer buffer (1.5 g Tris base, 7.2 g glycine, and 100 mL methanol made up to 500 mL with distilled water).
9. Block the membrane using NAP blocker for 1 h.
10. Incubate the membrane in 1:2 NAP blocker (in PBST) for 1 h.
11. Add primary antibody 1:500 dilution in 1:4 NAP blocker (*see* **Note 8**) for 90 min.
12. Wash with 1× PBST for 10 min × 6 times.
13. Add secondary antibody (HRP conjugate) 1:6000 dilution in 1:4 NAP blocker for 1 h.
14. Wash the membrane with 1× PBST for 5 min × 4 times.
15. Add substrate (TMB) 15 mL to cover the membrane till the blue color develops.

3.4 In Vitro *Cut Leaf and Plant-Level Bioefficacy Studies in the Transgenic and Wild-Type Pigeonpea Plants Against* Helicoverpa armigera

Efficacy of the transgenic *vis-à-vis* nontransgenic plants against *H. armigera* is generally assessed by in vitro cut leaf bioassay [9, 10] and plant level bioassay [11] (*see* **Note 9**).

In Vitro Cut Leaf Bioassay

1. Collect fully expanded trifoliate leaves from 45-to-60-day old transgenic and wild-type plants and cover the leaf petioles with cotton to maintain moisture in the bioassay plates.
2. Over each trifoliate leaf, release 10 neonate larvae. Maintain two replicates from each transgenic and wild type.
3. Record larval mortality, growth, and extent of damage on the plant tissue at 24, 48, 72, and 96 h after release of larvae.

4. Calculate the percent larval mortality by counting the number of dead larvae and percent leaf damage by the area of leaf damaged by the feeding larvae.

Plant-Level Bioassay

1. Raise both the transgenic and wild-type pigeon pea plants in earthen pots in the nethouse.
2. Do not apply any lepidopteran-specific insecticides or sprays during the entire duration of the experiment. Whenever the population of leaf hopper and mites increases, control with the application of specific insecticides.
3. Choose peak vegetative growth phase of plants for the release of larvae. Accordingly, on 45-day-old plants, release 10 neonate larvae of in vitro reared *H. armigera* (earlier collected from pigeon pea fields) onto the third leaf of each plant using a camel hair brush.
4. Cover the pot later with transparent plastic sheet to avoid larval escape. Observations are taken at an interval of 24 h for 7 days and assessed for a visual estimation of defoliation percentage.

4 Notes

1. Do not use more than 1-week-old *Agrobacterium* culture; always revive freshly from glycerol stock in culture plate and use.
2. Use the right stage of seedlings as shown in the figure as it is important for the transformation efficiency.
3. For tobacco leaf extract, stage (second or third last leaf) and color (greenish yellow) of the leaf is important to induce virulence in *Agrobacterium*.
4. Use only insulin syringe needle for the pricking, using bigger needles will damage the seedlings and reduce the recovery rate. During infection, maintain the agitation at 50 RPM. Increasing RPM will damage roots and reduce the transformation efficiency.
5. During the screening of putative transformants, stage of the seedling is important. Use just-emerged seedlings only, using old seedlings or very young seedlings (before radicle emergence) will introduce escapes.
6. It is better to use quartz sand instead of soilrite for growing seedlings after kanamycin screening, because growing in quartz sand will not damage the root during transfer.

7. Use topmost leaf for DNA extraction and fully expanded third leaf for expression analysis.
8. Try to use NAP blocker for blocking step in western blot to reduce background.
9. For *H. armigera* bioassay use 10 larvae if they are neonates, 3 to 5, 4–6-day-old larvae and use single larvae for bigger instars to avoid cannibalism.

Acknowledgments

The authors acknowledge financial support from ICAR-National Agricultural Science Fund (NFBSFARA/PB-2010) and ICAR-Network Project on Transgenic Crops (NPTC).

References

1. Bent AF (2000) Arabidopsis *in planta* transformation. Uses, mechanisms, and prospects for transformation of other species. Plant Physiol 124(4):1540–1547
2. Bechtold N, Pelletier G (1998) *In planta* Agrobacterium mediated transformation of adult *Arabidopsis thaliana* plants by vacuum infiltration. Methods Mol Biol 82:259–266
3. Rohini VK, Rao KS (2000) Transformation of peanut (*Arachis hypogaea* L.): a non-tissue culture based approach for generating transgenic plants. Plant Sci 150(1):41–49
4. Chee PP, Fober KA, Slightom JL (1989) Transformation of soybean (*Glycine max*) by infecting germinating seeds with *Agrobacterium tumefaciens*. Plant Physiol 91(3): 1212–1218
5. Sankara Rao K, Sreevathsa R, Sharma PD, Keshamma E, Udaya Kumar M (2008) *In planta* transformation of pigeon pea: a method to overcome recalcitrancy of the crop to regeneration *in vitro*. Physiol Mol Biol Plants 14: 321–328
6. Winans SC (1990) Transcriptional induction of an Agrobacterium regulatory gene at tandem promoters by plant-released phenolic compounds, phosphate starvation, and acidic growth media. J Bacteriol 172(5):2433–2438
7. Sambrook J, Fritsch EF, Maniatis T (1989) Molecular cloning: a laboratory manual, 2nd edn. Cold Spring Harbor Laboratory Press, Woodbury
8. Sambrook J, Russell DW (2006) SDS-polyacrylamide gel electrophoresis of proteins. CSH Protoc 1(4):pdb.prot4540
9. Ramu SV, Rohini S, Keshavareddy G, Gowri Neelima M, Shanmugam NB, Kumar ARV, Udayakumar M (2012) Expression of a synthetic cry1AcF gene in transgenic Pigeon pea confers resistance to *Helicoverpa armigera*. J Appl Entomol 136(9):675–687
10. Singh S, Kumar NR, Maniraj R, Lakshmikanth R, Rao KYS, Muralimohan N, Sreevathsa R (2018) Expression of Cry2Aa, a *Bacillus thuringiensis* insecticidal protein in transgenic pigeon pea confers resistance to gram pod borer, *Helicoverpa armigera*. Sci Rep 8(1):1–12
11. Ramkumar N, Rathinam M, Singh S, Kesiraju K, Muniyandi V, Singh NK, Sreevathsa R (2020) Assessment of Pigeonpea (*Cajanus cajan* L.) transgenics expressing Bt ICPs, Cry2Aa and Cry1AcF under nethouse containment implicated an effective control against herbivory by *Helicoverpa armigera* (Hübner). Pest Manag Sci 76(5):1902–1911

Chapter 18

CRISPR/Cas9-Mediated Genome Editing in the Management of Oriental Fruit Fly, *Bactrocera dorsalis* (Hendel) (Tephritidae: Diptera)

Chikmagalur Nagaraja Bhargava, Ashok Karuppannasamy, and Asokan Ramasamy

Abstract

The oriental fruit fly, *Bactrocera dorsalis* (Hendel), is a highly invasive polyphagous pest that causes significant damage to horticultural crops of global importance. Traditional management practices have not been effective in controlling this pest, and therefore, there is a need for alternative management strategies. CRISPR/Cas9-driven genome editing has been successfully used in a wide range of insects to induce site-specific, off-target minimized mutations that result in loss of function. This technique can be used to develop precision-guided sterile insect technique (pgSIT) and gene drive programs, which can be used for area-wide suppression of the pest. This chapter provides a brief overview of the workflow for RNP-based genome editing, which can be used to validate and establish gene function for large-scale gene drive programs aimed at combating this pest. The RNP, or ribonucleoprotein complex, comprises the sgRNA and Cas9 protein, which are microinjected into the G0 stage embryos for heritable editing of the target gene(s).

Key words CRISPR/Cas9, Oriental fruit fly, Genetic biocontrol, Microinjection, Genome editing

1 Introduction

Bactrocera dorsalis (Hendel), commonly referred to as the oriental fruit fly, is a species of tephritid fruit fly that is endemic to Southeast Asia. It is a notorious pest species that shows extreme polyphagia and causes financial depreciation of higher scale [1]. It was first recorded in Taiwan [2] and further distributed to Asia-Pacific regions [1]. It causes fruit yield losses and also results in rejection of exports due to quarantine regulation. They are considered as most economically important pest species in the world, attacking a more than 400 kinds of fruit and vegetables. However, avocado (*Persea americana*), papaya (*Carica papaya*), and mango (*Mangifera indica*) are the most common hosts [3]. Oriental fruit fly has a

Muthappa Senthil-Kumar and Hema Ramanna (eds.), *Plant Biotic Stress: Methods and Protocols*, Methods in Molecular Biology, vol. 2966, https://doi.org/10.1007/978-1-0716-4746-2_18,

Fig. 1 Damage symptoms of *B. dorsalis* in mango fruit. (Left) The gravid females of *B. dorsalis* oviposition in mango, (Right) The infested fruits start to rot due to feeding by maggots remaining inside the decaying fruit

short life cycle, which results in rapid population increase under favorable conditions. The adult fruit flies lay eggs within mature fruit and upon hatching larvae tunnels through the fruit, causing it to drop (Fig. 1). Due to its broader host range, it is even considered as a quarantine pest.

In the management of *B. dorsalis*, chemical control has been the main stay [4, 5], bait application technique (BAT) using protein bait mixed with insecticides and the male annihilation technique (MAT) using methyl eugenol (ME) traps [6, 7] have also been used as a monitoring tool and a component of integrated pest management. In addition to the above, area-wide pest management through mass release of sterile males (SIT) derived through irradiation is also widely followed in different countries [8, 9]. With the dawn of CRISPR/Cas-based genome editing, genetic biocontrol methods such as the precision guided sterile insect technique (pgSIT) and gene drive have gained importance and serious consideration for adoption into the area of pest control, owing to their technical feasibility, species specificity, environmental and ecological safety [10, 11]. Being a truly sexually reproducing species [12] with a high invasive potential, *B. dorsalis* is an attractive target for pest control by such genetic biocontrol methods. However, for these methods to be implemented, the biological trait(s) to be chosen and the genetic loci to be targeted in order to produce those traits need to be studied thoroughly and validated through in vivo mutagenesis.

CRISPR (clustered regularly interspaced short palindromic repeats) and the Cas9 (CRISPR associated protein 9) commonly known as CRISPR-Cas9 were discovered as an adaptive immune response by prokaryotes to infection by bacteriophages [13]. Presently programmable CRISPR/Cas9 is increasingly being used as a cutting-edge tool for gene editing in eukaryotes for various purposes [14, 15]. The natural CRISPR-Cas9 system in bacteria has two essential RNA components: mature CRISPR RNA (crRNA) and transactivating crRNA (tracrRNA). Combining these two RNA structures forms the synthetic guide RNA (sgRNA) [16]. These two RNAs have partial sequence complementarity and together form a well-defined two-RNA structure that directs Cas9 to target DNA and creates a double-stranded break (DSB) [17, 18]. Followed by DSB, NHEJ (Nonhomologus end Joining) repair mechanism will be initiated, which is error prone and thereby end up in imparting mutation.

From a pest management perspective, genetic control through CRISPR-Cas9-based genome editing holds great promise in developing an area-wide pest management strategy for the oriental fruit fly. However, creating a functional genetic control system requires the creation of sgRNA and Cas9 transgenic lines for males and females, respectively [19]. This process requires higher biosafety levels, more time for development and maintenance, and there is uncertainty regarding the loss of function of the target genes. Therefore, it is important to validate gene targets in an easy, nontransgenic way by employing the ribonucleoprotein (RNP) complex (sgRNA: Cas9 protein) to be injected into the G0 stage of the embryo of the target species. The CRISPR/Cas9-based genome-editing technique has emerged as the method of choice for manipulating specific sequences in insect genomes [20, 21] due to its technical simplicity, site specificity, and controllable/low off-target effects. This technique has been successfully carried out in *B. dorsalis* by embryo microinjection of Cas9-sgRNA reagents [22–26]. In this chapter, we describe the detailed methodology for Cas9-RNP complex-based nontransgenic genome editing in the oriental fruit fly.

2 Materials

2.1 Cloning and Characterization of Candidate Genes

2.1.1 RNA Isolation

1. TRIzol® Reagent.
2. Chloroform (100%).
3. Isopropanol (100%).
4. Ethanol (70%).
5. DNase enzyme.
6. Nuclease-free water.

2.1.2 cDNA Synthesis

1. RNA template (DNase-treated).
2. Reverse transcriptase (RT) buffer.
3. dNTPs.
4. oligoDT primer/random hexamer.
5. RNase inhibitor (RI).
6. RNase-free water.

2.1.3 PCR Amplification of the Candidate Gene

1. 10× *Taq* polymerase buffer.
2. dNTPs.
3. $MgCl_2$.
4. Template (cDNA).
5. Forward primer.
6. Reverse primer.
7. *Taq* polymerase.
8. 6× DNA loading dye—To prepare 10 mL of 6× DNA loading dye, weigh out 25 mg bromophenol blue. Transfer it to a 15-mL screw-capped graduated tube. Add 7.06 mL of 85% glycerol and 2.94 mL deionized/Milli-Q water.
9. 50× TAE buffer—dissolve 242 g Tris base in 400–500 mL of Milli-Q water, add 100 mL of 0.5 M EDTA, and 57.1 mL glacial acetic acid to it. Make up the volume with deionized water to 1000 mL. The pH of the final solution should be between 8.2 and 8.4.
10. Agarose gel—weigh the required mass of agarose in a given volume of 1× TAE to make gels of desired concentrations.
11. Any commercially available gel elution/PCR cleanup kit.

2.1.4 Cloning and Transformation of PCR Amplicon

1. PTZ57R/T TA cloning vector.
2. Ligase enzyme and buffer.
3. DH5α *E. coli* competent cells.
4. LB agar media—dissolve 40 g of any commercially available LB agar in 1 L of distilled water in a 2 L flask. Cover with foil and autoclave. Monitor the temperature as the solution cools. When the temperature reaches 50 °C, add 1 mL of ampicillin stock solution. Pour ~20 mL into each petri dish.
5. LB broth—dissolve 25 g in 1 L of distilled water. Mix well and pour into appropriate containers and sterilize in autoclave at 121 °C for 15 min.
6. Ampicillin antibiotic.
7. X-Gal (5-bromo-4-chloro-3-indolyl-beta-D-galacto-pyranoside).
8. 1 M IPTG (Isopropyl β-d-1-thiogalactopyranoside)
9. Any commercially available plasmid harvesting kit.

1. PCR reagents and cloning reagents.
2. *SacI, HindIII, and BbsI* restriction enzymes.
3. pBluescript plasmid.
4. pEn-C1.1 plasmid.

5. sgRNA DNA construct (generated by PCR).
6. Any commercially available in vitro transcription kit.
7. Phenol: chloroform: isoamyl alcohol (25:24:1).
8. Chloroform: isoamyl alcohol (24:1).
9. 3 M sodium acetate (pH 5.2)
10. Ethanol (100%).
11. Nuclease-free water.
12. Agarose gel (2%).

2.2 Identification of Off-Target Minimized sgRNAs and In Vitro Restriction Assay

2.2.1 In Vitro Synthesis of sgRNA and Purification

2.2.2 In Vitro Cleavage Assay

1. Purified DNA substrate.
2. Purified sgRNA.
3. Cas9 enzyme and buffer.
4. KCl.
5. Proteinase K.
6. Agarose gel (2%).

2.3 Microinjection of Cas9-sgRNA Ribonucleoprotien (RNP) Complex

1. Oviposition substrate.
2. Glass slide (7.5 × 2.5 cm).
3. Double-sided tape.
4. Oil.
5. Agar plates.
6. Freshly deposited eggs.
7. Purified sgRNA.
8. Cas9 enzyme and buffer.
9. Micromanipulator apparatus.
10. Microinjection needle.
11. Adult rearing cages with gravid females.

2.4 Phenotypic and Molecular Characterization of the Edits

1. Stereo binocular microscope fitted with camera.
2. Any commercially available insect DNA isolation kit.
3. DNA from mutant flies.
4. Mutation screening primers.
5. PCR reagents (dNTPs, $MgCl_2$, Buffer, *Taq* polymerase).
6. Any commercially available PCR cleanup kit.
7. Ligation and cloning reagents (*see* Subheading 2.1.4).

3 Methods

3.1 Cloning and Characterization of Candidate Genes

3.1.1 RNA Isolation and cDNA Conversion

1. Isolate total RNA by using TRIzol® Reagent (according to manufactures protocol) from desired tissue (*see* **Note 1**).
2. Check the integrity of the RNA by running it in agarose gel (1.5%).
3. Treat DNase if there is any DNA contamination.
4. Take 1 μg of RNA and convert it to cDNA by using any of the commercially available first-strand cDNA synthesis kit.

3.1.2 PCR Amplification of Target Gene

1. Add required reagents and template to PCR tubes.
2. Mix properly and centrifuge.
3. Amplify in thermo cycler based on primers melting temperature.
4. Resolve amplified product in 1% agarose gel by gel electrophoresis.
5. Perform PCR clean up with any commercially available kit.

3.1.3 Cloning of PCR Amplicon

1. The purified PCR products were ligated into pTZ57R/T vector using manufactures protocol (the ligation mix is incubated overnight at 4 °C).
2. Thaw a tube of DH5α *E. coli* competent cells in ice until the last ice crystals disappear.
3. Mix gently and carefully pipette 50 μL of cells into a transformation tube in ice.
4. Add 5–10 μL of ligation mix to the cell mixture.
5. Mix the contents carefully by flicking the tube 4–5 times. Do not vortex.
6. Incubate the mixture on ice for 30 min.
7. Heat shock is given exactly at 42 °C for 45 s.
8. Place on ice for 5 min.
9. Add 1 mL of LB broth into the reaction mixture.
10. Incubate at 37 °C for 60 min. Shake vigorously (250 rpm) or rotate.
11. Pellet down the cells by centrifuge at 7000×*g* for 5 min.
12. Mix the cells thoroughly by flicking the tube and inverting, then retain only 100 μL of LB broth by discarding the rest.
13. Spread X-Gal and IPTG (40 μL + 4 μL/plate) on LB plates (to perform blue/white screening) and allow it to dry.
14. Spread 50–100 μL of each dilution onto a selection plate and incubate overnight (14–16 h) at 37 °C.

15. Pick the transformed colony (white colony) and grow overnight (14–16 h) at 37 °C in a vial containing LB broth (5 mL) added with ampicillin as resistant marker.
16. Harvest the plasmid using any commercially available kit by following manufacturer's protocol.
17. 200–300 ng of harvested plasmid is sequenced in triplicates by Sanger sequencing.
18. The identity of gene can be confirmed by NCBI nucleotide BLAST.

3.2 Identification of Off-Target Minimized sgRNAs and In Vitro Restriction Assay

3.2.1 Designing and Synthesis of Single Guide RNA (sgRNA)

1. The cloned gene was annotated for intron/exon information using the mRNA sequence.
2. Select 17–20 nucleotides upstream of the PAM (*see* **Note 2**) within the genome using any of the online tools, namely CRISPOR (https://crispor.tefor.net/), CHOP-CHOP (https://chopchop.cbu.uib.no/), or IDT (https://www.idtdna.com).
3. Design the primers to amplify the Bbs1 site as well as sgRNA scaffold and terminator from pEn-C1.1 plasmid, the restriction sites of *SacI* and *HindIII* were tagged at 5′ end in forward and reverse primer, respectively.
4. Predigest the pBluescript plasmid with *SacI* and *HindIII* restriction sites and purify using PCR cleanup kit.
5. Perform the PCR to amplify using sgRNA Sac + Bbs1 F and SgRNA Scaff + HindIII R primers using pEn-C1.1 plasmid as template.
6. Purify the PCR products and digest with *SacI* and *HindIII* restriction sites and purify again.
7. Ligate these products into predigested *SacI* and *HindIII* pBluescript II SK plasmid.
8. Confirm the clones by sequencing. Then the confirmed plasmid will be digested with *BbsI* restriction site and use this plasmid to ligate the sgRNAs.
9. Add the linker sequence of ATTG and CCCG at 5′ region in forward and reverse primer, respectively.
10. Hybridize both forward and reverse oligonucleotides and ligate to the *BbsI* predigested pBluescript II SK plasmid.
11. Clone the ligated products and confirm the presence of sgRNA by sequencing.
12. Linearize the confirmed clones with HindIII restriction site, purify using PCR cleanup kit, and use this as template for in vitro transcription.

3.2.2 In Vitro Transcription of sgRNA

1. Take 2 μg of confirmed clones as template and assemble the reaction mixture (Tris-buffered NTP set, T7 RNA Polymerase, and RiboLock RNase Inhibitor) as per the kit manufacturer's protocol.
2. Mix, spin, and incubate at 37 °C overnight for transcription reaction.
3. Add 2 μL of DNase to the reaction mixture and incubate at 37 ° C for 1 h.
4. Transfer the mixture to 2 mL phase lock gel tube and add equal volume of phenol: chloroform: isoamyl alcohol (25:24:1), then mix for 15 s and centrifuge at 10,947×*g* for 5 min at 4 °C.
5. Transfer the top aqueous layer to fresh 2 mL phase lock gel tube and add equal volume of chloroform: isoamyl alcohol (24: 1) followed by centrifugation at 10,947×*g* for 5 min at 4 °C.
6. Again, transfer the top aqueous layer to new 1.5 mL tube and add two volume of ethanol followed by mixing and incubation at −20 °C for 1 h to precipitate the sgRNA.
7. Pellet down the gRNA by doing centrifuge at 10,947×*g* for 15 min at 4 °C.
8. Remove the supernatant and wash the pellet with 70% ethanol.
9. Centrifuge at 10,947×*g* for 5 min at 4 °C.
10. Airdry the pellet.
11. Resuspend the sgRNA with nuclease-free water and check the integrity of sgRNA in 2% agarose gel.

3.2.3 In Vitro Cleavage Assay

1. Design the primers to generate substrate, which is to be targeted by Cas9-sgRNA ribonucleoprotein (RNP) complex (*see* **Note 3**).
2. Perform the standard PCR with the specific primers, which is designed as described above to amplify the region in genomic DNA and purify through PCR cleanup kit.
3. Assemble sgRNA, cas9 enzyme, and cas9 buffer in room temperature (RNP complex) (*see* **Note 4**).
4. Incubate the above mixture at 25 °C for 30 min to get the RNP complex.
5. Add 100 ng of substrate DNA (generated through PCR) to RNP complex.
6. Incubate at 25 °C for 1 h.
7. Resolve in 2% agarose gel to confirm the in vitro restriction (Fig. 2).

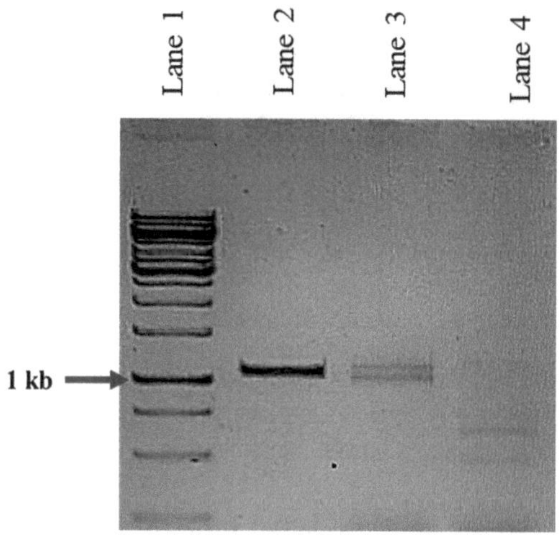

Fig. 2 In vitro restriction assay to test the efficiency of RNP complex. *Lane 1* 1 kb ladder; *Lane 2* DNA template; *Lane 3* DNA template + sgRNA1 + Cas9; *Lane* 4 DNA template + sgRNA2 + Cas9. Lane 3 and Lane 4 shows the in vitro restriction of DNA template. The restriction can be confirmed by visualizing the restricted fragments. Ex. Restriction with sgRNA1 gives two bands at 1056 and 54 bp (Lane 3), whereas in case of sgRNA2, restricted fragments can be seen at 670 and 440

3.3 Embryonic Editing of Target Genes Employing Cas9-sgRNA Ribonucleoprotien (RNP) Complex

3.3.1 Preparation of RNP Complex

1. Assemble sgRNA, cas9 enzyme, and cas9 buffer in room temperature.
2. Incubate at 25 °C for 30 min to get the RNP complex.

3.3.2 Collection of Eggs for Microinjection

1. Provide the oviposition substrate to the adult flies, which are maintained in insect rearing cages under laboratory conditions.
2. Collect the freshly laid eggs (*see* **Note 5**).
3. Arrange the eggs on a glass slide (Fig. 3).

3.3.3 Embryonic Microinjection

1. Load RNP complex to the injecting needle with the help of micro loader.
2. Inject the embryos using micromanipulator (*see* **Note 6**).
3. Place the glass slide in the agar plate and incubate the eggs at 25 °C until it hatches (*see* **Note 7**).
4. Transfer the hatched larvae to fresh banana and rear under controlled environmental conditions (25 ± 1 °C, 60–70%

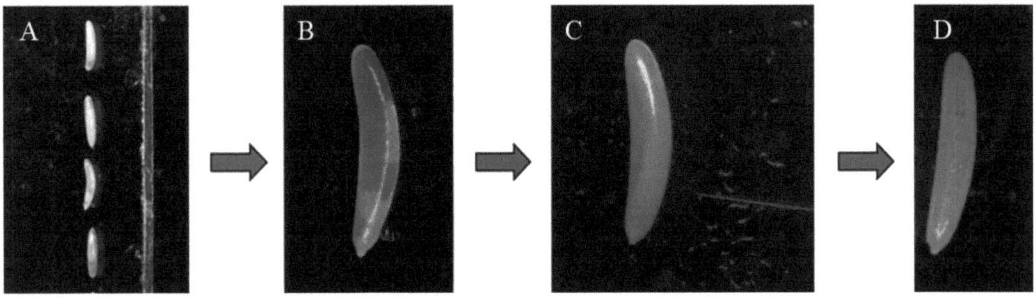

Fig. 3 Schematic representation of *B. dorsalis* embryonic microinjection. (**a**) Arrangement of *B. dorsalis* on glass slide; (**b**) single egg showing the polarity of egg; (**c**) microinjection needle loaded with RNP complex poking the embryo; (**d**) egg after the microinjection

relative humidity, and photoperiod of 14 h light:10 h dark cycle) (*see* **Note 8**).

3.4 Phenotypic and Molecular Characterization of the Edits

3.4.1 Phenotypic Characterization of the Edits

1. Visualize the visible phenotypic changes in edited flies by comparing it with wild flies.
2. Stereo binocular microscope can be used to visualize and capture the changes in the phenotypic characters of the edited flies.
3. If there are no observable phenotypic changes in the fly, different assays will be carried out according to the character of the subject gene to confirm the edit.

3.4.2 Molecular Characterization of the Edits

1. Isolate the genomic DNA from the adult flies after conducting the required assays.
2. Perform the PCR to amplify the guide region.
3. Confirm the amplification by resolving it in 1.2% agarose gel.
4. Purify the PCR product by using PCR cleanup kit.
5. Clone the PCR amplicon (refer Subheading 3.1.3).
6. Sequence the clones.
7. Analyze the sequence to know the nature of edit either manually or by using the online tools (*see* **Note 9**).

4 Notes

1. RNA isolation, as mentioned in Subheading 2.2.1, should be performed on different tissues and developmental stages where the candidate genes show maximum expression.
2. PAM site (Protospacer Adjacent Motif) is a short DNA sequence (5′-GGN-18nts-NGG-3′) that follows the targeted DNA region for cleavage by the CRISPR-Cas9 system.

3. RNA isolation, as mentioned in Subheading 2.2.1, should be performed on different tissues and developmental stages where the candidate genes show maximum expression.
4. PAM site (Protospacer Adjacent Motif) is a short DNA sequence (5′-GGN-18nts-NGG-3′) that follows the targeted DNA region for cleavage by the CRISPR-Cas9 system.
5. The sgRNA target sequence should be located asymmetrically within the amplicon sequence during primer design for substrate generation in in vitro cleavage assays. Cas9 and sgRNA reagents can be delivered using various methods, including Cas9 protein with sgRNA, Cas9 mRNA with sgRNA, and Cas9 plasmid with sgRNA.
6. The desiccation time after injection is primarily determined by the moisture content in the room. If the moisture level is low, the desiccation time should be reduced to prevent overdrying and reduced embryo hatching.
7. Proper incubation temperature must be maintained after injection, as lower temperatures can lead to longer hatching times.
8. Excess moisture or water droplets on the embryos after injection should be avoided.
9. Cas-Analyzer (http://www.rgenome.net/cas-analyzer/#!), ICE Synthego (https://ice.synthego.com/).

Acknowledgments

The authors profusely thank the Director, ICAR-IIHR, Bengaluru for the support and infrastructural facilities. This work was financially supported by the CABin project.

References

1. Clarke AR, Armstrong KF, Carmichael AE et al (2005) Invasive phytophagous pests arising through a recent tropical evolutionary radiation: the *Bactrocera dorsalis* complex of fruit flies. Annu Rev Entomol 50:293–319
2. Wan X, Liu Y, Zhang B (2012) Invasion history of the oriental fruit fly, *Bactrocera dorsalis*, in the Pacific-Asia region: two main invasion routes. PLoS One 7(5):e36176
3. USDA (2016) United States department of agriculture animal and plant health inspection service. Cooperative fruit fly emergency response triggers & guidelines. https://www.aphis.usda.gov/plant_health/plant_pest_info/fruit_flies
4. Huan LI, Zhang DJ, Xu YJ et al (2019) Invasion, expansion, and control of *Bactrocera dorsalis* (Hendel) in China. J Integr Agri 18(4):771–787
5. Abdullahi G, Obeng-Ofori D, Afreh-Nuamah K et al (2020) Acute and residual concentration-dependent toxicities of some selected insecticides to adult *Bactrocera invadens* drew, tsuruta and white (Diptera: Tephritidae). J Basic Appl Zool 81(1). https://doi.org/10.1186/s41936-020-00144-4
6. Stonehouse JM, Mumford JD, Verghese A et al (2007) Village-level area-wide fruit fly suppression in India: bait application and male annihilation at village level and farm level. Crop Prot 26:788–793
7. Vargas RI, Piñero JC, Leblanc L (2015) An overview of pest species of *Bactrocera* fruit flies (Diptera: Tephritidae) and the integration

of biopesticides with other biological approaches for their management with a focus on the Pacific region. Insects 6:297–318

8. Sutantawong M, Orankanok W, Enkerlin WR et al (2002) The sterile insect technique for control of the oriental fruit fly, *Bactrocera dorsalis* (Hendel), in mango orchards in Ratchaburi Province, Thailand. In: Proceedings, symposium: 6th international symposium on fruit flies of economic importance, pp 6–10
9. Orankanok W, Chinvinijkul S, Thanaphum S et al (2007) Area-wide integrated control of oriental fruit fly *Bactrocera dorsalis* and guava fruit fly *Bactrocera correcta* in Thailand. In: Area-wide control of insect pests. Springer, Dordrecht, pp 517–526
10. Alphey L, Benedict M, Bellini R (2010) Sterile-insect methods for control of mosquito-borne diseases: an analysis. J Vector Borne Dis 10(3): 295–311
11. Kapranas A, Collatz J, Michaelakis A (2022) Review of the role of sterile insect technique within biologically-based pest control – an appraisal of existing regulatory frameworks. Entomol Exp Appl 170(5):385–393
12. Gokhman VE, Kuznetsova VG (2018) Parthenogenesis in Hexapoda: holometabolous insects. J Zool Syst Evol Res 56(1):23–34
13. Jansen R, van Emben JDA, Gaastra W et al (2002) Identification of genes that are associated with DNA repeats in prokaryotes. Mol Microbiol 43:1565–1575
14. Cong L, Zhou R, Kuo YC et al (2012) Comprehensive interrogation of natural TALE DNA-binding modules and transcriptional repressor domains. Nat Commun 3(1):968
15. Knott GJ, Doudna JA (2018) CRISPR-Cas guides the future of genetic engineering. Science 361(6405):866–869
16. Bassett AR, Tibbit C, Ponting CP et al (2013) Highly efficient targeted mutagenesis of *Drosophila* with the CRISPR/Cas9 system. Cell Rep 4(1):220–228
17. Brouns SJ, Jore MM, Lundgren M, Westra ER, Slijkhuis RJ, Snijders AP, Dickman MJ, Makarova KS, Koonin EV, van der Oost J (2008) Small CRISPR RNAs guide antiviral defense in prokaryotes. Science 321:960–964
18. Deltcheva E, Chylinski K, Sharma CM et al (2011) CRISPR RNA maturation by trans-encoded small RNA and host factor RNase III. Nature 471(7340):602–607
19. Kandul NP, Liu J, Sanchez CHM et al (2019) Transforming insect population control with precision guided sterile males with demonstration in flies. Nat Commun 10(1):84
20. Ashok K, Bhargava CN, Babu KP et al (2023) First report on CRISPR/Cas9 mediated editing of the eye colour gene, tryptophan 2, 3-dioxygenase in eggplant shoot and fruit borer *Leucinodes orbonalis* Guenée (Lepidoptera: Crambidae). J Asia Pac Entomol 26(1): 102031
21. Gantz VM, Akbari OS (2018) Gene editing technologies and applications for insects. Curr Opin Insect Sci 28:66–72
22. Sim SB, Kauwe AN, Ruano RE et al (2019) The ABCs of CRISPR in Tephritidae: developing methods for inducing heritable mutations in the genera *Anastrepha*, *Bactrocera* and *Ceratitis*. Insect Mol Biol 28(2):277–289
23. Zhao S, Xing Z, Liu Z et al (2019) Efficient somatic and germline genome engineering of *Bactrocera dorsalis* by the CRISPR/Cas9 system. Pest Manag Sci 75(7):1921–1932
24. Bai X, Zeng T, Ni XY et al (2019) CRISPR/Cas9-mediated knockout of the eye pigmentation gene white leads to alterations in colour of head spots in the oriental fruit fly, *Bactrocera dorsalis*. Insect Mol Biol 28(6):837–849
25. Chen X, Lei Y, Li H et al (2021) CRISPR/Cas9 mutagenesis abolishes odorant-binding protein BdorOBP56f-2 and impairs the perception of methyl eugenol in *Bactrocera dorsalis* (Hendel). Insect Biochem Mol Biol 139: 103656
26. Paulo DF, Cha AY, Kauwe AN et al (2022) A unified protocol for CRISPR/Cas9-mediated gene knockout in tephritid fruit flies led to the recreation of white eye and white puparium phenotypes in the melon Fly. J Econ Entomol 115(6):2110–2115

Chapter 19

A Novel Approach Using SSR Markers for Identifying and Breeding Bruchid Pest-Resistant Blackgram (*Vigna mungo* L. Hepper) Genotypes

Ragul Subramaniyan, Manivannan Narayana, and Soundararajan Raga Palanisamy

Abstract

Molecular and biochemical techniques are beneficial in generating markers to investigate interesting genes in populations. SSRs, as genetic markers, provide significant advantages over traditional techniques of breeding and identifying essential traits. Under storage circumstances, the bruchid beetle *Callosobruchus* sp. can cause significant yield loss in blackgram. Hence, the identification of bruchid-resistant lines and breeding for resistance are important in blackgram. In this paper, the screening technique useful for the bruchid resistance breeding program and the use of SSR markers in identifying resistance are presented.

Key words Bruchid, Blackgram, Resistance breeding, Phenotyping, Genotyping, SSR

1 Introduction

Bruchid beetle (Chrysomelidae: Bruchinae) causes quantity and quality losses during storage in the tropics and subtropics. It is a well-known storage grain pest that can cause total seed destruction if proper care (drying, fumigation, pesticide application, etc.) is not performed throughout the storage period. These beetles primarily infest legumes in the field via pods and seeds, then move on to storage conditions. It can swiftly multiply and spread into uncontaminated storage lots [1].

Food legumes are infested by several Coleopteran species of *Callosobruchus* (Bruchidae) viz., *C. maculatus*, *C. chinensis*, *C. analis*, and *C. phaseolous*. Among them, cowpea weevils (*C. maculatus* F.) and azuki bean weevils (*C. chinensis* L.) are the most devastating storage pests of pulses during storage. In India, compared to other species, *C. maculatus* causes a yield loss of up to

Muthappa Senthil-Kumar and Hema Ramanna (eds.), *Plant Biotic Stress: Methods and Protocols*, Methods in Molecular Biology, vol. 2966, https://doi.org/10.1007/978-1-0716-4746-2_19,

80–90% in the blackgram crop [2]. To manage storage pests in legumes, pesticides and fumigants are utilized; nevertheless, a green strategy is needed to prevent health risks from chemicals [3].

Scientists face a difficult task in improving the genetic resistance of pulse crops to storage grain pests. Understanding the genetic mechanisms of resistance and the finding of genes or QTLs are critical for developing bruchid-resistant lines. An appropriate mapping population and a high-density genetic linkage map are required for efficient QTL discovery. Breeding programs use molecular markers to improve their efficiency and effectiveness. Because of their abundance, molecular markers are the most often used genetic markers.

SSRs have several advantages over other DNA markers and have been used to assess genetic diversity [4]. SSR has proven to be the most effective technology for identifying crop varieties. It holds a great deal of promise in genetic and breeding studies. Substantial progress has not been made in the creation of genomic resources in blackgram. Nevertheless, genetic maps were created utilizing genomic materials from azuki bean, common bean, and mungbean [5–7]. In general, SSR primers showed a low degree of polymorphism in cultivated blackgram.

In this paper, we show how SSR may be used to identify and breed bruchid resistance in blackgram. A standardized methodology for phenotyping, that is, bruchid pest screening procedures, has been developed. SSR genotyping procedures, lab techniques, and specialized formulations are all included. Important suggestions and reasons to remember are mentioned as notes.

2 Materials

2.1 Plant Material

1. Diverse parents with the desired trait (*see* **Note 1**).
2. Bruchid-susceptible blackgram and resistant blackgram genotype (e.g., TU 68) [8] (*see* **Note 2**) (*see* Fig. 1).

2.2 Pest Culture

1. Bruchid species *Calasobruchus maculatus* (F) (*see* **Note 3**).
2. Plastic bottles (50 g capacity).
3. Plastic petri plates (15 cm diameter).
4. Compound light microscope (10× magnification).

2.3 DNA Extraction

1. Extraction buffer (CTAB (2%)): 73 mL of double-distilled water, 10.0 mL of 1 M Tris-HCl (pH 8.0), made by dissolving 121.10 g of Tris-hydrochloride in 750 mL of distilled water and adjusting the volume to 1000 mL, 14.0 mL of 5 M NaCl, made by dissolving 292.2 g of NaCl in distilled water and adjusting the volume to 1000 mL, 2.0 mL of 0.5 M EDTA

a. Susceptible blackgram genotype   b. Resistant blackgram genotype

Fig. 1 Diverse blackgram genotypes for bruchid infestation (**a**) bruchid susceptible blackgram genotype and (**b**) bruchid-resistant blackgram genotype

(pH 8.0), made by dissolving 186.12 g of disodium tetra-acetic acid (EDTA) in 750 mL of distilled water and adjusting the volume to 1000 mL, 1.0 mL of 2-mercaptoethanol, and 2.0 g of CTAB. Mix all the components thoroughly to prepare the extraction buffer.

2. Phenol: Chloroform: isoamyl alcohol (25:24:1) (*see* **Note 4**).
3. Chloroform: isoamyl alcohol (24:1).
4. 100% ethanol
5. TE buffer (pH 8.0) (makeup volume—1000 mL): 10.0 mL of 1 M Tris - Hcl (pH 8.0) and 2.0 mL of 0.5 M EDTA (pH 8.0).
6. RNAse (*see* **Note 5**).

2.4 DNA Quantification

1. TBE 10× (Tris borate buffer) (makeup volume—1000 mL): 109 g of Tris base, 55 g of boric acid, 800 mL of double-distilled water, and 40 mL of 0.5 M EDTA (pH 8.0).
2. Ethidium bromide: 100 mg of ethidium bromide and 10 mL of double-distilled water.
3. 10× loading dye (for 100 mL): 50 mL of glycerol, 0.4 g of bromophenol blue, 0.4 g of xylene cyanol, and 50 mL of ultrapure water.

2.5 PCR Amplification

1. 10× *Taq* polymerase buffer.
2. 15 mM magnesium chloride.
3. 2 mM dNTP's (mixture of dATP, dCTP, dGTP, and dTTP).
4. *Taq* polymerase (Genei 5 μ/μL).
5. 10 μM SSR primers.
6. Template DNA.

7. Sterile distilled water.
8. Thermal cycler.

3 Methods

3.1 Identification of Male and Female Sex Among Bruchid Pest

1. Identify the *C. maculatus* or cowpea weevil (*see* **Note 6**) by reddish-brown coloration, with black and gray elytral markings with two black spots at center.
2. Select the female beetles with the abdomen extends out from the elytra, and also have two black spots in the last segment of the body. The males are smaller than the female.
3. Identify the male and female insects morphologically with two key traits such as,
   - (a) Presence of less dense setae on the ventral side of the 2,3,4 abdominal segments (sternites).
   - (b) Presence of serrate type of antenna in both male and female.

3.2 Bruchid Pest Purification and Multiplication

1. Identify 10 pairs of beetles, that is, male and female as per the key traits.
2. Collect and culture the beetles in 15–20 plastic bottles with 50 g of blackgram or greengram seeds to maintain them as a pure form of *Callasobruchus* species.
3. Allow the beetles for oviposition and leave it undisturbed for 30–35 days to give rise to the next generation.
4. Use the freshly emerged adults from the stock culture for bioassay (*see* Fig. 2).

3.3 Bioassay for Bruchid Resistance

1. Release five pairs of adults on 50 seeds of each genotype placed in 15 cm diameter plastic petri plates.
2. Allow the insects to remain in petri plates for 5 days for oviposition.
3. Remove the adults after 5 days from petri plates (*see* **Note 7**).
4. Observations were recorded on.
   - (a) 100-seed weight (g)
   - (b) Oviposition on 50 seeds.
   - (c) Mean number of eggs per seed [9].
   - (d) Developmental time (egg+larval+pupal stages) (days).
   - (e) Adult emergence.
   - (f) Mean developmental period (days) (*see* Eq. 1).
   - (g) Howe's Index of susceptibility [10] (*see* Eq. 2).

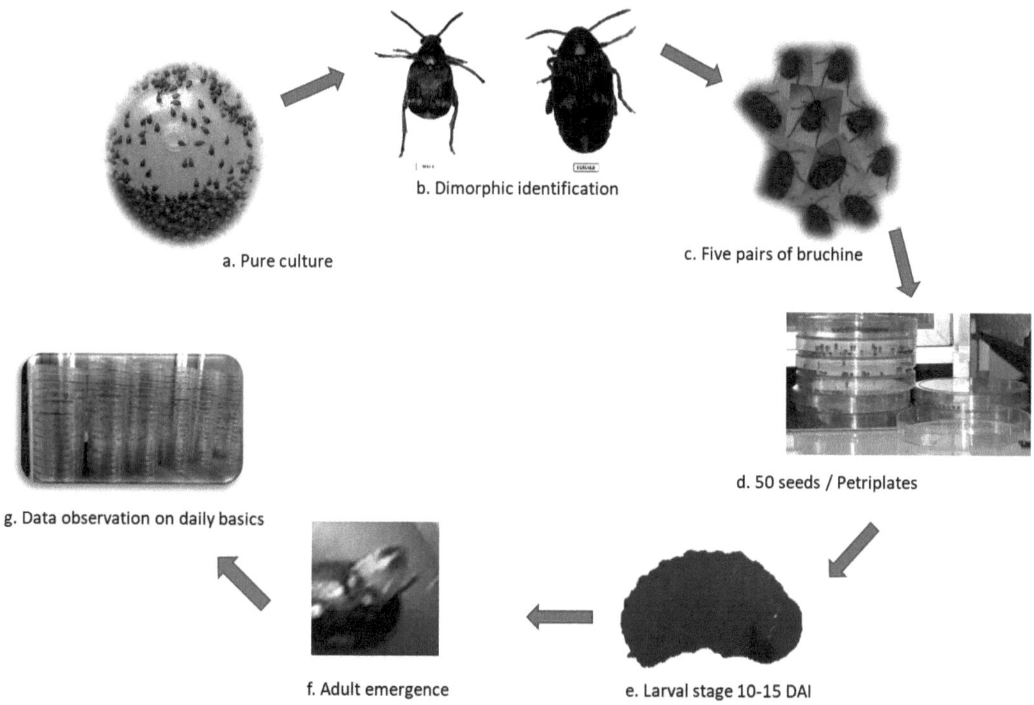

Fig. 2 Bruchid screening procedure including (**a**) pure culturing and mass multiplication of the beetles, (**b**) dimorphic identification of the beetles, (**c**) collect five pairs of beetles, (**d**) allow it to oviposit on the seeds of the genotypes, (**e**) larval stage will be upt o 10–15 DAI, (**f**) adult emergence will be observed after 25–30 DAI and (**g**) observe daily to collect data for bruchid screening

(h) Seed damage (%) (*see* Eq. 3).

(i) Seed weight loss (%) (*see* Eq. 4) (*see* **Note 8**).

5. The adults emerged were counted on a daily basis and removed from the petri plates. The experiments will be carried out up to 50 days after infestation (DAI) only (*see* **Note 9**).

3.4 Screening for Bruchid Resistance

Use the following essential traits to assess the level of resistance to bruchid infestation:

1. Observe the initial seed weight of the 50 seeds of all genotypes before adult release and record it as initial seed weight (g).
2. Count the total number of eggs laid on the seeds with a 10× magnification compound microscope.
3. Estimate the mean number of eggs by dividing the total number of eggs by the number of seeds.
4. Estimate the developmental time by counting the combined duration of egg, larval and pupal stages (*see* **Note 10**).
5. After 25 days of adult release, observe the adult emergence daily up to 50 days after infestation (DAI), and the cumulative

adult emergence can be calculated as the total number of adult emergence.

6. Use Eq. (1) to assess mean developmental period (days)

$$\text{MDP(days)} = \frac{d_1 a_1 + d_2 a_2 + d_3 a_3 + \ldots + d_n a_n}{\text{Number of adults emerged}} \quad (1)$$

Where

d_1—day at which the adults started emerging (first day).

a_1—number of adults emerged on the d_1th day.

7. Use Eq. (2) to assess Howe's index of susceptibility (IS) (*see* **Note 11**).

$$\text{Howe's index of susceptibility} = \text{Loge}\frac{F_1}{D} \times 100 \quad (2)$$

Where

F_1—the total number of emerging adults.

D—mean developmental period in days.

8. Observe the seed damage (%) using Eq. (3) at 50 DAI (*see* **Note 12**).

$$\text{Seed damage(\%)} = \frac{\text{Number of seeds damaged}}{\text{Total number of seeds}} \times 100 \quad (3)$$

9. Estimate the seed weight loss (%) using Eq. (4) at 50 DAI.

$$\text{Seed weight loss (\%)} = \frac{\text{Initial weight of 50 seeds} - \text{Final weight of 50 seeds}}{\text{Initial weight of 50 seeds}} \times 100 \quad (4)$$

3.5 DNA Extraction

Extract the genomic DNA samples using the CTAB method [11].

1. Collect young fresh leaves 10–20 days old and use them for DNA extraction.
2. Ground the sample with a pestle and mortar by using preheated (65 °C) CTAB buffer.
3. Add a quantity of 500 μL of CTAB buffer.
4. Transfer the extracted samples into Eppendorf tubes and incubate them in the water bath for 30 min at 65 °C.
5. After incubation, add 300 μL of phenol:chloroform:isoamyl alcohol (25:24:1) into the tubes, and invert it twice to mix and keep the tubes in a centrifuge for 10 min at 10,000 rpm.
6. Collect the supernatant in the fresh tubes, add 500 μL of chloroform:isoamyl alcohol (24:1), and invert it twice to mix.
7. Centrifuge the tubes for 5 min at 10000 rpm. Collect the aqueous layer into the new Eppendorf tubes.

8. Add 0.7 μL of 100% ethanol (stored at −20 °C) to each sample, invert to mix, and keep it overnight at 4 °C.
9. Centrifuge the samples at 8000 rpm for 10 min on the next day. Discard the supernatant from each sample and air dry the pellets settled in the bottom for 30 min.
10. Add a quantity of 50 μL of TE buffer into each sample and store overnight at 4 °C.
11. Further, add RNAse (3 μL) into each sample to exclude the RNA contamination from the sample and leave it at 65 °C for 30 min (*see* **Note 5**).

3.6 DNA Quantification

Check the quality and concentration of DNA before carrying out a PCR reaction.

1. Check the quality of the DNA using 0.8% agarose gel.
2. Prepare 0.8% agarose gel with 0.8 g agarose and heat it along with 100 mL 1× TBE buffer (*see* **Note 13**).
3. Add ethidium bromide (10 mg/mL) in a concentration of 10 μg/mL to the gel before it solidifies.
4. Cast the gel in a gel tray and keep it on a horizontal surface.
5. Place the comb in such a way that a 2 mm gap was maintained between the bottom of the gel and the comb tip.
6. Allow the gel to solidify.
7. After solidification, place the gel in the electrophoresis chamber and add 1× TBE buffer to fill the chamber to flood the surface of the gel.
8. Remove the comb carefully without damaging the wells of the gel.
9. Load the Lambda DNA with gel loading buffer into the wells of known concentrations (25, 50, and 100 ng/μL) to compare the DNA concentration.
10. Connect the apparatus to the power pack and allow the DNA to run through the gel at 80 V for 1 h (*see* **Note 14**).
11. After electrophoresis, remove the gel from the chamber and view it under UV illumination and photograph using a gel documentation unit (M/s Medicare, India; Model: Gelstain 4× Advanced).
12. Then dilute the DNA accordingly using TE buffer to the desired concentration.

3.7 Assessment of Parental Polymorphism

1. Involve all SSR primers (*see* **Note 15**) to identify the polymorphism between the susceptible and the resistant parents.
2. Identify polymorphic SSR markers between parents among the amplified markers.

3. Perform genotyping with polymorphic SSR markers among the F_2 mapping population of the cross between the susceptible and the resistant parents.

3.8 PCR Amplification

1. Check the DNA concentration of each sample as it should be approximately 5–10 ng/μL.
2. Prepare the polymerase chain reaction (PCR) mixtures, with 2 μL of 10 ng/μL template DNA of each sample, and add 1.0 μL of 10× Taq buffer + $MgCl_2$ (15 mM).
3. Add 1.0 μL of dNTPs (2 mM) and 0.5 μL each of forward and reverse SSR primers (10 μM).
4. Add 0.3 μL of Taq polymerase (3 IU) and 4.7 μL of sterile double-distilled water in the PCR tube.
5. Place the PCR tube in a thermocycler (M/s Eppendorf, Germany; Model: AG 6325) for DNA amplification.
6. Follow the PCR reaction protocols with 94 °C for 4 min of initial denaturation, followed by 30 cycles of 94 °C for 30 s for denaturation, 55 °C for 45 s for annealing (*see* **Note 16**), 72 °C for 1 min for extension, with a final extension step of 72 °C for 20 min.
7. Load the PCR products onto a 3% agarose gel and document the image using GELSTAIN 4× advanced gel documentation unit (M/s Medicare, India). (Prepare the 3% agarose gel by mixing 3 g agarose and 100 mL 1× TBE buffer and placing it in the oven for heating. The solution is kept in liquid form by keeping it at 45 °C.)
8. Add ethidium bromide (10 mg/mL) was added at the concentration of 0.5 μg/mL before it solidifies. (**The remaining steps will be the same as from** Subheadings 3.6, **stpes 4–11**).
9. Mark the base pair details of the amplified DNA for all genotypes from the mapping population for amplified SSR markers.
10. Score the banding pattern, as "A" for the parent 1 type, "B" for the parent 2 type, and "H" for the heterozygous type.

3.9 Linkage Map and QTL Analysis

1. Prepare the data as per the genotype with consecutive SSR markers in the Excel sheet.
2. Perform the linkage analysis using QTL IciMapping software [12].
3. Select a minimum LOD threshold of 3.0 and a maximum distance of 50 cM for the construction of linkage groups.
4. Calculate the map distance in centimorgan (cM) values using the Kosambi mapping function.
5. After linkage map analysis, perform the QTL analysis using QGene 4.4.0 software [13].

6. Use composite interval mapping (CIM) with automatic cofactor selection. Identify the LOD threshold significance for each QTL with 1000 runs of a permutation test at $P = 0.05$.
7. Utilize the QTL information obtained from the QTL analysis to frame the breeding program for the desired trait improvement.

4 Notes

1. The selection of parents is the first step in the production of the mapping population. Parents selected to develop a mapping population should have sufficient variation for the trait of interest both at the DNA sequence level and at phenotypic level.
2. MDU 1 is a popular high-yielding blackgram variety grown in Tamil Nadu. This variety was developed at Agriculture College and Research Institute, Tamil Nadu Agricultural University, Madurai, India, and released in 2014 for commercial cultivation in October–November season of Tamil Nadu, India. But this variety (MDU 1) is highly susceptible to bruchid (*Callosobruchus maculatus*). TU 68 is a genotype developed from *Vigna mungo* var. *sylvestris* by Bhabha Atomic Research Centre, Trombay, India. It is less susceptible toward bruchid infestation. TU 68 can be selected as a male parent and crossed with a female parent MDU 1. The F_2 generation derived from the cross can be used as a mapping population for bruchid resistance.
3. In India, compared to other species, *C. maculatus* is the most predominant species. It causes a yield loss of up to 80–90% in the blackgram crop.
4. The phenol is highly reactive toward light. Hence, the stock solution has to be stored in an amber color bottle and placed in the refrigerator.
5. RNAse can be used to remove the RNA contamination in the extracted DNA.
6. The bruchid beetle is not a true weevil, as it lacks the "snout" of a true weevil.
7. The quantity of seeds involved for the screening and insect pair has been increased to provide maximum pressure load on the genotype under screening to obtain good results. After the oviposition, adults are to be removed from the petri plates. Otherwise, confusion may arise while counting the adult emergence rate.

8. Seed damage and seed weight loss are estimated 50 days after infestation (DAI).
9. The emerged adults are removed from the petri plates daily after the examination so as to avoid secondary infestation. Though the adults are removed daily, there may be a chance to get secondary infestation as the single pair of adults can lay eggs within 1 day after emergence. Hence, the whole experiment has to be stopped by 50 DAI to avoid the second cycle of adult emergence.
10. The time taken for the first adult emergence on the genotypes from the date of adult release.
11. The genotypes were classified as Resistant (R) (0.0–2.5), Moderately resistant (MR) (2.6–5.0), Moderately susceptible (MS) (5.1–7.5), Susceptible (S) (7.6–10.0), and Highly susceptible (HS) (>10.0).
12. The genotypes were classified as Highly resistant (HR) (0–10%), Resistant (R) (10.1–20%), Moderately resistant (MR) (20.1–40%), Susceptible (S) (40.1–80%), and Highly susceptible (HS) (80.1–100%).
13. For the working solution (1× TBE buffer), the stock solution was diluted 10 times with double-distilled water.
14. DNA is negatively charged and it runs from black to red.
15. All the SSR primers on the entire 11 chromosomes are to be used for full coverage. The SSR primers and their sequence information can be utilized from the related *Vigna* sp., viz., mungbean maps [14–18] and adzuki bean map [19, 20] as the SSR markers are cross-species amplification in nature.
16. The annealing temperature is 55 °C in this reaction, but should be empirically determined for each set of primers using the formula (2) (#A and T nucleotides) + (4)(#G and C nucleotides).

References

1. Dasbak M, Echezona B, Asiegbu J (2009) Pigeon pea grain physical characteristics and resistance to attack by the bruchid storage pest. Int Agrophys 23(1):19–26
2. Soundararajan R, Geetha S, Chitra N, Dinakaran D (2013) Resistance in Vigna mungo var silvestris against Bruchids, Callosobruchus maculatus (F.). Ann Plant Prot Sci 21(2): 279–282
3. Sharma S, Thakur D (2014) Studies on the varietal preference of *Callosobruchusmaculatus* on soybean genotypes. Asian J Biol Sci 7(5): 233–237
4. Morgante M, Olivieri AM (1993) PCR amplified microsatellites as markers in plant genetics. Plant J 3(1):175–182
5. Chaitieng B, Kaga A, Tomooka N, Isemura T, Kuroda Y, Vaughan DA (2006) Development of a black gram (*Vigna mungo* (L.) Hepper) linkage map and its comparison with an adzuki bean (*Vigna angularis* (Willd.) Ohwi and Ohashi) linkage map. Theor Appl Genet 113: 1261–1269
6. Gupta SK, Souframanien J, Gopalakrishna T (2008) Construction of a genetic linkage map of black gram. *Vigna mungo* (L.) Hepper,

based on molecular markers and comparative studies. Genome 51:628–637

7. Isemura T, Kaga A, Tabata S, Somta P, Srinives P, Shimizu T, Jo U, Vaughan DA, Tomooka N (2012) Construction of a genetic linkage map and genetic analysis of domestication related traits in mungbean (*Vigna radiata*). PLoS One 7(8):e41304. https://doi.org/10.1371/journal.pone.0041304
8. Ragul S, Manivannan N, Iyanar K, Ganapathy N, Karthikeyan G (2022) Screening and biochemical analysis on blackgram genotypes for resistance against storage pest bruchine (*Callosobruchus maculatus* (F.)). Legum Res 45(3):371–378. https://doi.org/10.18805/LR-4528
9. Sewsaran R, Khan A, Stone R, Jhon K (2019) Resistance screening of 14 *Cajanus cajan* (L.) Millsp, cultivars to *Callasobruchus maculatus* (F.) (Coleopters: Bruchidae). J Stored Prod Res 82:67–72
10. Mensah G (1986) Infestation potential of *Callosobruchus maculatus* (Fab.)(Coleoptera: Bruchidae) on cowpea cultivars stored under subtropical conditions. Int J Trop Insect Sci 7(6):781–784
11. Lodhi MA, Ye GN, Weeden NF, Reisch BI (1994) A simple and efficient method for DNA extraction from grapevine cultivars and *Vitis* species. Plant Mol Biol Report 12:6–13
12. Wang J, Li H, Zhang L, Meng L (2016) Users' manual of QTL IciMapping. The Quantitative Genetics Group, Institute of Crop Science, Chinese Academy of Agricultural Sciences (CAAS) and Genetic Resources Program, International Maize and Wheat Improvement Center (CIMMYT), Apdo, Beijing and Mexico
13. Joehanes R, Nelson JC (2008) QGene 4.0, an extensible JavaQTL-analysis platform. Bioinformatics 24:2788–2789
14. Gwag JG, Chung WK, Chung HK, Lee JH, Ma KH, Dixit A, Park YJ, Cho EG, Kim TS, Lee SH (2006) Characterization of new microsatellite markers in mungbean, *Vigna radiata* (L.). Mol Ecol Notes 6:1132–1134
15. Somta P, Musch W, Kongsamai B, Chanprame S, Nakasathien S, Toojinda T, Sorajjapinun W, Seehalak W, Tragoonrung S, Srinives P (2008) New microsatellite markers isolated from mungbean (*Vigna radiata* (L.) Wilczek). Mol Ecol Resour 8(5):1155–1157
16. Seehalak W, Somta P, Musch W, Srinives P (2009) Microsatellite markers for mungbean developed from sequence database. Mol Ecol Resour 9:862–864
17. Tangphatsornruang S, Somta P, Uthaipaisanwong P, Chanprasert J, Sangsrakru D, Seehalak W, Sommanas W, Tragoonrung S, Srinives P (2009) Characterization of microsatellites and gene contents from genome shotgun sequences of mungbean (*Vigna radiata* (L.) Wilczek). BMC Plant Biol 9:137
18. Chotechung S, Somta P, Chen J, Yimran T, Chen X, Srinives P (2016) A gene encoding a polygalacturonase-inhibiting protein (PGIP) is a candidate gene for bruchid (Coleoptera: Bruchidae) resistance in mungbean (*Vigna radiata*). Theor Appl Genet 129:1673–1683. https://doi.org/10.1007/s00122-016-2731-1
19. Wang XW, Kaga A, Tomooka N, Vaughan D (2004) The development of SSR markers by a new method in plants and their application to gene flow studies in azuki bean (*Vigna angularis* Willd.) Ohwi & Ohashi. Theor Appl Genet 109:352–360
20. Chankaew S, Isemura T, Isobe S, Kaga A, Tomooka N, Somta P, Hirakawa H, Shirasawa K, Vaughan DA, Srinives P (2014) Detection of genome donor species of neglected tetraploid crop *Vigna reflexo-pilosa* (cre'ole bean), and genetic structure of diploid species based on newly developed EST-SSR markers from azuki bean (*Vigna angularis*). PLoS One 9(8):e104990. https://doi.org/10.1371/journal.pone.0104990

Chapter 20

RNAi-Based Approach as a Promising Tool for Management of Leafhopper (*Amrasca biguttula biguttula*) in Cotton

K. Shankarganesh, V. Gowtham, K. P. Raghavendra, and Joy Das

Abstract

Leafhoppers pose a severe threat to cotton crops, causing substantial yield losses. The use of synthetic insecticides for pest control has drawbacks such as resistance development, environmental harm, and health risks. In this context, RNA interference (RNAi)-based strategies have emerged as an alternative for leafhopper management. RNAi involves introducing double-stranded RNA molecules (dsRNAs) that target vital genes, disrupting their function and causing reduced fitness or death. Specific target genes are identified and corresponding dsRNA molecules are produced and delivered to leafhoppers through various methods. This chapter presents a case study of *Amrasca biguttula*, showcasing the potential of RNAi in cotton leafhopper control. Despite delivery and cost challenges, RNAi offers precise, safe, and eco-friendly benefits, making it a valuable tool for managing this significant cotton pest. The chapter outlines techniques for mass rearing leafhoppers, gene selection, RNA isolation, cDNA synthesis, primer design, PCR amplification, and *in vitro* transcription for dsRNA synthesis. It also includes a protocol for RNAi diet bioassay for bio-efficacy testing and molecular characterization of dsRNA-treated insects. Proper handling and use of RNase-free reagents are emphasized throughout these processes. Overall, we present an RNAi protocol as a valuable tool for controlling leafhopper infestations in cotton research experiments.

Key words Cotton, *Amrasca biguttula biguttula*, dsRNA synthesis, RNAi, Microinjection and diet incorporation bioassay, qRT-PCR

1 Introduction

Leafhoppers represent the major sucking pest of cotton. The occurrence of this pest can be observed throughout crop growth resulting in up to 50% yield loss [9]. The indiscriminate use of synthetic insecticides for the management of this pest could lead to multiple issues such as resistance development, resurgence of minor pests, insecticide residue in the final product, health hazards for applicators, and contamination of the agroecosystem. As a result, alternative strategies like RNAi-based approaches have garnered attention. This method involves introducing small RNA molecules, majorly in the form of double-stranded RNA (dsRNA), specifically designed

Muthappa Senthil-Kumar and Hema Ramanna (eds.), *Plant Biotic Stress: Methods and Protocols*, Methods in Molecular Biology, vol. 2966, https://doi.org/10.1007/978-1-0716-4746-2_20,

to target and silence crucial genes responsible for the pest's survival or reproduction. These dsRNA molecules are subsequently processed to produce small interfering RNAs (siRNAs) that complementarily bind to the messenger RNA (mRNA) of the target gene, preventing its translation into functional proteins [1]. This interference disrupts the normal functioning of essential genes, ultimately leading to the death or reduced fitness of a target pest [4, 5].

1.1 RNAi for Effective Pest Control in Cotton

To develop an RNAi-based approach for controlling leafhoppers in cotton, specific genes vital for the pest's survival or reproduction are to be identified. Corresponding double-stranded RNAs (dsRNAs) to be synthesized and delivered to the leafhoppers through topical application, ingestion, or genetically modified plants expressing the target gene-specific dsRNAs. Upon consumption or absorption of the dsRNAs, the leafhoppers trigger the RNAi mechanism within their cells, leading to the silencing of the targeted genes and subsequent pest control. A previous study investigated the potential of RNAi to control the leafhopper *Amrasca biguttula* in cotton [6]. In this study, the transcriptome analysis of the cotton leafhopper revealed the presence of a robust RNA interference (RNAi) machinery, with 37 transcripts associated with this process. Core RNAi machinery genes (*Dicer-2*, *Ago-2*, *Piwi*, and *Staufen*) were validated through dsRNA feeding studies. Additionally, the transcriptome data helped to identify potential RNAi targets and genes related to insecticide detoxification, particularly the CYP450 family [6]. These findings provide valuable insights into the insect's ability to develop resistance to insecticides and highlights potential targets for pest control strategies.

1.2 Potential Benefits of Using RNAi to Control Leafhoppers

The use of RNAi for controlling leafhoppers offers various potential benefits. Firstly, RNAi is considered a safe and environmentally friendly technology. It allows targeting-specific genes crucial for the leafhopper's survival. Moreover, RNAi holds the promise of developing more effective insecticides. Nevertheless, some challenges must be addressed before widespread implementation. These challenges include devising efficient delivery methods for dsRNA to cotton plants, ensuring the safety of RNAi products for humans and the environment, and achieving cost-effectiveness. Despite these obstacles, the substantial advantages of RNAi for controlling leafhoppers make it a valuable tool in managing this significant cotton pest.

2 Materials

2.1 Mass Culture of Cotton Leafhopper

1. *A. biguttula biguttula*.
2. Cotton plant.

3. Aluminum-framed cages.
4. Growth chamber.

2.2 Target Genes for RNAi in Cotton Leafhopper (See Note 1)

2.3 RNAi of Candidate Genes

2.3.1 RNA Isolation

1. TRIzol Reagent [40% w/w phenol (saturated at pH 4.3), 1 M guanidine thiocyanate, 1 M ammonium thiocyanate, 0.1 M sodium acetate buffer (pH 5.0), 5% w/w glycerol].
2. Diethyl pyrocarbonate (DEPC)-treated water.
3. Chloroform.
4. Isopropanol.
5. 75% ethanol
6. Nuclease-free water.
7. RNase-free pipettes and tubes.

2.3.2 cDNA Synthesis

1. RNA sample(s) with known concentration.
2. Oligo dT (deoxythymidine) or random primers (depending on the specific application).
3. Reverse transcriptase enzyme.
4. Nuclease-free water.
5. Buffer(s) for reverse transcription.
6. Dithiothreitol (DTT) or other reducing agents.
7. RNase inhibitor (e.g., RNAsin).
8. Thermal cycler or a suitable apparatus for the reaction.

2.3.3 Primer Design for Target-Gene-Specific dsRNA Synthesis

1. Select target gene sequence for double-stranded RNA (dsRNA) synthesis.
2. Snapdragon double-stranded RNA (dsRNA) design tool (*see* **Note 2**).
3. T7 polymerase promoter sequence can be incorporated into both forward and reverse primers (*see* **Note 3**).
4. An Illustrative example of a gene-specific primer sets designed for PCR amplification of templates for dsRNA synthesis. For the target gene (e.g., *Snf7*), the primer sequences are as follows: the **forward primer** is 5′ TAATACGACTCACTA TAGGGXXXXXXXXXXXXXXXXXXX 3′, and the **reverse primer** is 5′ TAATACGACTCACTA TAGGGXXXXXXXXXXXXXXXXXXX 3′.

2.3.4 Amplification of the Candidate Gene for dsRNA Synthesis

1. 10× *Taq* polymerase buffer.
2. dNTPs.
3. $MgCl_2$.
4. Template (cDNA).
5. Forward primer and reverse primer.
6. *Taq* polymerase.
7. 6× loading dye [30% glycerol (v/v), 0.25% bromophenol blue (w/v), and 0.25% xylene cyanol FF (w/v)]
8. 50× TAE buffer [242 g Tris base in double-distilled H_2O; 57.1 mL glacial acetic acid; 100 mL 0.5 M EDTA solution (pH 8.0) and make up the volume to 1 L]
9. Agarose gel (weigh the required mass of agarose in a given volume of 1× TAE to make gels of desired concentrations).
10. Any commercially available gel elution/PCR cleanup kit.

2.3.5 In Vitro Transcription for dsRNA Synthesis

1. T7 DNA-dependent RNA polymerase (T7 pol) enzyme nuclease-free water for volume makeup (precalculated).
2. Template (5 μg).
3. T7 polymerase buffer (1× diluted).
4. Dithiothreitol (DTT) (10 mM).
5. Nucleoside triphosphate (NTP) mix (5 mM).
6. Inorganic pyrophosphatase (PPase) (0.01 units/μL).
7. RNase inhibitor (0.01 units/μL).
8. T7 polymerase (0.01 units/μL).

2.4 Insect Bioassay

2.4.1 Diet Incorporate Assay

1. dsRNA (for selected target genes viz. *dsSnf7*, *dsAQP1*, *dsVAT-Pase,* and *dsIAP*).
2. Diet ingredients [7].
3. 1.5 mL tubes.
4. Muslin cloth.
5. Parafilm.

2.4.2 Synthetic Diet Composition (in mg/10 mL)

5.0 mg of L-cysteine, 2.0 mg of glycine, 1.0 mg of nicotinic acid, 500.0 mg of sucrose, 50.0 mg of K2HPO4, 0.04 mg of ZnCl2, 0.25 mg of thiamine HCl, 0.25 mg of Vitamin B6, 2.0 mg of Becosules capsule powder, and 2.0 mg of Green food dye. These components are dissolved in the appropriate volume to prepare the solution.

2.5 Molecular Characterization of the dsRNA-Affected Insect

1. RNA isolation (refer Subheading 2.3.1).
2. cDNA synthesis (refer Subheading 2.3.2).

3. qRT-PCR amplification of the candidate genes (viz. *Snf7*, *AQP1*, *VATPase,* and *IAP*) along with a normalizer (stable housekeeping gene/s**) (*see* **Note 3**).

3 Methods

3.1 Mass Culturing

1. Collect *A. biguttula biguttula* nymphs from the field.
2. Maintain a healthy colony of leafhoppers on cotton plants within aluminum-framed cages covered with a fine voile-type nylon mesh.
3. Place the cages in a growth chamber with conditions set at 65–70% relative humidity, a 14:10 h light and dark photoperiod, and a temperature of 27 ± 2 °C.
4. Regularly replace the cotton plants to ensure a continuous supply of fresh food for the insects and to maintain the availability of the leafhopper culture throughout the study period.

3.2 RNAi of Candidate Genes

3.2.1 RNA Isolation

1. Collect the insect tissue (50–100 mg of tissue) to be used for RNA isolation.
2. Add 1 mL of TRIzol reagent to each sample.
3. Incubate the samples at room temperature for 5 min.
4. Add 250 μL of chloroform to each sample and shake vigorously for 15 s.
5. Centrifuge the samples at 12,000×*g* for 15 min at 4 °C.
6. Carefully transfer the upper aqueous phase to a new tube.
7. Add 0.5 mL of isopropanol to each sample and mix gently.
8. Centrifuge the samples at 12,000×*g* for 15 min at 4 °C.
9. Carefully remove the supernatant and discard.
10. Add 1 mL of 75% ethanol to each sample and mix gently.
11. Centrifuge the samples at 12,000×*g* for 5 min at 4 °C.
12. Carefully remove the supernatant and discard.
13. Allow the pellets to airdry for 10 min.
14. Resuspend the pellets in RNase-free water.
15. Check the integrity of the RNA by running it in agarose gel (1.5%).
16. Quantity the RNA samples using nano-spectrophotometry.
17. Treat DNase if there is any DNA contamination.
18. Take 1 μg of RNA and synthesize cDNA.

3.2.2 cDNA Synthesis

The cDNA synthesis protocol is a step-by-step procedure used to convert RNA molecules into complementary DNA (cDNA) using reverse transcriptase enzyme. Here is a general outline of a typical cDNA synthesis protocol:

1. Ensure the quality of the RNA sample by using spectrophotometry or other methods.
2. Prepare the cDNA synthesis reaction mix: This will include the RNA sample (usually 500 ng to 1 μg per reaction), appropriate reverse transcription buffer, oligo dT or random primers (based on the experimental needs), dithiothreitol (DTT) or another reducing agent to maintain the enzyme's activity, RNase inhibitor to prevent RNA degradation, nuclease-free water to bring the final volume to the desired level in a sterile, nuclease-free microcentrifuge tube (*see* **Note 5**).
3. Incubate the reaction mix in a thermal cycler or a suitable apparatus capable of controlled temperature. Typically, the reaction is carried out at 42–55 °C for 30 min to 1 h.
4. After the incubation, heat inactivate the reverse transcriptase enzyme at a higher temperature (e.g., 70–75 °C) for 5–10 min to stop the reaction.
5. Store the synthesized cDNA at −20 °C or −80 °C for short-term or long-term storage, respectively.

3.2.3 PCR Amplification of Target Gene Using Primers Tailed with T7 RNA Polymerase Promotor Region

1. Add template (cDNA), 10× *Taq* polymerase buffer, dNTPs, $MgCl_2$, forward primer and reverse primer, and *Taq* polymerase to PCR tubes.
2. Mix properly and centrifuge.
3. Amplify in thermo cycler based on primers melting temperature.
4. RT-PCR using the following cycling conditions: 94 °C for 2 min, followed by 35 cycles at 94 °C for 30 s, 56 °C for 30 s, and 72 °C for 45 s and finally extension step for 4 min at 72 °C.
5. Resolve amplified product in 1% agarose gel by gel electrophoresis.
6. Perform PCR cleanup with any commercially available kit.

3.3 In Vitro Transcription Protocol

1. Add the aforementioned components to a microcentrifuge tube in the following order: nuclease-free water for volume makeup (precalculated), template (5 μg), T7 polymerase buffer (1× diluted), dithiothreitol (DTT) (10 mM), nucleoside triphosphate (NTP) mix (5 mM), inorganic pyrophosphatase (PPase) (0.01 units/μL), RNase inhibitor (0.01 units/μL), and T7 polymerase (0.01 units/μL) (*see* **Note 6**).

2. Mix the reaction mixture gently by pipetting up and down.
3. Incubate the reaction mixture in a circulating water bath at 37 ° C for 6 h to allow for dsRNA synthesis.
4. After the incubation period, quantify the resultant dsRNA using a nano-spectrophotometer.
5. Store the dsRNA at −20 °C until further use.

3.4 Delivery of dsRNA for Insect RNAi (Diet Method for RNAi-Based Silencing and Testing)

1. Mix dsRNA at a concentration of 500 ng/μL for respective target genes.
2. Prepare the diet by incorporating the dsRNA mixture.
3. Synthetic diet composition for delivery of dsRNA in cotton leafhopper through membrane feeding [8].
4. Cut a 1/4th portion from the bottom side of a 1.5 mL tube.
5. Cover the cut portion of the tube with muslin cloth.
6. Place the diet incorporated with dsRNA between two layers of stretched parafilm in the tube.
7. Release three biological replicates of thirty *A. biguttula biguttula* insects released into tubes each containing 5 insects.
8. Observe the insects 48 h after release and record the mortality.
9. Compare the mortality of the insects treated with *dsAQP*, *dsIAP*, *dsVATPase*, *dsSnf7* to those treated with *dsGaEPSPS* (dsRNA for nonhost gene used as control).

3.5 Molecular Characterization

1. Isolate the total RNA from the adult leafhopper from both untreated and dsRNA treated insect groups (a pool of 4-5 insects from each group can be used).
2. Use 500 nanogram RNA from each samples (untreated and dsRNA treated groups) for cDNA synthesis (refer section 3.2.2) and subjected to quantitative real-time reverse-transcription PCR (qRT-PCR or qPCR) for gene expression studies following standard protocol [3].
3. Calculate the relative expression of the target genes with respect to the normalizer (*see* **Note 4**) using ΔΔCT method [2], in comparison with *dsGaEPSPS*-fed control insects, after normalization with appropriate housekeeping gene [8].

4 Notes

1. Before delving into the design of primers, it is crucial to select the appropriate target genes for RNAi. Researchers need to identify genes that play critical roles in essential biological processes, such as development, reproduction, or immunity. These genes are considered ideal targets as their silencing can

have a significant impact on the survival and reproductive capabilities of leafhoppers. As for instance, selection of specific genes as targets for RNAi in leafhoppers along with this a nonhost/nontarget gene example *EPSPS* from *Gossypium arboreum* (*GaEPSPS*) [3] could be used as control. Some of the candidate target genes and their encoded proteins, along with their functions, are as follows: Snf7 encodes a vacuolar sorting protein, which is involved in the endosomal sorting complex required for transport (ESCRT) pathway. The ESCRT machinery plays a critical role in directing protein trafficking and membrane remodeling within cells. AQP1 encodes an aquaporin, which facilitates the transport of water and small solutes across cell membranes. It regulates water movement and excretion, allowing insects to adapt to changing environmental conditions. VATPase encodes a vacuolar-type ATPase, essential for various cellular processes, including protein degradation, vesicular trafficking, and ion transport. Lastly, IOP encodes an inhibitor of apoptosis, promoting cell survival and playing crucial roles in tissue development, immune response, and cellular homeostasis.

2. The design of appropriate primers is a critical step in the successful application of double-stranded RNA (dsRNA) for RNA interference (RNAi) experiments. The design of primers for dsRNA, while avoiding off-target effects, can be facilitated by utilizing online tools like the Snapdragon double-stranded RNA (dsRNA) design tool (https://www.flyrnai.org/cgi-bin/RNAi_find_primers.pl). Primers for the target genes can be designed, and an additional T7 polymerase promoter sequence can be incorporated into both forward and reverse primers to enable dsRNA synthesis [3].
3. Bold italicised nucleotides represent T7 RNA polymerase promoter site. "X" represents any nucleotide specific for the target gene.
4. Some commonly used housekeeping genes (as normalizer for gene expression studies) in insects include actin, which is a cytoskeletal protein involved in maintaining cell structure and facilitating intracellular transport. It is known for its stable expression and is widely used as a reference gene in many organisms. GAPDH (glyceraldehyde 3-phosphate dehydrogenase) enzyme involved in glycolysis, the metabolic pathway that generates energy from glucose. It is a commonly used housekeeping gene due to its constitutive expression and relatively high abundance. EF1α (elongation factor 1-alpha) essential protein involved in translation during protein synthesis. It is considered a stable reference gene and is frequently used in gene expression studies. RPL (ribosomal protein L) are components of the ribosomes, the cellular machinery responsible

for protein synthesis. Specific ribosomal proteins, such as RPL, are often used as reference genes due to their constant expression levels. Similar to RPL, ribosomal protein S (RPS) is another component of the ribosome that is used as a housekeeping gene in various studies.

5. The specifics of the protocol may vary based on the reverse transcription kit or enzyme used, the nature of the RNA tem plate, and the application for which the cDNA will be used (e.g., PCR, qPCR, and gene expression analysis). Always follow the manufacturer's instructions and optimize the reaction conditions if needed.
6. To avoid nuclease contamination during dsRNA synthesis, it is essential to use nuclease-free reagents, and add RNase inhibitors to protect RNA from degradation. Utilize RNAse-free tubes and tips, treat surfaces with RNAse-inactivating agents, and avoid freeze-thaw cycles. Store RNA at −80 °C until experimental use.

Acknowledgments

The authors gratefully acknowledge the support from Director, ICAR-CICR, Nagpur and Head, ICAR-CICR, Regional Station, Coimbatore.

References

1. Fire A, Xu S, Montgomery M et al (1998) Potent and specific genetic interference by double-stranded RNA in *Caenorhabditis elegans*. Nature 391:806–811. https://doi.org/10.1038/35888
2. Livak KJ, Schmittgen TD (2001) Analysis of relative gene expression data using real-time quantitative PCR and the 2– ΔΔCT method. Methods 25(4):402–408
3. Das J, Kumar R, Shah V, Raghavendra KP, Sharma AK (2023) Identification and functional characterisation of N-acetylglucosamine kinase from *Helicoverpa armigera* divulge its potential role in growth and development via UDP-GlcNAc salvage pathway. Int J Biol Macromol 242:124674
4. Das J, Kumar R, Shah V, Sharma AK (2023) Simple cost-effective larval injection method for dsRNA delivery to induce RNAi response in *Helicoverpa armigera* (Hübner). J Appl Entomol 147(4):289–298
5. Das J, Kumar R, Shah V, Sharma AK (2022) Functional characterization of chitin synthesis pathway genes, HaAGM and HaUAP, reveal their crucial roles in ecdysis and survival of *Helicoverpa armigera* (Hübner). Pestic Biochem Physiol 188:105273
6. Gupta M, Singh S, Kaur G, Pandher S, Kaur N, Goel N, Kaur R, Rathore P (2021) Transcriptome analysis unravels RNAi pathways genes and putative expansion of CYP450 gene family in cotton leafhopper *Amrasca biguttula* (Ishida). Mol Biol Rep 48(5):4383–4396
7. Koyama K (1988) Artificial rearing and nutritional physiology of the planthoppers and leafhoppers (Hemiptera: Delphacidae and Deltocephalidae) on a holidic diet. JARQ 22(1):20–27
8. Singh S, Gupta M, Pandher S, Kaur G, Rathore P, Palli SR (2018) Selection of housekeeping genes and demonstration of RNAi in cotton leafhopper, *Amrasca biguttula biguttula* (Ishida). PLoS One 13(1):e0191116
9. Manivannan A, Kanjana D, Dharajothi B, Meena B (2020) Evaluation of resistance in cotton genotypes against leafhoppers *Amrasca biguttula biguttula* (Ishida), (Homoptera: Cicadellidae). Int J Trop Insect Sci 41:2409–2420

Chapter 21

Germplasm Evaluation in Chickpea for Resistance to Plant Parasitic Nematodes

Zakaullah Khan, Bharat H. Gawade, Gayacharan, and V. Celia Chalam

Abstract

Cicer arietinum L. (chickpea) is the second most widely cultivated food legume crop after common bean worldwide. Chickpea is an important source of dietary protein for human beings. It also helps in improvement of soil fertility by adding atmospheric N_2 (ca. 60 kg/ha) through biological nitrogen fixation. Plant parasitic nematodes (PPNs) are one of the important biotic factors, which affects the crop production and productivity worldwide. PPNs cause severe yield losses estimated to the tune of US\$100–173 billion worldwide. PPNs not only feed on host plants, but also responsible for inducing wounds that enables entry of secondary soilborne pathogens. PPNs species causing significant economic damage in chickpea include root-knot nematodes (*Meloidogyne* spp.), cyst nematode (*Heterodera ciceri*), root-lesion nematode (*Pratylenchus thornei*), and reniform nematode (*Rotylenchulus reniformis*). Identification and utilization of inherent resistance in host plants along with the agronomic management practices is considered as the best effective approach to control PPNs infection in crop fields. Here, we discuss the methodology for evaluation of resistance to various important plant parasitic nematodes infecting chickpea.

Key words Plant-parasitic nematodes, *Meloidogyne*, *Heterodera*, *Pratylenchus*, *Rotylenchulus*, Resistant screening

1 Introduction

Chickpea (*Cicer arietinum* L.), an annual grain legume crop grown on larger area worldwide is an important crop for food and nutritional security due to its nutritional value. Besides, chickpea being a legume crop, it fixes atmospheric nitrogen, thus increasing soil fertility. Also, this crop is infected by various pests and pathogens among which plant-parasitic nematode are important constraints.

There are more than 100 species of PPNs, which infect the chickpea and can cause up to 14% losses at the global level [1–3]; however, only a few predominant species are known to cause economic damage to chickpea crops throughout the world, which include root-knot nematodes (*Meloidogyne artiellia, M. arenaria, M. javanica, M. incognita*), cyst nematode (*Heterodera ciceri*),

Muthappa Senthil-Kumar and Hema Ramanna (eds.), *Plant Biotic Stress: Methods and Protocols*, Methods in Molecular Biology, vol. 2966, https://doi.org/10.1007/978-1-0716-4746-2_21,

root-lesion nematode (*Pratylenchus thornei*), and reniform nematode (*Rotylenchulus reniformis*). Here, we discuss the methodology for evaluation of resistance to various important plant parasitic nematodes infecting chickpea.

1.1 Root-Knot Nematodes

Meloidogyne spp. (RKN) are one of the most polyphagous nematodes with wide host range and distribution and highly damaging genera of PPNs [4]. The annual global economic losses due to PPNs are estimated to approximately US$170 billion, in which RKNs are the major contributors [4]. Till date around 100 species under *Meloidogyne* genus are reported worldwide.

1.2 Cyst Nematodes

The chickpea cyst nematode, *H. ciceri,* is one of the important damaging nematodes of chickpea that can cause significant damage to crop. Mature cysts are adhered on root surface (Fig. 1), which contains eggs. Among 10,000 chickpea accessions of Kabuli and Desi type chickpea evaluated against *H. ciceri* by International Center for Agricultural Research in the Dry Areas (ICARDA) in collaboration with the Institute for Sustainable Plant Protection, Italy, none of the accessions were found immune, except a few moderately resistant lines [5, 6]. However, some sources were identified in wild *Cicer* germplasm [6, 7]. The screening experiments on wild *Cicer* germplasm indicates that among nine annual *Cicer* species *C. bijugum* has comparatively better resistance against

Fig 1 Cysts of *Heterodera ciceri* attached on the surface of chickpea roots. (Source: ICARDA)

the cyst nematode infection. In a study, eight annuals wild *Cicer* species with varying accession numbers were screened against the cyst nematode but resistance was only observed in *C. bijugum* [7]. This indicates that *C. bijugum* species can be further explored for a new sources and utilization of such accessions in chickpea cultivar development. However, resistance to *H. ciceri* were also discovered in some other species such as *C. reticulatum* (one accession), and *C. pinnatifidum* (six accessions) [6]. The wild species, particularly species which are easily crossable with cultivated species, are also being utilized in breeding programs [8].

1.3 Root-Lesion Nematode

Pratylenchus spp. are root migratory endoparasites that cause heavy damage to root cortical cells during excessive feeding and migration [9], which caused necrotic lesions on roots (Fig. 2). They feed on the root content with the help of a stylet and migrate intracellularly to withdraw nutrients from the cytoplasm of the root cortex cells [10]. found a moderate level of resistance to *P. thornei* in accessions from both *C. echinosperum* and *C. reticulatum*, as well as accessions from *C. bijugum*. In a wild germplasm screening consisting of 133 accessions of *C. reticulatum* and 41 accessions of *C. echinospermum*, total 13 resistant accessions, 9 and 4, respectively, from these two species were identified [11].

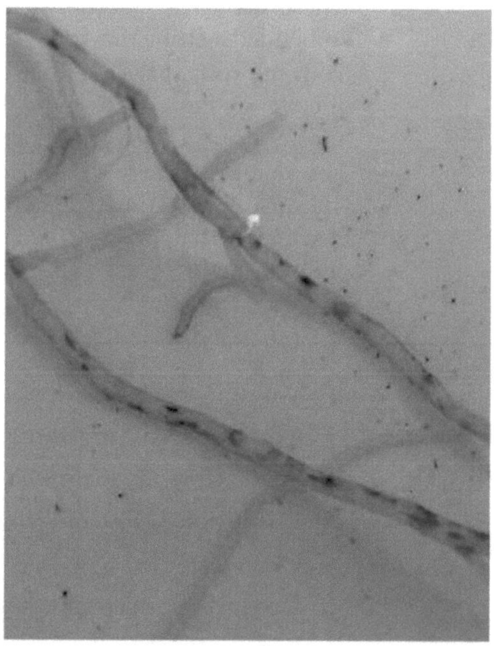

Fig 2 Necrotic lesions caused by *Pratylenchus thornei* on chickpea roots (Source: DPIRD). Methods for screening of chickpea against root lesion nematode, *Pratylenchus thornei*

Fig 3 Stained reniform nematode female attached to the roots. (Figure courtesy: CABI Digital Library)

1.4 Reniform Nematode

Reniform nematode, *Rotylenchulus reniformis*, is the most economically important species, and is an important pathogen of chickpea in Egypt, Ghana, and India [12, 13]. The term "reniform" refers to the kidney-shaped body form of the sedentary mature female protruding from infected roots. Reniform nematodes (*Rotylenchulus* spp.) are semi-endoparasites of roots and occur commonly in tropical and subtropical regions. Infected plants are distributed in patches and show pale green leaves, stunted growth, and early senescence. The anterior part of the nematode body remains embedded within the root, whereas the posterior end protrudes from the root surface and swells during maturation of the reproductive system to assume a kidney-like shape (Fig. 3). Mature females deposit between 60 and 200 eggs into a gelatinous matrix. The damage threshold for *R. reniformis* in greenhouse grown chickpeas was estimated at 0.5 to 1.0 nematodes per g of soil [14, 15], with up to 80% yield reduction occurring at 10 nematodes per g of soil [15].

2 Materials

2.1 Common Materials and Instruments

1. Chickpea seeds (test accessions and susceptible check varieties).
2. Sterilized sandy loam soil.
3. Plastic pots (15- cm diameter).
4. Clean water.
5. Droppers or micropipettes (for inoculation).
6. Forceps (for handling egg masses, cysts, females, etc.)
7. Watering setup (watering can or irrigation system).

8. Stereo zoom microscope (for counting nematodes, observing root lesions, cysts, and eggs).
9. Tools for uprooting plants (trowel or small shovel).
10. Petri dishes/cavity blocks.
11. Measuring cylinder/pipettes (for accurate measurement of suspension).
12. Glass slides and coverslips (for stained root observation).
13. Knife or blade (for cutting roots into 10 cm pieces).
14. Glass beakers or jars (for incubation).
15. Permanent marker and plant labels (for pot identification).
16. Fine brush or needle (for separating and counting egg masses on roots).

2.2 Pure Culture, Egg Collection, and Evaluation for Root-Knot Nematodes

1. Infected roots of chickpea, brinjal, tomato, okra, chili (for initial nematode source).
2. Brinjal or tomato seeds (host plant for nematode multiplication).
3. BOD incubator (28 °C, for hatching J2s).
4. Wire mesh and tissue paper (for egg hatching setup).
5. Susceptible check variety (e.g., tomato or okra).
6. Hand lens or magnifier (for preliminary observation of root galls).
7. Ruler or digital caliper (to measure gall size if needed).
8. Dissecting tray or root observation tray.

2.3 Evaluation for Cyst Nematodes

1. Infected chickpea root samples containing cysts.
2. 60-mesh sieve (for isolating cysts from soil samples).

2.4 Evaluation for Root-Lesion Nematode (*Pratylenchus thornei*)

1. Infected roots of chrysanthemum (source of *P. thornei*).
2. Scissors or scalpel (to cut roots into 1 cm pieces).
3. Root staining solution (e.g., acid fuchsin or lactophenol cotton blue).
4. Ruler or measuring scale (to measure necrotic lesion length).

2.5 Screening of Chickpea Germplasm Against Reniform Nematode

1. Infected cowpea roots with egg masses of *R. reniformis*.
2. Nematode suspension containing 100 immature females and 100 males per plant.
3. 60-mesh sieve (for collecting detached females from soil).

3 Methods

3.1 Pure Culture, Egg Collection, and Evaluation for Root-Knot Nematodes

3.1.1 Preparation of Pure Culture of Root-Knot Nematodes

1. Collect infected roots of crops such as chickpea, brinjal, tomato, okra, chili etc. from field.
2. Select the single and unmerged root galls containing single egg mass adhered to it. Collect single egg mass in separate cavity blocks and allow second-stage juvenile nematodes to hatch. Also collect the female from respective galls and label it. Use female and few juveniles for morphological and molecular identification. Remaining juveniles will be used for inoculation and multiplication.
3. Raise the plants of brinjal or tomato in pots filled with sterilized soil.
4. Inoculate the J2s in separate plants near the roots of these plants and plug the holes by sterilized dry soil. Label each pot accordingly and irrigate the plants on regular basis.
5. After 6–7 weeks of inoculation, harvest the plant for observation. Slightly water the plants and invert the pot and take out the plants along with roots.
6. Wash the soil adhered to roots by clean water, collect the egg masses using forceps, and keep it for hatching.
7. Reconfirm using females and juveniles and confirmed population can further be used for inoculation and multiplication as discussed above.

3.1.2 Collection of Egg Masses and Isolation of Juveniles

1. Wash plant roots and handpick egg masses using sharp forceps into a cavity block or glass petri dish.
2. Place egg masses in a BOD incubator at 28 °C for hatching.
3. Alternatively, place egg masses on tissue paper supported by wire mesh over water in a petri dish.
4. Collect freshly hatched J2s daily from the water suspension using a narrow beaker.
5. Let the suspension sit undisturbed for 1 h, then gently remove the upper water with a dropper.
6. Mix the remaining suspension gently and take 1 mL into a counting dish.
7. Count the number of nematodes per mL under a stereo zoom microscope three times and calculate the average.
8. Calculate the suspension volume needed to maintain 2 J2s per gram of soil.
9. Remove 2–3 cm of soil near the chickpea root zone, apply the calculated suspension, and cover with dry soil.

3.1.3 Preliminary Evaluation

1. Fill the pots with mixture of soil, sand, compost in the ratio of 2:1:1 [use completely randomized design (CRD) design Fig. 4].
2. Sow 3–4 seeds per pot and thin it to single plant after 10 days.
3. Inoculate freshly hatched J2s near the root zone of plants @ 2 J2s per g of soil. Remove upper 2–3 cm layer of soil around stem for inoculation and pour the suspension around root zone and cover it by dry soil.
4. Always keep one susceptible check as control treatment and inoculate with same procedure to check the viability of inoculum as well as for comparison of results.
5. Uproot the plants gently after 6–7 weeks of inoculation as discussed above.
6. Count the roots galls, egg masses per plant and eggs per egg mass. Designate host as per the galling index (*see* **Note 1**).

3.1.4 Confirmatory Evaluation

1. Accessions with less than 10 root galls/plant were selected for confirmatory evaluation.

Fig 4 Pot experiment for evaluation of chickpea germplasm against nematodes

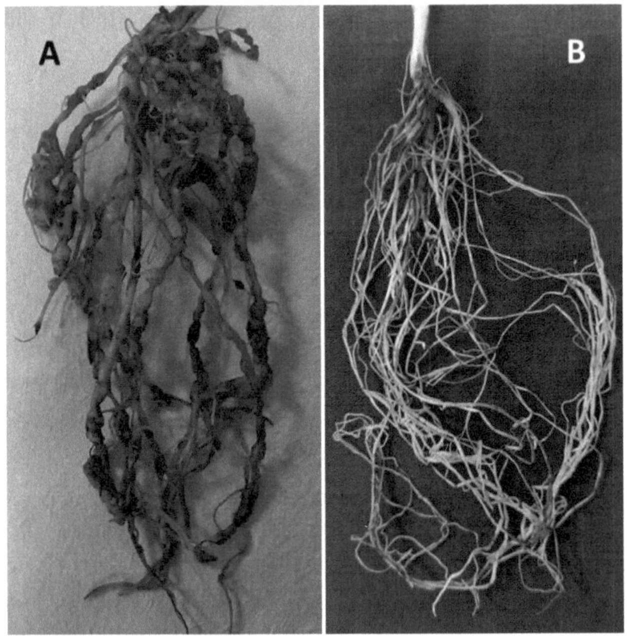

Fig 5 Reaction of root-knot nematode, *Meloidogyne incognita* infection in chickpea roots. (**a**) Susceptible accession and (**b**) resistant accessions

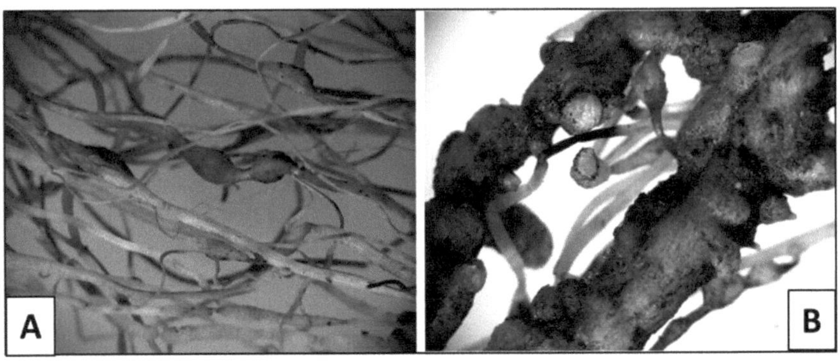

Fig 6 Microphotogarphs of root galls induced by root-knot nematode on roots of chickpea. (**a**) Highly susceptible accession, with entire root covered with galls and (**b**) resistant accession, with only a few small galls

2. Seeds were sown in pots and inoculated @ 2 J2s, 3 J2s, and 4 J2s per g of soil.
3. Number of galls/plant and number of egg masses/plant were counted to confirm the host reaction.
4. Accessions with consistent resistance reaction (less than 10 root galls per plant) should be rated as resistant (Figs. 5 and 6).

3.2 Method of Evolution for Cyst Nematodes

1. Collect the cysts from infected roots of chickpea maintained in net house/greenhouse.
2. Incubate the cysts for 7 days in water at room temperature for hatching.
3. Sow 5 seeds of each accession of chickpea in 15-cm-diameter plastic pots containing 1 kg sterilized sandy loam soil. Maintain only one healthy plant per pot while thin out the rest plants.
4. Keep one susceptible variety as check for comparison of results.
5. 15–20 days after sowing each plant was inoculated with one second-stage juvenile (J2) per gram of soil. Remove top 1 cm of soil around roots and dispense nematode suspension and then cover with soil.
6. Uproot plants from pots after 6 weeks of inoculation for observations. Count the number of cysts attached to roots and present in soil (*see* **Note 2**).

3.3 Method of Evolution for Root-Lesion Nematode

1. Collect the roots from *P. thornei*-infected plants such as chrysanthemum maintained in glasshouse conditions.
2. Cut the roots in small pieces about 1 cm in length and dip it in fresh clean water to allow nematode movement in water.
3. Sow 5 seeds of each accession of chickpea in 15-cm-diameter plastic pots containing 1 kg sterilized sandy loam soil. Maintain only one healthy plant per pot while thin out the rest plants.
4. Keep one susceptible variety as check for comparison of results and confirmation of results.
5. After 2 weeks of sowing, each plant should be inoculated with mixed stages of nematodes such as juveniles and adults @1000 nematodes/pot. Remove top 1 cm of soil around roots and dispense nematode suspension and then cover with soil.
6. Uproot the plants from pots after 6–7 weeks of inoculation for observations. Observe the root lesions induced on roots under microscope. Cut the roots into 10 cm pieces and cut open to observe the necrotic area and measure the length of necrotic area and convert it into percentage (*see* **Note 3**).
7. Stain the roots for observation on nematodes inside the roots (*see* **Note 4**).

3.4 Methods for Screening of Chickpea Germplasm against Rotylenchulus reniformis

1. Collect nematode egg masses from infected roots of cowpea maintained in net house/greenhouse.
2. Collect the hatched nematodes and incubate them for 10 days in water at room temperature.
3. Sow the chickpea seeds in 15-cm-diameter plastic pots containing sterilized sandy loam soil. Each pot should be sown with 5 seeds and thin it to one plant per pot after 10 days of sowing.

Always kept one susceptible variety as check for comparison of results.

4. After 15–20 days of sowing, inoculate each plant with 100 immature females and males each per plant. Dispense the nematodes in 5 mL water around the rhizosphere.
5. Uproot the plants from pots after 15 days of inoculation for observations.
6. Count the number females attached to the roots. Females of the reniform nematodes get easily dislodged from chickpea roots. Wash the soil and collect 60-mesh catch and count the number of detached females. Add the number of females from soil to females attached to the roots and use total number for host rating (*see* **Note 5**).

4 Notes

1. Root-knot nematodes develop root galls as major symptom of infection; therefore, number of root galls was used as the criteria for assessment of resistance and host designation was done based on those numbers. Number of root galls/plant were recorded for each replication of accessions and based on mean values, host designation was given according to standard scale followed by All India Coordinated Research Project on Nematodes. Plant resistance to pests is categorized based on the number of galls or egg masses. No galls or egg masses indicate high resistance, while 1–10 galls/egg masses show resistance, 11–30 galls/egg masses represent moderate resistance, 31–100 galls/egg masses indicate susceptibility, and 101 or more galls/egg masses reflect high susceptibility. This scale is useful for evaluating the level of resistance or vulnerability of plants to pest infestations.
2. The resistance rating for plants based on cyst count is as follows: No cysts per plant indicates high resistance, 1–5 cysts per plant shows resistance, 6–25 cysts per plant represents moderate resistance, 26–50 cysts per plant indicates susceptibility, and more than 50 cysts per plant reflects high susceptibility.
3. The Necrosis Index is used to assess plant resistance based on root condition and nematode presence. No necrosis and no nematodes in the roots indicates resistance, whereas up to 25% necrosis and 1–10 nematodes per 5 g of roots indicates moderate resistance, up to 50% necrosis and 1–10 nematodes per 5 g of roots represents susceptibility, and with more than 50% necrosis and 11–25 nematodes per 5 g of roots indicates high susceptibility.

4. Roots can be stained by sodium-hypochlorite-acid-fuschin method [16]. The nematode stages can be seen under stereo zoom microscope and counted.
5. The resistance rating based on the number of females per plant is as follows: No females per plant indicates high resistance, 1–10 females per plant indicates resistance, 11–20 females per plant represents moderate resistance, 21 30 females per plant indicates susceptibility, and more than 30 females per plant reflects high susceptibility.

Acknowledgments

Authors duly acknowledge the funding support from the Department of Biotechnology, Ministry of Science and Technology, Government of India, New Delhi, India (SAN No. 102/IFD/SAN/1504/2020-21). Authors gratefully acknowledge the ICAR-National Bureau of Plant Genetic Resources, New Delhi; DBT-National Institute of Plant Genomic resources, New Delhi; and International Crops Research Institute for the Semi-Arid Tropics, Hyderabad, India for providing necessary support.

Conflict of Interest The authors declare that the study was conducted in the absence of any commercial or financial relationships that could be construed as a potential conflict of interest.

Author Contributions ZK conceptualized the area addressed in the manuscript. ZK and BHK collected the literature and wrote on their respective area. G and VCC made scientific improvement and editing. All the authors checked the final draft manuscript and gave their approval.

References

1. Zwart RS, Thudi M, Channale S, Manchikatla PK, Varshney RK, Thompson JP (2019) Resistance to plant-parasitic nematodes in chickpea: current status and future perspectives. Front Plant Sci 10:966
2. Nene YL, Sheila VK, Sharma SB (1996) A world list of chickpea and pigeonpea pathogens. International Crops Research Institute for Semi-Arid Tropics, Patancheru
3. Sikora RA, Claudius-Cole B, Sikora EJ (2018) Nematode parasites of food legumes. In: Sikora R, Coyne D, Hallmann J, Timper P (eds) Plant parasitic nematodes in subtropical and tropical agriculture, 3rd edn. CABI, Boston, pp 290–345
4. Elling AA (2013) Major emerging problems with minor *Meloidogyne* species. Phytopathology 103:1092–1102
5. Di Vito M, Greco N (1988) The relationship bet ween initial population densities of *Meloidogyne artiellia* and yield of winter and spring chickpea. Nematol Mediterr 16(2):163–166
6. Di Vito M, Singh KB, Greco N, Saxena MC (1996) Sources of resistance to cyst nematode in cultivated and wild *Cicer* species. Genet Resour Crop Evol 43:103–107
7. Singh KB, Ocampo B, Robertson LD (1998) Diversity for abiotic and biotic stress resistance

in the wild annual Cicer species. Genet Resour Crop Evol 45(1):9–17

8. Malhotra RS, Greco N, Di Vito M et al (2008) Registration of FLIP 2005-8C and FLIP 2005-9C, chickpea germplasm lines resistant to chickpea cyst nematode. J Plant Regist 2:65–66
9. Castillo P, Mora-Rodriguez MP, Navas-Cortes JA, Jimenez-Diaz RM (1998) Interactions between *Pratylenchus thornei* and *Fusarium oxysporum* f. sp. *ciceris* on chickpea. Phytopathology 88:836–844
10. Thompson JP, Reen RA, Clewett TG et al (2011) Hybridization of Australian chickpea cultivars with wild *Cicer* spp. increases resistance to root-lesion nematodes (*Pratylenchus thornei* and *P. neglectus*). Australas Plant Pathol 40:601–611
11. Reen RA, Mumford MH, Thompson JP (2019) Novel sources of resistance to root-lesion nematode (*Pratylenchus thornei*) in a new collection of wild *Cicer* species (*C. reticulatum* and *C. echinospermum*) to improve resistance in cultivated chickpea C. arietinum. Phytopathology 109:1270–1279
12. Ali SS (1995) Nematode problems in chickpea. Indian Institute of Pulses Research, Kanpur
13. Oteifa BA (1987) Nematode problems of winter season cereals and food legume crops in the Mediterranean region. In: Saxena MC, Sikora RA, Srivastava JP (eds) Nematodes parasitic to cereals and legumes in temperate semi-arid regions, Proceedings of a workshop held at Larnaca, Cyprus, 1–5 March. ICARDA, Syria, pp 199–209
14. Darekar KS, Jagdale GB (1987) Pathogenicity to *Rotylenchulus reniformis* to chickpea. Curr Res Rep Mahatma Phule Agrc Univ 3:89–91
15. Mahapatra BC, Pahdi NN (1986) Pathogenicity and control of *Rotylenchulus reniformis* on *Cicer arietinum*. Nematol Mediterr 14:287–290
16. Byrd DW Jr, Kirkpatrick T, Barker KR (1983) An improved technique for clearing and staining plant tissue for detection of nematodes. J Nematol 14:142–143

Chapter 22

Novel Method for Rapid Screening of Chickpea for Combined Dry Root Rot Disease and Osmotic Stress

Shubhashish Ranjan, Chaitali Narendra Chavan, and Muthappa Senthil-Kumar

Abstract

Chickpea (*Cicer arietinum* L.), confronts substantial challenges from the emerging pathogenic fungus *Macrophomina phaseolina* (Tassi) Goid, causing dry root rot (DRR) disease. Chickpea plants severely affected by combined DRR and drought stress. Currently sick plot and sick pot method are utilized for germplasm screening to identify tolerant genotypes. These methods are time-consuming; therefore, we propose a novel methodology for the rapid screening of chickpea under combined DRR and osmotic stress conditions. This chapter introduces an adept high-throughput phenotyping methodology, conducted within controlled laboratory conditions, aiming to investigate the interaction between osmotic stress and DRR disease in chickpea crops. The methodology employs an innovative pouch technique for screening combined stress, providing a streamlined temporal investigation process and precise control over stress parameters. The incorporation of polyethylene glycol (PEG) enables the simultaneous imposition of osmotic stress alongside pathogen infection, making the methodology versatile for studying combined stress scenarios. This approach fills a gap in concurrent stress imposition techniques, enhancing germplasm screening by identifying genotypes with varying susceptibility and resistance levels. Thus, we suggest use of high-throughput phenotyping in combination genome-wide association study (GWAS) can take combined stress resistance breeding in chickpea at next level to combat food security and climate change.

Key words Chickpea, DRR, Combined DRR osmotic stress phenotyping, GWAS and *Macrophomina phaseolina*

1 Introduction

Chickpea (*Cicer arietinum*), an annual legume in global agriculture and nutrition, encounters challenges from both abiotic and biotic stressors that affect productivity [1–3]. The emerging fungal pathogen *M. phaseolina*, causing DRR in chickpea, poses a substantial threat, exhibiting symptoms such as scattered drying of plants in the field, drooping petioles, and compromised lateral and taproots, ultimately leading to severe yield losses, particularly in regions with suboptimal growing conditions [4]. Additionally, this pathogen, a

Muthappa Senthil-Kumar and Hema Ramanna (eds.), *Plant Biotic Stress: Methods and Protocols*, Methods in Molecular Biology, vol. 2966, https://doi.org/10.1007/978-1-0716-4746-2_22,

member of the Botryosphaeriaceae family, demonstrates a broad host range, infecting over 500 plant species, emphasizing its significant implications for chickpea crops [4, 5]. Concurrently, drought or osmotic stress serves as a major constraint, contributing to a global yield reduction of 40–50% in chickpea crops [6, 7]. Osmotic stress during the reproductive stage leads to a considerable decrease in plant yield and related traits [8]. Studies on DRR suggest that osmotic stress influences the interaction between the DRR pathogen and chickpea, exacerbating the severity of DRR disease [9]. The combined osmotic stress and DRR pathogen infection proves to be particularly detrimental to the overall plant performance, resulting in a significant decline in yield. Reductions in yield are more pronounced under the combined stress, surpassing the impact observed under sole pathogen stress [1, 3, 10]. This combined stress condition also leads to significant reductions in photosynthetic rate, transpiration, stomatal conductance, and overall plant growth [9]. Notably, osmotic stress exacerbates DRR, amplifying disease incidence and severity regardless of soil type, geography, and genotype [11, 12]. These findings underscore the importance of understanding and mitigating the impact of combined stress in chickpea cultivation, forming the basis for the high-throughput phenotyping methodology developed in this chapter.

The evolving climatic conditions anticipated to affect the severity of DRR across various crops, thereby raising concerns about cultivating crops in changing environmental condition. This scenario necessitates a more in-depth investigation into the molecular basis of resistance against combined stress and the pathogenicity of *M. phaseolina*. Such insights can be harnessed to formulate more effective control strategies [13]. In addressing the complex dynamics of combined stress, particularly the combat interplay between osmotic stress and DRR in chickpea crops. High-throughput phenotyping studies emerge as crucial factor to understand. Understanding the diverse responses of different genotypes to varying degrees of damage becomes essential for advancements in crop development initiatives. Currently, chickpea screening relies on field-based sick plot and sick pot methodologies, which are time-intensive, laborious, and require substantial infrastructure. Screening a large number of germplasm lines under combined stress conditions within a short time frame is not feasible by using traditional methods. This chapter primarily focuses on establishing a robust methodology for high-throughput phenotyping, specifically directed at assessing the impact of combined stress on chickpea crops. This methodological foundation serves as a crucial stepping-stone toward comprehending the intricate interactions involved in combined stress, offering invaluable insights for future strides in chickpea cultivation and stress resilience. Furthermore, the integration of the high-throughput phenotyping approach with GWAS and developed databases like the Stress Combinations and

Their Interactions in Plants (SCIP) Database [3] offers opportunities for comprehensive research. This integrated strategy streamlines the process of identifying critical genetic and phenotypic markers essential for enhancing crop resilience when confronted with multiple stressors in climate change.

Here phenotyping methodology for combined stress of DRR and osmotic stress simultaneously. This approach, beneficial for high-throughput phenotyping of chickpea germplasm under combined stress, contributes significantly to advancing our understanding of crop responses and resilience under combined stress in chickpea.

2 Materials

2.1 Seed Material

Chickpea seeds [seedbank of International Crop Research Institute for Semi-Arid Tropics (ICRISAT) Patancheru, Hyderabad, Telangana, India].

2.2 Pathogen

Culture of DRR causing pathogen *Macrophomina phaseolina* (ITCC 8635, NCBI GenBank ID MH509971.1) (*see* **Note 1**).

2.3 Media and Chemical Reagents

1. Potato dextrose agar (PDA) (*see* **Note 2**).
2. Potato dextrose broth (PDB).
3. 2% Sodium hypochlorite.
4. Polyethylene glycol (mw 20,000).
5. Soilrite mix.

2.4 General Requirements

1. Petri plates for culturing fungus.
2. Falcon tubes (50 mL).
3. Grow bags.
4. 500 mL conical flask
5. 500 mL laboratory bottle
6. Nylon mesh.
7. Petri plates.
8. Forceps.
9. Autoclaved RO water.
10. Polythene ziplock bags (5 × 6 in.).
11. Plastic tray.
12. Wooden sticks.
13. Paint brush.
14. Scalpel.
15. Test tubes.

16. Test tube stands.
17. Milli-Q (MQ) and reverse osmosis (RO) water.

2.5 Instruments

1. Laminar airflow.
2. Incubator (28 °C).
3. Shaker incubator (28 °C).
4. EPSON root scanner.
5. Weighing balance.
6. Electrolyte conductivity (EC) meter.

2.6 Software

RhizoVision [14], WinRHIZO (Regent Instruments Inc., Canada), ImageJ [15], or other similar image analysis tools (*see* **Note 3**).

3 Methods

3.1 Seedling Cultivation

1. Chickpea seeds (40) per genotype wash with autoclaved RO water under laminar airflow, repeating the process 3 to 4 times.
2. Surface sterilize the seeds using 2% sodium hypochlorite, shaken intermittently for 3 min, and rinsed with autoclaved RO water 3–4 times to remove traces of sodium hypochlorite.
3. Fill autoclaved soilrite into grow bags, sow surface-sterilized seeds in the bags and maintain at 25 °C with a 16-h light and 8-h dark photoperiod, and a light intensity of 150 μmol/m^2/s in growth room or chamber (Fig. 1).
4. Uproot the plants 7 days after sowing. Gently wash the plants to remove soil particles without damaging the roots, and place in a tray with autoclaved RO water.

3.2 Fungal Inoculum and 10% PEG Preparation

1. Prepare 200 mL of PDB media in a 500 mL conical flask and autoclave.
2. Inoculate the medium with a 1 mm agar plug from a 5-to-6-day-old actively growing *M. phaseolina* fungal agar plate.
3. Incubate the conical flasks at 28 °C in a shaker for 7 days at 180 rpm in darkness.
4. Filter fungal mycelia and microsclerotia using nylon mesh after 7 days (*see* **Note 4**).
5. Weigh 10 g of PEG with a molecular weight of 20,000 (*see* **Note 5**).

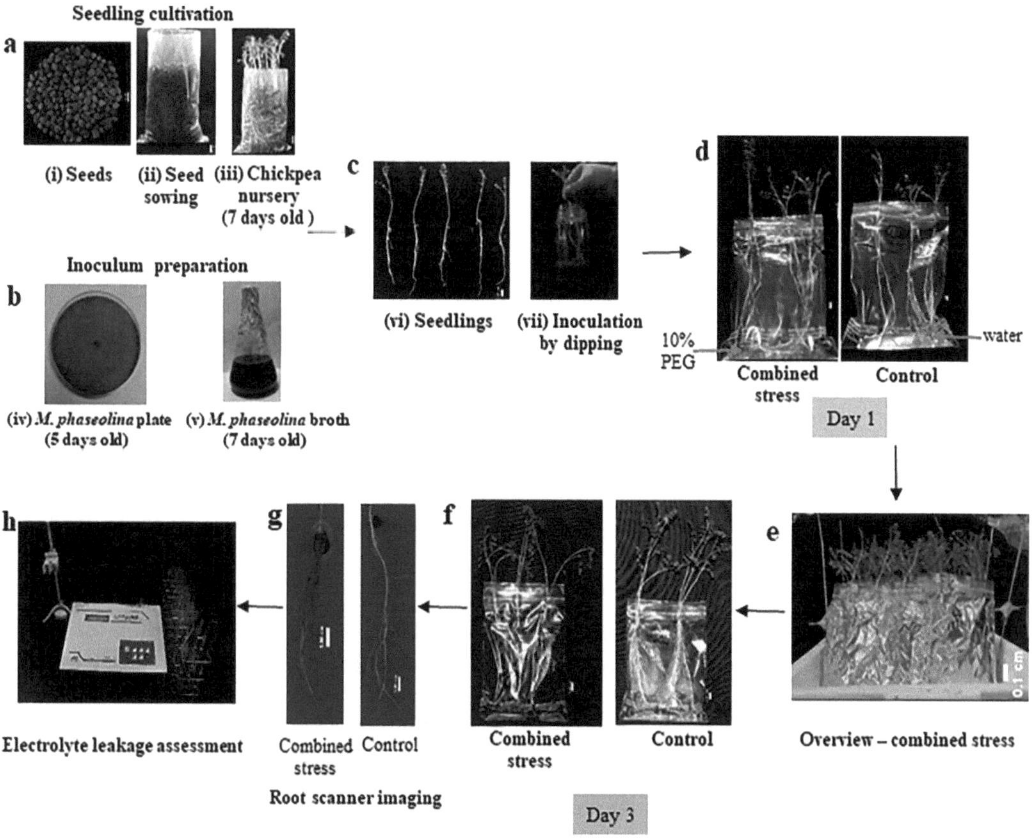

Fig. 1 Experimental workflow for imposition of combined DRR and osmotic stress in chickpea. (**a**) Seedling cultivation—(*i*) chickpea seeds, (*ii*) seeds sown in a bag containing soilrite mix for germination, and (*iii*) 7-day-old seedlings. (**b**) Fungal inoculum preparation—(*iv*) 5 day-old *M. phaseolina* plate for broth inoculation and (*v*) broth culture of *M. phaseolina* for root inoculation. (**c**) Dipping method—(*vi*) washed chickpea seedlings used for fungal inoculation. (*vii*) Inoculation of fungus on chickpea roots by dipping method. (**d**) Fungus inoculated and control seedlings placed in pouch containing 10% PEG and water, respectively. (**e**) Experimental setup in a tray, maintained in the infection room for 3 days. (**f**) Combined stress and control seedlings after 3 days after inoculation. (**g**) Root scanner images of combined stress and control seedling after 3 days of infection. (**h**) Electrolyte leakage assessment by using an EC meter

6. Dissolve 10 g of PEG 20,000 mw in 60 mL of MQ water in laboratory bottle. Adjust the final volume to 100 mL with MQ water and autoclave.
7. Store the 10% PEG solution at 4 °C or room temperature.

3.3 Pouch Preparation

1. Fill 20 mL of autoclaved RO water in bags for control treatment.
2. Fill 20 mL of 10% PEG solution in bags for stress imposition (Fig. 1) (*see* **Notes 6** and 7).

3.4 Fungus Inoculation by Dipping

1. Filter the 7-day-old fungal culture using a nylon mesh with a 1 mm diameter.
2. Dissolve the washed mycelia in autoclaved RO water within an autoclaved petri dish.
3. Dip the plant roots into the inoculum for 1 min, moving intermittently to ensure even mycelia attachment (Fig. 1) (*see* **Note 8**).

3.5 Combined Stress Imposition

1. Place the control, osmotic stress, and combined stress (fungus-inoculated roots in 10% PEG medium) in separate bags, using different bags for each genotype.
2. Place plants in prepared respective bags with partial root immersion in water and 10% PEG solution.
3. Cover all bags with aluminum foil to ensure darkness. Leave the cut part of the bag open for air circulation to prevent excess moisture inside.
4. Arrange the bags in a tray with wooden sticks for support. Maintain the setup at 25 °C for 3 days in the infection room under 150 μmol/m^2/s light intensity, 60–70% relative humidity (Fig. 1) (*see* **Note 9**).

3.6 Assay for Quantification in Combined Stress

3.6.1 Root Cleaning

1. Remove all plants from the bags after 3 days of incubation.
2. Wash the roots thoroughly in water to eliminate fungal mycelium and PEG from combined and osmotic stress samples.

3.6.2 Root Scanner Imaging and Scoring

1. Visually observe the roots for necrosis, tap and lateral root damage or loss.
2. Impute the observation in provided range of manual score (Table 1) for each plant of genotype.
3. Calculate average score of genotype for manual score.
4. After manual scoring, capture root image using root scanner.
5. Assess these root images using Image J or WinRhizo software, assigning scores on a scale from 0 to 9 to reflect the severity of necrosis in the root region. Additionally, conduct manual scoring, also on a scale from 0 to 9 (as shown in Table 1), for the analysis of infected or diseased roots.
6. To calculate the total area of necrosis from root scanner image, use the following formula:

$$\text{Area of necrosis (\%)} = \left(\frac{\text{Area covered by specific score}}{\text{Total root area}}\right) \times 100$$

7. Calculate average necrosis area from all samples of a genotype to obtain total area of necrosis.

Table 1
Disease severity for combined DRR under osmotic stress of chickpea

Rating	Symptoms	Disease reaction
1	No infection on roots	Immune
>1 and <3	Few small lesions on infected roots, only little effect of drought and disease appears on roots	Resistant
>3 and <5	Small and clear lesions on roots; new and lateral roots infected	Moderately resistant
>5 and <7	Lesions on roots; many new roots generally have lesions, discoloration of roots, lateral and tap roots, root started drying	Susceptible
>7	Roots are fully infected and completely discolored; sometimes complete roots dried	Highly susceptible

3.6.3 Electrolyte Leakage Assessment

1. After imaging, roots were cut into small pieces using a scalpel, treating each root as a replicate.
2. Collected roots for each genotype under control, osmotic, and combined stress conditions.
3. Weigh 100 mg of roots from each replicate into labeled test tubes and arrange the test tubes properly in a test tube stand.
4. Add 15 mL of MQ water to each test tube and wash for 30 min.
5. After 30 min, remove the water and add 15 mL of fresh water to each test tube to measure electrolyte leakage (*see* **Note 10**).
6. Record the EC using an EC meter immediately within 30 min after adding water (ECa), after 4 h of rehydration (ECb), and ultimately after autoclaving to measure total leakage (ECt).
7. Express the measured electrolyte leakage as a percentage electrolyte leakage using the formula.

$$\mathrm{EC}\ (\%) = \left[\frac{(\mathrm{ECb} - \mathrm{ECa})}{\mathrm{ECt}}\right] \times 100$$

3.6.4 Genotype Screening

1. Sow seeds of chickpea genotypes in a soilrite-containing bag.
2. Uproot the 7-day-old sown plants.
3. Wash the plants and keep in tray by labeling genotype name.
4. Prepare the experimental setup (refer to Subheadings 3.2, 3.3, and 3.4) for combined and osmotic stress imposition.
5. Remove the 3 days postinoculated combined and osmotic stress plant and wash.

6. Measure the manual score (Table 1) and record the image for area of necrosis scoring.
7. Calculate area of necrosis to obtain total area of necrosis.
8. Measure EC from the root samples and compute in the EC formula (*see* **Notes 11** and **12**).
9. Obtain all readings and analyze for combined and osmotic stress.

4 Notes

1. The necrotrophic fungus present in the soil is derived from the roots of a diseased plant harvested from a field in NIPGR, New Delhi (28.5307 °N, 77.1665 °E). A pure culture is successfully cultivated using the hyphal tip technique (Irulappan and Senthil-Kumar 2021).
2. It is preferable to use PDA and PDB from Becton, Dickinson and Company, Sparks, MD, USA for the experiment.
3. RhizoVision, WinRHIZO, Image J, or other similar tools used to analyze necrotic lesions from infected root images for high-throughput phenotyping. These tools help quantify and evaluate the extent of root damage and providing detailed information on the impact of infections.
4. Ensure thorough washing of mycelia with autoclaved RO water to remove all media and toxins.
5. At PEG concentrations above 15%, the roots may become either dry or soggy, depending on the genotype. Therefore, 10% PEG concentration recommended for optimal root phenotype.
6. PEG facilitates the imposition of osmotic stress by offering adjustable concentrations, thus allowing for customization according to specific experimental requirements. This adaptability in concentration selection ensures precise control over the intensity of osmotic stress applied, thereby enhancing the flexibility and effectiveness of experimental procedures.
7. Cut bags slightly on both sides of upper part to facilitate proper air circulation and preventing root sogginess due to excess moisture.
8. After loading fungal inoculum, place seedling roots on tissue paper to absorb excess water from the inoculum, thereby preventing sogginess in the roots.
9. Optimize environmental conditions are necessary for optimal disease expression and to obtain phenotype. High temperature (above 30 °C) and high light intensity may cause drying of

roots, whereas lower temperature (below 22 °C) and high relative humidity may cause sogginess in roots. At 20 °C and below temperature, no fungal infection may be observe.

10. Maintain a washing time of 30 min to ensure accurate observation of the EC trend within the same genotype root samples. A shorter washing time may result in unclear EC readings. Additionally, calibrate the EC meter regularly to ensure precise measurements for proper experimentation.
11. Using this protocol, 25 genotypes/week/m^2 can be screen. The responses of genotypes categorized as resistant or susceptible across all three methods can be used to classify them as either resistant or susceptible to combined DRR osmotic stress.
12. Chickpea genotypes ICC4958 and ICC11088 (BG212) recommended as check genotypes for combined stress experiments. ICC4958 is moderately resistant, with a manual score of 4.5, an area of necrosis of 37%, and an EC of 25.5%. On the other hand, ICC11088 is highly susceptible, with a manual score of 8.4, an area of necrosis of 55.02%, and an EC of 45%. These genotypes can serve as benchmarks for evaluating stress responses using this protocol.

Acknowledgments

Combined stress projects at the MS-K lab are supported by the National Institute of Plant Genome Research core funding and partly by the Department of Biotechnology (DBT), Government of India (No. BT/Ag/Network/Chickpea/2019–20), under the mission program on "Characterization of genetic resources". Council of Scientific and Industrial Research (CSIR) senior research fellowship [CSIR File no.–09/0803(13718)/2022-EMR-I] to S.R. We thank to Komal Vitthalrao Mali and Athimoolam Durgadevi for their technical input during the experiment.

References

1. Chilakala AR, Mali KV, Irulappan V et al (2022) Combined drought and heat stress influences the root water relation and determine the dry root rot disease development under field conditions: a study using contrasting chickpea genotypes. Front Plant Sci 13: 890551. https://doi.org/10.3389/fpls.2022.890551
2. Ranjan S, Mirchandani R, Senthil-Kumar M (2023) Abiotic stress impact on the interaction between Macrophomina phaseolina and crop plants. Plant Physiol Rep 29:1–10. https://doi.org/10.1007/s40502-023-00753-5
3. Priya P, Patil M, Pandey P et al (2023) Stress combinations and their interactions in plants database: a one-stop resource on combined stress responses in plants. Plant J 116:1097–1117. https://doi.org/10.1111/tpj.16497
4. Mirchandani R, Irulappan V, Chilakala AR, Senthil-Kumar M (2023) Dry root rot disease: current status and future implications for chickpea production. Proc Natl Acad Sci India

Sect B Biol Sci 93:791–800. https://doi.org/10.1007/s40011-023-01451-w

5. Gupta GK, Sharma SK, Ramteke R (2012) Biology, epidemiology and management of the pathogenic fungus *Macrophomina phaseolina* (Tassi) Goid with special reference to charcoal rot of soybean (*Glycine max* (L.) Merrill). J Phytopathol 160:167–180. https://doi.org/10.1111/j.1439-0434.2012.01884.x
6. Ahmad F, Gaur PM, Croser JS (2005) Chickpea (Cicer arietinum L.). In: Singh RJ, Jauhar PP (eds) Genetic resources, chromosome engineering, and crop improvement. Taylor & Francis, London, pp 229–267. ISBN 978-0-8493-1430-87
7. Chilakala AR, Pandey P, Durgadevi A et al (2023) Drought attenuates plant responses to multiple rhizospheric pathogens: a study on a dry root rot-associated disease complex in chickpea fields. Field Crop Res 298:108965. https://doi.org/10.1016/j.fcr.2023.108965
8. Sachdeva S, Bharadwaj C, Patil BS et al (2022) Agronomic performance of chickpea affected by drought stress at different growth stages. Agronomy 12:995. https://doi.org/10.3390/agronomy12050995
9. Sinha R, Irulappan V, Mohan-Raju B et al (2019) Impact of osmotic stress on simultaneously occurring pathogen infection in field-grown chickpea. Sci Rep 9:5577. https://doi.org/10.1038/s41598-019-41463-z
10. Rai A, Irulappan V, Senthil-Kumar M (2022) Dry root rot of chickpea: a disease favored by drought. Plant Dis 106:346–356. https://doi.org/10.1094/PDIS-07-21-1410-FE
11. Sharma M, Pande S (2013) Unravelling effects of temperature and soil moisture stress response on development of dry root rot [Rhizoctonia bataticola (Taub.)] Butler in chickpea. Am J Plant Sci 4:584–589. https://doi.org/10.4236/ajps.2013.43076
12. Irulappan V, Kandpal M, Saini K et al (2022) Drought stress exacerbates fungal colonization and endodermal invasion and dampens defense responses to increase dry root rot in chickpea. Mol Plant Microbe Interact 35:583–591. https://doi.org/10.1094/MPMI-07-21-0195-FI
13. Ranjan S, Mirchandani R, Senthil-Kumar M (2023) Dry root rot in chickpea: a perspective on disease resistance breeding strategies. In: Jha UC, Nayyar H, Sharma KD et al (eds) Diseases in legume crops: next generation breeding approaches for resistant legume crops. Springer Nature, Singapore, pp 85–98. https://doi.org/10.1007/978-981-99-3358-7_4
14. Seethepalli A, Dhakal K, Griffiths M et al (2021) RhizoVision explorer: open-source software for root image analysis and measurement standardization. AoB PLANTS 13: plab056. https://doi.org/10.1093/aobpla/plab056
15. Schneider C, Rasband W, Eliceiri K (2012) NIH image to ImageJ: 25 years of image analysis. Nat Methods 9:671–675. https://doi.org/10.1038/nmeth.2089

Index

A

B

C

D

E

F

Muthappa Senthil-Kumar and Hema Ramanna (eds.), *Plant Biotic Stress: Methods and Protocols*, Methods in Molecular Biology, vol. 2966, https://doi.org/10.1007/978-1-0716-4746-2,

G

H

I

K

L

M

N

O

P

Q

R

S

T

U

V

W

X

Y

MIX
Papier aus verantwortungsvollen Quellen
Paper from responsible sources
FSC® C105338

If you have any concerns about our products,
you can contact us on
ProductSafety@springernature.com

In case Publisher is established outside the EU,
the EU authorized representative is:
Springer Nature Customer Service Center GmbH
Europaplatz 3, 69115 Heidelberg, Germany

Printed by Libri Plureos GmbH
in Hamburg, Germany